Springer Series in
MATERIALS SCIENCE 65

Springer Series in

MATERIALS SCIENCE

Editors: R. Hull R. M. Osgood, Jr. J. Parisi

The Springer Series in Materials Science covers the complete spectrum of materials physics, including fundamental principles, physical properties, materials theory and design. Recognizing the increasing importance of materials science in future device technologies, the book titles in this series reflect the state-of-the-art in understanding and controlling the structure and properties of all important classes of materials.

61 **Fatigue in Ferroelectric Ceramics and Related Issues**
By D.C. Lupascu

62 **Epitaxy**
Physical Principles and Technical Implementation
By M.A. Herman, W. Richter, and H. Sitter

63 **Fundamentals of Ion-Irradiated Polymers**
By D. Fink

64 **Morphology Control of Materials and Nanoparticles**
Advanced Materials Processing and Characterization
Editors: Y. Waseda and A. Muramatsu

65 **Transport Processes in Ion-Irradiated Polymers**
By D. Fink

66 **Multiphased Ceramic Materials**
Processing and Potential
Editors: W.-H. Tuan and J.-K. Guo

67 **Nondestructive Materials Characterization**
With Applications to Aerospace Materials
Editors: N.G.H. Meyendorf, P.B. Nagy, and S.I. Rokhlin

68 **Diffraction Analysis of the Microstructure of Materials**
Editors: E.J. Mittemeijer and P. Scardi

69 **Chemical-Mechanical Planarization of Semiconductor Materials**
Editor: M.R. Oliver

70 **Applications of the Isotopic Effect in Solids**
By V.G. Plekhanov

71 **Dissipative Phenomena in Condensed Matter**
Some Applications
By S. Dattagupta and S. Puri

72 **Predictive Simulation of Semiconductor Processing**
Status and Challenges
Editors: J. Dabrowski and E.R. Weber

73 **SiC Power Materials**
Devices and Applications
Editor: Z.C. Feng

74 **Plastic Deformation in Nanocrystalline Materials**
By M.Yu. Gutkin and I.A. Ovid'ko

75 **Wafer Bonding**
Applications and Technology
Editors: M. Alexe and U. Gösele

76 **Spirally Anisotropic Composites**
By G.E. Freger, V.N. Kestelman, and D.G. Freger

77 **Impurities Confined in Quantum Structures**
By P.O. Holtz and Q. Zhao

78 **Macromolecular Nanostructured Materials**
Editors: N. Ueyama and A. Harada

Volumes 10–60 are listed at the end of the book.

D. Fink (Ed.)

Transport Processes in Ion-Irradiated Polymers

With 107 Figures and 15 Tables

Springer

Dr. Dietmar Fink
HMI Berlin, Abt. SF4, Glienicker Str. 100
14109 Berlin, Germany

Series Editors:

Professor Robert Hull
University of Virginia
Dept. of Materials Science and Engineering
Thornton Hall
Charlottesville, VA 22903-2442, USA

Professor R. M. Osgood, Jr.
Microelectronics Science Laboratory
Department of Electrical Engineering
Columbia University
Seeley W. Mudd Building
New York, NY 10027, USA

Professor Jürgen Parisi
Universität Oldenburg, Fachbereich Physik
Abt. Energie- und Halbleiterforschung
Carl-von-Ossietzky-Strasse 9–11
26129 Oldenburg, Germany

Professor Hans Warlimont
Institut für Festkörper-
und Werkstofforschung,
Helmholtzstrasse 20
01069 Dresden, Germany

ISSN 0933-033X

ISBN 3-540-20943-3 Springer Berlin Heidelberg New York

Library of Congress Control Number: 2004104058

Springer is a part of Springer Science+Business Media.

springeronline.com

Printed in Germany

Final processing by PTP-Berlin Protago-TeX-Production GmbH, Germany
Cover concept: eStudio Calamar Steinen
Cover production: *design & production* GmbH, Heidelberg

Printed on acid-free paper SPIN: 10948033 57/3141/YU 5 4 3 2 1 0

Preface

The ion irradiation of polymers (which had been treated in Volume 1 of this book) is frequently followed by transport processes, either of the implanted ions themselves or of the activated polymer chains, of radiochemical reaction products inside the irradiated zone, or of foreign atoms, ions or molecules that penetrate into the polymer after the irradiation has taken place (these entities will be called "penetrants" from now on). Transport processes in a polymer can be described in terms of diffusion, permeation, and/or solution phenomena, but there are also cases when, e.g., viscous flow through pores plays a role. There exist many elaborate theories on transport processes in unirradiated polymers, the most important ones of which will be briefly summarized in this volume in Chap. 1. Transport processes in irradiated polymers are much less well known, but nevertheless the presently available data base already allows one to obtain a rough understanding of the processes occurring. This will be summarized in the subsequent two chapters. As it has turned out that the formation of tracks by energetic heavy ions leads to many peculiarities of the transport processes in polymers, Chap. 3 will be devoted to these cases, while Chap. 2 rather treats the penetrant migration in low-energy ion-irradiated polymers. Finally, Chap. 4 deals with ion-track etching. The latter is of importance as quite a number of technical applications have emerged from the use of etched ion tracks in polymers.

Ion irradiation of polymers has found applications already since about 1950. This concerned especially low-energy ion beams (Chap. 5) that were used in electronics for semiconductor doping through polymeric photoresist masks. In the meantime, the ion-beam modification of polymeric surfaces for medical applications to improve the material's biocompatibility has also found major use, and, furthermore, tribological and optical applications of low-energy ion beams have become more familar.

In order to prepare high-energy irradiated polymers for applications, frequently additional manipulation steps have to be undertaken, e.g. etching, doping with penetrants, deposition of nanoclusters, nanowires, or nanotubules within the ion tracks, or deposition of additional structures on the polymeric surfaces. Whereas these manipulations will be treated in Chap. 6, the last chapter will be devoted to the emerging applications themselves.

For high-energy ion-beam irradiation of polymers, the major field is found in the production of microporous foils. These etch-track filters have found a number of well-established applications, and more are currently under discussion or development.

For all applications of ion-irradiated polymers it is of fundamental importance to prevent the inherent degradation of their properties due to both the constituent and penetrant motion at ambient temperature, with possibly detrimental oxidative or corrosive reactions. Since this would otherwise restrict the applicability of polymers to inexpensive throwaway products with limited lifetimes and to low-temperature devices only, the continuous search for new polymer systems with good thermal stability even after their irradiation is of central importance.

This volume as well as the preceeding one on the fundamentals of ion-irradiated polymers are thought to challenge the reader to contribute himself to the yet open questions, to deepen our insight and to promote the technological applicability of this fascinating field.

Thanks to everybody who contributed to this book directly or indirectly by numerous discussions at conferences, meetings, etc. Apart from giving special credit to all the coauthors of this book, I want to acknowledge thankfully the continuous interest and help of Prof. Lewis T. Chadderton, Dr. Manfred Müller, and Dr. Jiri Vacík over the many years during which the idea of that book emerged, and which helped clarifying many findings that appeared to be a riddle in the moment of their discovery. We are also obliged to Dr. Ricarda Klett, Dr. Arne Schmoldt, Prof. Kamal K. Dvivedi, Dr. Svarnali Ghosh, Mr. Prashant Alegaonkar and Mr. Alexander Petrov who obtained quite a number of the results compiled in this book, partly in the frame of their PhD or Postdoc works. Many other important personal communications are mentioned in the references. We thank to Prof. Pedro Grande for his kind help in the chapter on stopping powers and ranges, to Prof. Jochen P. Biersack and to Dr. Amita Chandra for their critical reading, and to Dr. Mahalakshmi Iyer for her technical help in the last chapter of this book. Thanks also to the coauthors for their help in the chapters in which they are not explicitely named. Thanks also to all Publishing Companies who allowed the reproduction of figures in this book.

We are also obliged to the organizations that enabled this work by giving us financial support, especially to the DAAD and CAPES in the frame of the German-Brazil agreement "POBRAL". I also must not forget to mention the continuous support by my home institute, the Hahn-Meitner-Institute Berlin. Last not least I want to thank my good friends all over the world – especially to the families Chadderton, Stange-Ferreira, and Berdinsky, and also to my own family who always supported me with constant encouragement, friendship and love, and thus cared for the warm and good atmosphere necessary to write this book.

Berlin, February 2004 *Dietmar Fink*

Contents

Part I Transport Processes in Polymers

List of Contributors

Pawel Yu. Apel
Joint Institute of Nuclear Research
Dubna
Flerov Lab. of Nuclear Reactions
141980 Dubna, Russia
apel@cv.jinr.dubna.ru

Moni Behar
Instituto de Física
Universidade Federal do Rio Grande do Sul
Av. Bento Gonçalves 9500, C.P.15051
91501-970 Porto Alegre, R.S., Brazil
behar@if.ufrgs.br

Dietmar Fink
Abt. SF4
Hahn-Meitner-Institut Berlin
Glienicker Str. 100
D-14109 Berlin, Germany
fink@hmi.de

Vladimir Hnatowicz
Nuclear Physics Institute
25068 Řež, Czech Republic
hnatowicz@ujf.cas.cz

R.H. Iyer
Nuclear Recycle Group
Bhabha Atomic Research Centre
Trombay, Mumbai 400 085, India
rhiyer@magnum.barc.ernet.in

Ricardo M. Papaléo
Instituto de Física
Pontificada Universidade Catolica do Rio Grande do Sul
Av. Ipiranga 6681, C.P.1429
91619-900 Porto Alegre, R.S., Brazil
papaleo@if.ufrgs.br

Part I

Transport Processes in Polymers

1 Transport Processes: Fundamentals

D. Fink and M. Behar

In spite of the existence of many theories on the transport processes in polymers, one should always keep in mind that, due to the complexity of this type of matter, this field is by far not yet perfectly established. Though the multitude of presently available models will always allow the researcher to find one that might be adequate for his problem, it will not be a priori clear to him that his choice was unambiguous. As was shown by Perlmutter and Scrosati [1], estimates of the diffusion coefficent D from some given experimental data may vary within a factor 10^5 "for the same dopant–polymer combination using the same measurement technique in the same laboratory, differing only in regard to whether the (applied theoretical) model allows for homogeneous chemical reactions" – or for other peculiarities. Furthermore, not all measurements done on the same polymer specie really refer to the same system. Diffusion in polymers is frequently influenced in an uncontrollable way by the presence of – in general unspecified – additives like free-radical inhibitors, plasticizers, fillers, and eventually also residual solvents. For example, polymers containing additions of inhibitors may suppress irradiation effects that are present in the pure materials. Fillers that are incompatible with the polymer will initiate void formation at the interface, which leads to an increase for the free volume of the system and consequently, to an increase in the degree of permeability. Antioxidants will prevent corrosion processes otherwise present. Finally, occasionally structural or compositional inhomogeneities show up in different areas of the same foil, due to insufficiencies in the production processes.

This situation, already difficult for unirradiated polymers, becomes even worse for ion-irradiated polymers where there is much less known about the parameters governing the polymer modifications. Astonishingly, in spite of the multitude of models that polymer science offers to describe transport effects, almost none of them was ever applied to ion-irradiated polymers, apart from the simplest Fickian interpretation schemes. This chapter is therefore also intended to give the reader an overview about the different approaches to describe transport processes in pristine polymers, to encourage him to apply these models – if suitable – also to irradiated polymeric matter.

Anyhow, the presently available experimental results on polymer irradiation enable one already to come at least to a general understanding of the processes occurring. In the first part of this volume, we shall therefore summarize first the knowledge about transport processes in pristine polymers, then describe what is already known for irradiated ones, and finally summarize the experimentally achieved data base.

The Different Regimes of Penetrant Flow through Matter

Suppose penetrants with radius r_p and average distance λ_p to each other move through matter with some embedded free volume, with the penetrant's mean free flight path being λ. The free volume may either be of intrinsic nature such as microvoids as in polymers, or pores might have been artificially introduced by, e.g., irradiation-induced chain scissioning and degassing, and/or by chemical attack such as etching. The penetrants can be single atoms or ions, or molecules (with radius r_p in the nm range), macromolecules ($r_\mathrm{p} \approx 10$–$100\,\mathrm{nm}$), colloids ($r_\mathrm{p} \approx 100$–$1000\,\mathrm{nm}$), or larger particles with radius up to some $10\,\mu\mathrm{m}$. Then the diffusion velocity of the penetrant specie along the free volume is limited by two types of resistances to diffusion, namely the molecule–molecule collisions and molecule–wall collisions. Several regimes can be distinguished [2] by the so-called Knudsen number N_k, which is the ratio of mean free path to mean pore diameter: $N_\mathrm{k} = \lambda/r_\mathrm{p}$ in which the motion is thought to take place.

For $\lambda \gg r_\mathrm{p}$ one deals with capillaric transport. In the case of $\lambda \gg \lambda_\mathrm{p}$ one encounters the Knudsen flow, and for $\lambda \ll \lambda_\mathrm{p}$ the viscous flow regime. The first case is characteristic of gases in pores, and the second case of liquids in pores. In the first case the flow is reciprocal to the viscosity of the gas, in the second case the flow is reciprocal to the square root of the molecular mass of the diffusing species. Whereas in the first case, the penetrant velocity is constant all over the pore, the penetrant velocity drops to zero at the pore walls in the second case. That is, in the Knudsen regime, it is the resistance due to molecule–molecule collisions that is extremely low, whereas in the first case the resistance due to molecule–wall collisions is extremely low. In the transition between these two extremes, the so-called slip flow case ($\lambda \approx \lambda_\mathrm{p}$), both resistance mechanisms play an important role.

If the pore diameter is decreased to the point at which the molecular size is no longer negligible as compared with the pore radius ($2r_p \approx \lambda_\mathrm{p}$), molecular-sieving conditions occur, until the penetration mechanism gives rise to the configurational diffusion. One deals here with the molecular flux regime, where $\lambda \ll \lambda_\mathrm{p}$. There exists a correlation between pore diameter, particle size, and diffusion coefficient.

This brief overview makes it clear that by simple variation of the pore geometry diffusive flow merges into a viscous one, or vice versa. In order to find out from observation of penetrant flow which regime dominates in a given case, one can examine the different parametric dependences. Subsequently we shall first treat the case of diffusion, and then that of viscous flow.

1.1 Diffusion in Polymers: Overview

Diffusional motion of penetrants occurs due to random motions of individual molecules. The driving force behind the transport process that involves sorption, diffusion and permeation is a concentration gradient of the penetrant in the polymer. Then a net transport of the penetrant occurs in the direction of decreasing concentration, which is described by Fick's first law. This transport is determined by the free volume in between the polymer chains and by their segmental mobility. The latter depends on the geometry and the length of the polymeric chains, i.e., on the molecular weight. As polymer molecular weight increases, the number of chain ends – which are sites for permeant molecules to be sorbed – decreases. The transport properties are also determined by the interactions of the polymer chains with each other, which depends on the extent of unsaturation, the degree of cross-linking, the degree of crystallinity and the nature of the substituents. Unsaturation in polymer chains provides for segmental mobility, whereas saturation tends to restrict it. Bulky functional groups in the side chains decrease the diffusivity more strongly than substitution of these groups in the polymer backbone.

The mobility can be further influenced by various factors such as the transient or permanent sorption at reactive sites, internal or external electric fields, pressure gradients, e.g., due to osmosis, capillarity, and by phase transitions from glassy to rubbery, or from sol to gel, with associated swelling. Furthermore, the glass transition temperature of polymers has a profound influence on transport properties, insofar as polymers with low glass transition temperatures possess greater segmental mobility and have higher diffusivity. This makes the description of diffusion in polymers much more complicated than in, e.g., metallic or semiconducting crystals. In fact this is why a multitude of polymeric diffusional models has been developed [3].

Phase transitions determine the physical state of a given polymer that influences the penetrant diffusivity. Below a certain temperature T_g, the *glass transition temperature*, polymers are said to be in the *glassy* state, and the phase above T_g is called the *rubbery* phase. In glassy polymers, the relaxation time upon structural changes is finite, so that stress buildup becomes an important effect and diffusion is slow. Glassy polymers are not in a state of true thermodynamic equilibrium. By contrast, in rubbery polymers the relaxation takes place nearly instantaneously, the diffusion is faster, and marked memory effects do not exist. Consequently, a change in temperature causes immediate adjustment of small penentrant molecules in rubbers to an equilibrium state, whereas the polymeric motion is not sufficiently rapid to completely homogenize the penetrant's environment in glassy polymers. Therefore penetrant molecules can thus potentially sit in holes or irregular cavities in glassy polymers with very different intrinsic diffusional mobilities.

Due to their structural alignment, the crystalline domains are more compact than the noncrystalline, *"amorphous"* regions. Consequently, the diffusion and solubility of penetrants in crystalline phases is usually negligible or at least substantially reduced, in comparison with amorphous polymers. Presumably, the ordered zones serve as tortuous impediments to diffusion. Hence, for the discussion of penetrant diffusion it is important to know how the structure of both pristine and irradiated polymers changes upon annealing.

Another point where polymeric diffusion differs from that in other media is the depth dependence of the diffusion coefficient. Whereas the motion of polymeric chains is strongly hindered in the bulk due to the presence of the other molecules, they move practically undisturbed when crawling along the polymeric surface. The surface being readily accessible for polymer chains up to a depth corresponding to their chain length, one observes an increase in the selfdiffusion coefficient above the bulk value from that depth on, up to the surface. This surface-vicinity effect leads, e.g., to changes in T_g (as determined by, e.g., DSC) when going from thick polymer foils to very thin ones [4].

The nature of the penetrant–polymer interaction determines the type of penetrant diffusion mechanism. It is drastically different for the glassy and the rubbery states. An important point in the description of penetrant diffusion in polymers is the degree of penetrant–polymer interaction. Small noncondensable gas molecules or water in hydrophobic polymers usually migrate without significant interaction between them and the polymer matrix. In these cases the solubility is low and approximately obeys Henry's law, i.e., the equilibrium concentration in the polymer is directly proportional to the external partial pressure of the penetrant. Here, the penetrant diffusion is almost independent of its concentration. However, the coupling of penetrant diffusion and host polymer dynamics increases with the size of the penetrant molecules. Arrhenius plots of large molecules are strongly nonlinear in the rubbery state [5]. A simple universal correlation between the diameter d of a penetrant molecule and its activation energy for diffusion ΔE does not exist; however, for small nearly interaction-free gas molecules often the relation $\Delta E \sim d^2$ holds well [5].

The penetrant–polymer interaction is much stronger for organic vapors and liquids as for water in hydrophilic polymers. As a consequence, very high sorption levels, strong plasticization effects and non-Fickian transport are observed. Even in the Fickian range the diffusivity is a strong, often approximately exponential function of the concentration. For penetrants such as water and methanol that can self-associate, clustering takes place at higher concentrations, which leads to concentration-dependent diffusion. Finally, even the motion of the most noble metal atoms is always governed by strong non-negligible interactions with the polymer host, and reactive metals often do not show any observable diffusion in polymers [5]. A lot of literature is available, e.g., on the influence of finite solute concentrations, swelling, plasticization, excess volume, structural relaxation, penetrant–polymer interaction, and other aspects of penetrant transport in unirradiated polymers. The reader is referred here to [5], and to the reviews recommended there. In this book, only those topics are mentioned that might be of importance in connection with diffusion in irradiated polymeric matter.

According to [3, 6, 7], diffusion in polymers can be classified into three categories:

1. *Case I, or Fickian diffusion.* Here the rate of diffusion is much smaller than that of relaxation due to mechanical, structural, etc., modes of the polymer–penetrant system. Sorption equilibrium is rapidly established, leading to time-independent boundary conditions that do not depend on swelling kinetics. For microscopic descriptions, two groups of models have been developed, the molecular and the free-volume theories.
2. *Case II, or non-Fickian diffusion.* Here the rate of diffusion is much faster than that of slow relaxation and rearrangement processes. Sorption processes are complicated and strongly dependent on swelling kinetics.
3. *Anomalous diffusion.* Here diffusion and relaxation rates are comparable. Penetrant sorption and transport are affected by the presence of pre-existing microvoids within the polymer matrix.

In general, penetrants in rubbery polymers (i.e., $T > T_\mathrm{g}$) diffuse exclusively according to a Case I model. By contrast, in glassy polymers (i.e., $T < T_\mathrm{g}$), all three above-mentioned diffusion cases can be found. The same system that shows regular mobility above T_g may show irregularities when T is decreased below T_g.

1.2 Case I Diffusion

1.2.1 Regular Fickian Diffusion

One-Dimensional Diffusion in an Infinite Medium

In this case, the penetrants are known to behave according to Fick's first and second laws [8, 9]:

$$J = -D\frac{\partial C_\mathrm{f}}{\partial x}, \tag{1.1a}$$

$$\frac{\partial C_\mathrm{f}}{\partial t} = \frac{\partial}{\partial x}\left(D\frac{\partial C_\mathrm{f}}{\partial x}\right). \tag{1.1b}$$

Here J is the flux, and C_f is the concentration of free solid, liquid or gaseous penetrant atoms or molecules that can diffuse in the material, and D is their diffusion coefficient. We restrict the discussion here to the one-dimensional case: $C_\mathrm{f} = C_\mathrm{f}(x,t)$, and we assume further homogeneity and isotropicity of the polymer. (If this is not given, (1.1a) would have to be replaced by a matrix equation, with D becoming a diffusion tensor. For details in this case see [6].)

Two major groups of models – molecular and free-volume models – have been developed for the description of small penetrant molecules in rubbery polymers. *Molecular models* (see [3] and [10] and the references cited therein, especially [11]) involve specific motions of the penetrating molecule and of the polymer chains relative to each other. It is assumed that microcavities of different sizes are continuously created and annihilated due to thermal random movements of polymer segments. The penetrants are assumed to occupy holes (total volume V_{FH}) in the polymer, in which they vibrate thermally, and occasionally are allowed to jump into a neighboring site.

The various molecular models differ essentially in the way the activation energies E of diffusion are related to the jump processes and the dynamics of the polymer chains. Special attention has been given to the way in which jump processes occur. For example, in the model of DiBenedetto [12] it was assumed that the van der Waals bonds between parallel polymer segments are transiently activated, i.e., partially broken up by a diffusing small penetrant molecule to allow its passage towards another position. Other models assume also that transient transport perpendicularly to the molecular chains is possible [11]. Frequently, proportionality or subproportionality is found between E and the square of the collision diameter σ of the penetrant. Consistency has been found between the theories and measurements for the penetration of gases, aromatic compounds, dyes, and organic solvents in various polymers. For self-diffusion of linear chains, it is assumed that the transport proceeds via *reptation*, i.e., that the single chains diffuse by crawling within a virtual tube defined by the topological constraints provided by the other molecules. As a result of its entanglement the diffusivity D of a polymer chain exhibits a strong nonlinear dependence on the number of friction-causing monomers (i.e. also on its own molecular weight) according to $D \sim N^{-\lambda}$ with $1 \leq \lambda \leq 2$. The exponent λ depends on the degree of entanglement and approaches the value of 2 for chains longer than a few hundred monomers. On the other hand, the diffusion coefficient of a molecule diffusing by this mechanism is independent of the matrix molecular weight [13]. By contrast, isolated polymer chains in good solvents arange themselves as coils and diffuse like rigid spheres of the coil radius in a manner consistent with the Stokes–Einstein equation.

Free-volume models (see [3] and [10] and the references cited therein, especially [14]) originate from statistical-mechanical considerations. They assume that, additionally to the free volume V_{FH} present at 0 K, more free volume $V_{\mathrm{FS}}(T)$ is created by the polymer's thermal expansion. The overall free volume is governed by parameters such as the temperature, pressure, strain, presence of already-incorporated molecules, or radiation damage. One of the major outcomes of free-volume theories is the possibility to compare mean cavity volumes and densities for different penetrants in different polymers. The free-volume theories are further capable of describing well the pressure and temperature dependence of diffusion and permeability coefficients. Please note that apart from V_{FH} and $V_{\mathrm{FS}}(T)$, there might exist additional voids in the polymer arising from water and/or solvent driven off during subsequent curing steps. In an example [15], the volume fraction of these latter voids was determined to be of the order of 7×10^{-4}.

The diffusion process in *glassy polymers* is more complex than that in rubbery polymers due to the nonuniformity of sizes of "frozen-in" microcavities in this case. For this, a *dual sorption model* (see [3] and [10], and the references cited therein) has been developed where two mechanisms take place during penetrant dissolution: the dissolution in pre-existing microcavities, and in those formed during dissolution by the penetrant equilibrium pressure (i.e., loosely speaking, the penetrants themselves create at least part of the cavities that they require).

Molecular-dynamics simulations indicate that the diffusion in glassy polymers is a hopping process, where even in the absence of strong attractive forces between penetrant and polymer the penetrant has to overcome an activation barrier to jump into a free-volume hole (defined by a local minimum in potential energy). On the time scale of the simulations the holes proved to fluctuate in size and shape, but did not move. Accordingly, D is described by the product of the jump frequency Γ, the square of the jump length a_0 that describes the distance between the penetrant's original position and the neighboring site, and the number Φ of available end positions of that jump, and f is a correlation factor:

$$D = f\,\Phi\,a_0^2\,\Gamma. \tag{1.2a}$$

In an idealized perfect and rigid lattice when penetrants can only move if they are small and energetic enough to fit in between the lattice elements (this is termed "interstitial diffusion" in crystalline media), then $f = 1$ holds. Larger impurities can move in a rigid lattice only if there is some free space (vacancies) available within which the penetrant can be accommodated. In this case, Γ is given by the product of the jump frequency v of the impurities in the given diffusion direction, the probability $w = \exp(-E_\mathrm{v}/kT)$ that there exists a vacancy at the place where the penetrant jumps to, and the probability that the jump was successful (i.e., that the penetrant does not jump back):

$$\Gamma = vw\,\exp(-E_\mathrm{m}/kT) = v\,\exp[-(E_\mathrm{v} + E_\mathrm{m})/kT]. \tag{1.3}$$

Equation (1.3) holds strictly only for self-diffusion; in the case of arbitrary penetrants there have to be included additionally also volume and valence effects of the penetrants [8]. However, for practical application, these considerations are frequently already sufficient to describe the diffusion coefficient by the so-called Arrhenius relation:

$$D = D_0\,\exp(-E/kT), \tag{1.2b}$$

with D_0 being a constant, which is temperature independent in a given interval $T_1 < T < T_2$. The relation (1.2b) implies the constancy of E with temperature. This is, however, not given, e.g., for simple gases or noble metals in polymers. Here E decreases with increasing temperature, and therefore the Arrhenius relation: $\{\ln(D)$ vs. $1/kT\}$ yields curved plots in this case, which become steeper at lower temperatures.

In any experiment, a concentration distribution $C(x)$ is not measured continuously but discretely at equidistant points x_i (channels). The measured value of $C(x_i)$ is therefore given by the convolution of the true value with a discrete observation function. This yields the average values

$$\overline{C(x_i)} = (\Delta x)^{-1} \int_a^b C(x)\,\mathrm{d}x, \tag{1.4a}$$

with $a = x_i - \Delta x/2$, $b = x_i + \Delta x/2$, and

$$C(x) = \sum_i K_i \partial(x - x_i) = \sum_i \overline{C(x_i)} \Delta x \partial(x - x_i). \tag{1.4b}$$

That is, all particles that in reality are distributed within the channel $a \leq x \leq b$, are now united at the position x_i. This necessary convolution procedure motivates the subsequent evaluation [16].

The general solution of (1.1b) for an infinite medium with initial concentration $C_\mathrm{f}(x, 0)$ is:

$$C(x,t) = D \int_{-\infty}^{+\infty} C(\xi, 0) G(x - \xi, \sigma)\,\mathrm{d}\xi, \tag{1.5a}$$

with the Poisson nucleus:

$$G(x - \xi, \sigma) = (2\pi\sigma)^{-1/2} \exp[-(x - \xi)^2/(2\sigma^2)]. \tag{1.5b}$$

For an initial concentration

$$C_\mathrm{f}(x, 0) = \sum_i K_i\, \delta(x - x_i) \tag{1.6}$$

it follows that

$$C(x,t) = \sum_i K_i\, G(x - x_i, \sigma), \tag{1.7a}$$

with $\sigma = (2Dt)^{1/2}$ describing the straggling of this distribution. It is related to the full width half maximum (FWHM) of the distribution via:

$$\text{FWHM } = 2\sigma \sqrt{-2 \ln(1/2)} = 2.354\sigma. \tag{1.8}$$

Equation (1.7a) can be written more simply as:

$$C(x,t) = (2\pi\sigma)^{-1/2} \exp[-(x - x_0)^2/(2\sigma^2 t)]. \tag{1.7b}$$

In this and the subsequent considerations the concentration distributions are always thought to be normalized. In the case of n subsequent diffusion steps the stragglings of the individual steps add geometrically: $\sigma^2_\mathrm{tot} = \sum_n \sigma^2_n$.

1.2.2 Some Special Cases and Peculiarities

Whereas many systems follow Fickian diffusion for $T \gg T_g$ (T_g = glass transition temperature), for $T \leq T_g$ in the same system occasionally anomalies show up. If in the rubbery phase near T_g the microcavities are in perfect equilibrium, one should expect that upon freezing below T_g this equilibrium is more or less maintained. In this case the Arrhenius curves should smoothly pass the glass transition temperature without any change. However, in non-equilibrium cases the thermal expansion coefficent might rapidly change at T_g, which then leads to discontinuities in the slopes of the Arrhenius plots at $T = T_g$ [14]. The exact difference in the activation energies in these cases depends on the size of the penetrants.

Organic Vapors

Strong polymer–penetrant interaction occurs with many organic penetrant molecules at $T > T_g$, and hence the diffusion coefficients are strongly concentration dependent. Therefore (1.1b) becomes [17]:

$$\frac{\partial C_f}{\partial t} = D\frac{\partial D(C_f)(\partial C_f/\partial x)}{\partial x}. \tag{1.9a}$$

As this equation usually cannot be solved analytically, another form is commonly used [17]:

$$\frac{\partial C_f}{\partial t} = D(C_f)\frac{\partial C_f^2}{\partial x^2} + \frac{\partial D(C_f)}{\partial x}\left(\frac{\partial C_f}{\partial x}\right)^2. \tag{1.9b}$$

An even simpler way is to approximate $D(C_f)$ by: $D = D_0(1 + k\,C_f)$ in Fick's equation, with D_0 and k being appropriate constants. This equation can be solved analytically, see [18]. By contrast, the diffusion coefficient of simple gases in polymers is independent of concentration. This results from the low solubility of these gases in polymers due to weak gas–polymer interaction [10].

Aqueous Solutions

In the case of water permeation into less hydrophilic polymers, the concentration dependence of the diffusion coefficient can be explained by the polar interaction of water molecules with themselves, which lets them cluster [10]. Clustering also takes place when ionic salts are dissolved in water, as the ions are then surrounded by a hydrate shell. The thickness of the shell of these polar water molecules depends on the salt concentration. Taking into account that both the interaction of water molecules with themselves as the interaction of water with ions leads to clustering, one arrives at the understanding that ionic penetrants will, in general, float alongside with the water, rather than being retarded. Only, e.g., applied chemical potentials may lead to the buildup of salt concentration gradients in the solution penetrating polymeric matter.

If an ionic compound is dissolved in water, diffusing anions and cations may have different drift velocities, thus eventually resulting in the buildup of internal local potential gradients. Unless an external voltage is applied, the latter will, however, in general, counteract the migration of the faster component, so that the overall charge equilibrium is approximately maintained. In fact, the measured transference numbers describing the differences in anionic and cationic mobilities for different polymeric systems reveal that usually, both mobilities differ from each other by less than a factor 2 [19].

Metal Deposition onto and Metal Implantation into Polymers in Vacuum

The cohesive energy of metals is typically two orders of magnitude higher than the cohesive energy of polymers, and the interaction between metals of low reactivity and polymers is generally weak in comparison to the strong metal–metal binding forces. Therefore metals of low reactivity, when attached to a polymer surface, do not wet untreated polymer surfaces and exhibit a Volmer–Weber growth mode, i.e., they form clusters on the polymer surface that finally coalesce at high metal coverages. The condensation coefficient C, defined as the fraction of those atoms of a metal currently impinging onto a polymer surface that is really adsorbed, decreases with temperature and depends sensitively on the polymer specie. Whereas C is 0.95 for condensation of Ag on PI at ambient temperature, it drops to ~ 0.4 at 400°C. For PC: $C = 0.12$, for PS: $C = 0.11$, and for Teflon: $C \approx 0.002$ at 20°C. C does not depend on the fluence of impinging metal atoms, but it roughly follows the trend of the surface energies of the polymers (PI, PC: $\sim 45\,\mathrm{mJ\,m^{-2}}$; Teflon: $15.7\,\mathrm{mJ\,m^{-2}}$). Metals of higher reactivity have condensation coefficients near unity [20].

Also the solubility of low-reactivity metals in polymers is generally very low so that equilibrium thermodynamics predicts almost no intermixing at metal/polymer interfaces. However, in contrast to these expectations low-reactivity metals such as Pt, Au, Ag, Cu, or Pd do diffuse into polymers and even form clusters. Noble-metal diffusion has even been found to occur at room temperature (e.g., for Ag in trimethylcyclohexane PC [21]). Only more reactive metals such as Cr and Ti appear to form relatively sharp and thermally stable interfaces exhibiting no significant intermixing in contact with polymers [5, 22].

At $T < T_g$ only a small fraction of noble metals diffuses within polymers by a uniform mechanism with high velocity, while the overwhelming majority migrates with strongly reduced mobility due to cluster formation. Noble-metal cluster formation takes place already at room temperature. For $T > T_g$, the metal mobility is greatly enhanced; it appears to be largely dominated by polymeric self-diffusion [5,22]. Thermally activated Cu is capable of diffusing deep into polymers such as PI. However, this diffusion is not a regular one but is governed by clustering, the clusters growing with increasing temperature. No metal clustering was ever observed for reactive metals, such as Cr and Ti in polymers.

For the diffusion of metal atoms from evaporated metallic layers on polymers into the polymeric bulk, the preparation conditions of the metal layer have been found to be of extreme importance. For chemically interaction-free (i.e., noble) metals evaporated continuously and at very low rates onto a polymer held at high temperature, there is a great chance to insert individual metal atoms into the polymer that subsequently show a long-range mobility and distribute in a large volume at low concentration. Under normal evaporation conditions the metals (and also Si and Ge) cluster on the polymer surface, i.e., are nonwetting. In exceptional cases (Se, Sn, In, and Ag in styrene-rich styrene hexamethacrylate copolymer), subsurface clusters are formed a few tens of nm below the polymer surface. If, by contrast, the metal deposition takes place rapidly (e.g., by flash evaporation), the metal atoms neither enter the polymer nor cluster on the polymer surface but just form a homogeneous film, even at elevated substrate temperatures [5, 22].

The Arrhenius plots of noble metals in polymers below T_g are characterized by convex curves that are very steep at low temperatures and flatten towards T_g, i.e., the activation energy for these metals decreases with increasing temperature. They can be described well by the Williams–Landel–Ferry equation [23]:

$$\ln\left(\frac{D(T)}{D(T_g)}\right) = \frac{A\xi(T - T_g)}{B + T - T_g}, \tag{1.10}$$

where ξ is a coupling parameter between the polymer and penetrant diffusion (full coupling signifies: $\xi = 1$). This means that free-volume models are adequate to describe the motion of dispersed noble metal atoms in polymers. The diffusion coefficients of Cu in PI are of the order of 10^{-18} to 10^{-14} cm^2 s^{-1} between 20 and 400°C, respectively; the corresponding values for Ag in PI range from 10^{-17} to 10^{-13} cm^2 s^{-1} between 300 and 500°C, respectively [5, 22].

By contrast, metal diffusion in BPA-PC shows a regular behavior with activation energies of 1.02 eV (Ag) and 1.13 eV (Au), respectively. These are much larger than the activation energies of light molecules of comparable size (N_2, CO_2, etc.) in PC, which are of the order of only 0.36–0.44 eV, as they essentially do not interact with each other [5, 22]. Obviously, even diffusion of noble metals is mainly controlled by metal–polymer interactions rather than by size and free-volume effects, and the noble-metal transport is accompanied by a strong local immobilization of the polymer chains near the metal atoms.

Whereas Au and Pd do not undergo any chemical reaction with polymers, for Ag and Cu pure physical and only weak chemical interactions have been considered. Ni is more reactive than Cu [5, 22]. Clear signs of strong chemical reactions were found for highly reactive metals, i.e., for Co, the transition metals Cr and Ti, the rare-earth element Ce, and Al, involving the formation of new compounds. Whereas in the case of the alkali metals K and Cs in PMDA-ODA, polymer reduction (with concomitant metal ionization) was shown to be the responsible interaction mechanism, such electron-transfer mechanisms could be ruled out for Cr and Ti. In this case the polyimide appears to be disrupted, and the formation of oxidic, nitridic and carbidic compounds is initiated. Nevertheless, these reactions do not necessarily imply that the metal atoms are strongly bound to the polymer chains.

For chemically reactive metals such as Cr or Ti, some mobility is observed immediately after they are brought into contact with the polymer [5,22]. The diffusion is governed by thermally activated bond breaking, which signifies a very low mobility. Diffusion in, e.g., PI comes to rest after migration of less than one nm, apparently due to strong localized chemical reactions with the polymer [5,22]. These metals act as efficient diffusion barriers [5]. The alkali metals K and Cs do not cluster but exhibit high single-ionic mobility in PI, once the latter has been reduced by them. Similarly, high diffusion has also often been found for Li ions implanted into polymers [24]. In spite of all these individual peculiarities, one might say as a rule of thumb that, the more noble a metal, the higher is its tendency to migrate and to cluster at a later stage.

Thermal annealing of ion-implantated Cu in PI at elevated fluences leads first (460–525 K) to a profile narrowing due to self-gettering of Cu (resulting in clustering), and thereafter (525–623 K) again to a profile broadening due to cluster migration. Strinkingly different values for the activation energy of migrating clusters have been found, ranging up to 1.9 eV; however, without knowledge of the individual cluster sizes these values are not very meaningful [5,22]. Isolated metal clusters in polymers are very stable against dissociation of atoms, and therefore do not show Ostwald ripening. Therefore, even very small particles remain next to very large ones.

There is a marked difference between the behavior of metals deposited onto polymers and of polymers deposited onto metals, concerning the aggregation of metallic precipitates and their diffusion away from the interface, due to the influence of the polymer's solvent in the latter case [25].

The above discussion essentially treated the behavior of metal atoms that were evaporated onto polymer surfaces in vacuum. Instead, one also can deposit metals on polymers from the liquid phase, by making use of galvanic or chemical deposition processes. As it seems that in these cases, precipitated metal atoms do not penetrate into the polymers but only show surface-bonding reactions, and as this case is more application oriented, it will be treated separately in Sects. 6.2.2 and 6.2.3.

Heat Diffusion

Please remember that Fick's second law not only holds for diffusing matter but also for diffusing heat. If, e.g., the heat dissipation after high-fluence ion impact is of interest, (1.1a) can be rewritten as [26]:

$$\frac{\partial T}{\partial t} = \frac{\partial}{\partial x}\left(\frac{\lambda}{c\rho}\frac{\partial T}{\partial x}\right), \tag{1.1c}$$

with c being the specific heat, λ being the heat conductivity, and ρ being the densisty of the material of interest. This implies that all mathematical solutions derived above can be overtaken directly. In connection with ion–polymer interaction, heat conductivity problems play important roles in the ion-track formation ("thermal-spike" model), and in the question of heat dissipation from irradiated samples to avoid pyrolytic effects by overheating.

1.2.3 One-Dimensional Diffusion in a Semi-infinite Medium

In the more realistic case of penetrants in a semi-infinite polymeric medium the solution of the diffusion equation depends on the boundary condition. Two special cases are most important: (a) $C(x=0,t) \approx 0$, which holds for volatile, and (b) $[dC(x,t)/dx]_{x=0} = 0$, which holds for nonvolatile penetrants in polymers. (In the case of polymers in a liquid, boundary condition (a) can be understood as the leaching of a soluble penetrant into the surrounding liquid, whereas the condition (b) describes an unsoluble penetrant.) In the first case where volatile penetrants can leave the sample through its surface, the original solution (1.7b) of a diffusing delta function in an infinite medium now turns into [23, 27, 28]:

$$C(x,t) = (2\pi\sigma)^{-1/2}\{\exp[-(x-x_0)^2/(2\sigma^2 t)] \\ - \exp[-(x+x_0)^2/(2\sigma^2 t)]\}, \tag{1.11a}$$

and for nonvolatile impurities:

$$C(x,t) = (2\pi\sigma)^{-1/2}\{\exp[-(x-x_0)^2/(2\sigma^2 t)] \\ + \exp[-(x+x_0)^2/(2\sigma^2 t)]\} \tag{1.11b}$$

holds. This implies that in the case of volatile impurities the residual total concentration in the sample becomes [28]:

$$C_{\text{tot}}(t) = \int_0^\infty C(x,t)\,\mathrm{d}t = \operatorname{erfc}[x_0/(2\sigma^2 t)], \tag{1.12}$$

whereas for diffusing nonvolatile impurities their total concentration does, of course, not change. The maximum concentration of volatile impurities decreases according to [28]:

$$C_{\max}(t) = (2\pi\sigma)^{-1/2}\{1 + \exp[-x_0^2/\sigma^2 t]\}, \tag{1.13a}$$

and that of nonvolatile impurities follows the relation:

$$C_{\max}(t) = (2\pi\sigma)^{-1/2}\{1 - \exp[-x_0^2/\sigma^2 t]\}. \tag{1.13b}$$

Of course, the given cases are somewhat idealized, and in reality one may encounter deviations. For example, it may occur that the surface is less permeable for a volatile penetrant, so that the surface boundary condition is better described by $0 \leq C(0,t) \leq C_{\text{regular}}(0,t)$, with $C_{\text{regular}}(0,t)$ being the concentration at $x = 0$ for an infinite medium. As a specific case, for $x(0,t) = aC_{\text{reg}}(0,t)$ with $0 < a = \text{const.} < 1$, (1.11a) is modified so that its second exponential term has to be multiplied by the factor $(1-a)$.

1.2.4 One-Dimensional Diffusion in a Thin Foil

In the case of penetrant diffusion from both sides into a thin foil with thickness d, it is convenient to look at the fractional penetrant uptake $M_\mathrm{t}/M_\mathrm{max}$ where M_t and M_max refer to the cumulative penetrant masses $M_\mathrm{t}/M_\mathrm{max}$ desorbed from the foil at times t and $t \to \infty$, respectively. In the general case [6]:

$$\frac{M_\mathrm{t}}{M_\infty} = 1 - \sum_{n=0}^{\infty} \frac{8}{n^2(2n+1)^2} \exp\left[- \frac{\pi^2(2n+1)^2}{4d^2} D\,t \right]. \tag{1.14}$$

A number of simplifications of (1.14) have been proposed, one of which is [30, 31]:

$$\frac{M_\mathrm{t}}{M_\infty} = 1 - \exp\left(\frac{2t}{\tau}\right)\left(\frac{2t}{\tau} + 1\right), \tag{1.15}$$

with τ being a characteristic time, after which 59.4% of the penetrant has been absorbed. With this notation, the diffusion coefficient can be derived from the experiment by the simple relation $D = d^2/(4\tau)$. In this way, diffusion coefficients of numerous penetrants in (pristine and irradiated) polymers were determined [31]. For a similar approach see [29].

Also, the reverse case, the desorption of matter from thin polymeric foils, can be described by the approach (1.12). This takes place, e.g., when absorbed penetrants are to be desorbed again, by putting the polmer foil into a solvent. Another case is given whenever incorporated volatile components degas from a polymer foil inside a vacuum system. An important application of this is the analysis of volatile reaction products after ion impact by means of mass spectrometry, to allow for the determination of the different possible radiochemical reactions taking place. Suppose that a polymer foil with area F is irradiated inside a vacuum vessel with volume V, to which a pump with pumping speed P [gas molecules/s] is attached. Then the cumulative desorbed gas mass M_t as given by (1.15) is proportional to the integral of the pressure increase inside the vacuum chamber:

$$M_\mathrm{t} = Nw = wPc \int_0^t p(t)\,\mathrm{d}t, \tag{1.16a}$$

with N being the total number of desorbed molecules, w being the weight of one molecule, p being the pressure in the vacuum system, and c being a proportionality factor, given by the ratio of the number of molecules per unit volume over the ambient pressure (i.e., 760 Torr). Hence D can be derived via integration of the curve $p(t)$ vs. t, making use of (1.15) and (1.16a). Interestingly, (1.16a) is independent of F. For very low gas-diffusion coefficients, hence very low desorption rates M_t, one can switch off the pump to enhance the sensitivity of the measurement. In this case,

$$M_\mathrm{t} = Nw = wFVc \int_0^t p(t)\,\mathrm{d}t. \tag{1.16b}$$

The desorption occurs first from the polymer surface and then, progressively, from the whole polymer with thickness d. Assuming the validity of Fick's law, the time delay τ before a stable flow is reached, can be used to estimate the penetrant's diffusion coefficient via: $D = (d/2)^2/\tau$.

1.2.5 Permeation through a Membrane

Another frequently occurring case is that of the diffusive permeation through a thin foil with thickness d from one side to the other. In this case, the foils are frequently referred to as membranes. A membrane enables separation of two phases, or of two constituents in the same phase. It allows or limits very specifically the transfer of different chemical species from one medium to another [32]. The penetrant-transfer process through homogeneous membranes, or permeation, may be described as follows: the penetrant is absorbed on the membrane, then is solubilized in it, then diffuses in the polymer, and finally is released on the opposite interface. Generally, it is assumed that the solubilization follows Henry's law and the diffusion process obeys Fick's first and second laws. (Henry's law, $C_0 = S\,p_0$, signifies that the volume concentration C_0 in the foil surface region of the permeating specie (e.g., a gas) is proportional to the pressure difference p_0 to the neighboring phase (e.g., the gas), where the proportionality factor S is denoted as the solubility coefficient.) Other, more complex models exist to describe these phenomena [33].

For the permeation through a membrane, the correlation of the total amount of penetrant $Q(t) = \int_0^t J(t')\,\mathrm{d}t'$ permeated after the time t with the diffusion coefficient allows the determination of the latter. For this one plots $Q(t)$ vs. time t. The slope $(\mathrm{d}Q/\mathrm{d}t)_{t\to\infty}$ of this plot yields the steady-state penetrant flux J through the foil.

The product of D and S is called the permeation coefficient or permeability: $P = DS$. All the three quantities P, D, and S follow the Arrhenius relation (i.e., proportionality with $\exp(-E_{P,D,S}/kT)$, with different activation energies $E_{P,D,S}$ [32]. In the stationary state (i.e., under equilibrium conditions) the diffusion flow is constant and the diffusion coefficient is independent of concentration. Then the permeant flow J through the membrane is $J = Pp_0/d$ [34]. Experimentally, the permeability can be derived from a quasistationary state of gas flow through the membrane by: $P = (p_2/p_1)S_\mathrm{v}d/F$, where p_1 and p_2 are the pressures on both sides of the foil of area F, and S_v is the pumping speed of the pump that maintains the pressure difference [35]. The permeation rate increases for gases that induce swelling and plasticization. The lower the density or the higher the free volume, the higher is the permeability. Cross-linking, crystallinity and orientation decrease P, and therefore only amorphous thermolastic polymers are useful for gas separation. Furthermore, the transport of gases through glassy polymers differs from that through rubbery polymers due to an additional mode of sorption that is available in glassy polymers, which is the filling of Langmuir-type microvoids or regions of localized free volume present in the glassy state.

The diffusion coefficient is usually derived from the time lag τ between the onset of permeation and arrival of the penetrant at the other membrane side after which the flux has reached about 62% of its maximum value [36]: $D = d^2/(6\tau)$. If a number of different permeation curves are obtained with the same foil but for varying concentrations C_0, then the concentration-dependent diffusion coefficient follows from $D(C_0) = (\mathrm{d}J/\mathrm{d}C)d$ [10]. For a discussion of closer details of the different approaches to determine D, see [34].

For the transport of a material in a solvent through a membrane within a time t, the physicochemical model of Danesi et al. [37] gives the correlation:

$$\ln(C_t/C_0) = -(F\varepsilon/V)P_{\mathrm{c}}t, \tag{1.17}$$

where C_0 and C_t are the concentrations of the material in the solvent at times 0 and t, respectively, F is the membrane area, ε the membrane porosity, V the permeating volume, and P_{c} the permeability coefficient [$\mathrm{cm\,s^{-1}}$]. Furthermore, the flux J of matter through an etched track of length l is related to a so-called partition coefficient K of a penetrating molecule between the feed solution and the membrane via:

$$J = -D\,K\,C_{\mathrm{feed}}/l, \tag{1.18}$$

with C_{feed} being the concentration of the permeant molecule in the feed solution.

This relation is important as it helps to explain the occasionally anti-intuitive findings for the transport of solvents through structures in etched tracks. For example, toluene in aqueous feed penetrates better through Au tubules with chemisorbed hydrophobe thiole layers within etched tracks, the longer the chain length of the thiol applied. Taking into account that the partition coefficient K of toluene (and therefore also J) between the solution and the membrane increases with the thiol chain length, this effect becomes understandable, as the toluene preferentially migrates through the thiol layer [38]. In general, it is the interplay between the partitioning and diffusional terms in (1.18) that determines the migration through narrow pores.

Elastomers such as polyolefine-based microporous membranes are usually relatively permeable to gases and hydrophobic molecules. Hydrogels are permeable to water and water-soluble molecules.

A case relevant for practical application in environmental protection is the diffusion of the radioactive gas radon through membranes. Here, a different quantity, the permeation coefficient P [$\mathrm{cm^2\,s^{-1}}$] describing this specific transport process was introduced by:

$$P = \frac{\lambda\,V_2\,d\,A^{\infty}_{v_2}}{F(A_{v_1} - A^{\infty}_{v_2})}, \tag{1.19}$$

with d being the membrane thickness, V_1 being the upstream volume, V_2 the downstream volume, A_{v1} and A_{v2} being the radon activity concentrations in the compartments upstream and downstream of the membrane, respectively, and λ being the decay constant of the radioactive nuclide [39]. For unperturbed foils, one should expect P to be constant. However, it was found that for irradiated PVC foils P decreases with time, i.e., that the foils become more impermeable due to cross-linking.

In fact, ion-beam modification of membranes enables one to tailor not only their permeability but also the selectivity. Escoubes et al. [40] report that irradiation of polyimide membranes with 2-MeV α particles leads to an increase in free volume, hence facilitating the permeation of larger molecules, whereas irradiation with 170-keV N^+ ions essentially prevents the permeation of larger molecules due to cross-linking-induced compaction (i.e., free-volume reduction). This means that different membrane permeability/selectivity ratios can be tailored by the choice of different projectiles – in this case light low-energetic ions.

The opposite behavior can also be achieved, namely that polymer foils act as efficient diffusion barriers against diffusion of large molecules. This can be realized by ion-irradiation-induced cross-linking that is known to reduce the free volume to near zero, and this inhibits the permeability.

In the case of a porous material of thickness l where a gas penetrates via Knudsen diffusion (e.g., a polymer with etched tracks), the gas flux J [$\mathrm{mol\,s^{-1}}$] is related to the pore radius r [cm] via:

$$J = 4/3(2\pi/\mathrm{MRT})^{1/2}(r^3\Delta P/l), \tag{1.20}$$

with ΔP being the pressure difference [$\mathrm{dynes\ cm^{-2}}$], M being the molecular weight of the gas, R being the gas constant [$\mathrm{erg\,K^{-1}\,mol^{-1}}$], l the track length [cm], and T being the absolute temperature [K] [41].

One should keep in mind that when considering the permeation through objects like microporous foils, the total flux of matter through them consists of the sum of both the flux through the pores and through the matrix itself, which in the case of polymers is frequently non-negligible [42].

1.2.6 Case I Diffusion with Drift

In the case that a drift is overlaid the regular diffusion – e.g., in the case of ionic penetrants in an external electric field – then the basic diffusion equation (1.1b) becomes:

$$\frac{\partial C_f}{\partial t} = \frac{\partial}{\partial x}\left(D\frac{\partial C_\mathrm{f}}{\partial x}\right) - v\frac{\partial C_\mathrm{f}}{\partial x} \tag{1.21}$$

for the simplest case that the drift velocity v is constant and independent of x. According to Smoluchowski [43] the solution of this second-order differential equation can be obtained by the substitution:

$$C(x,t) = M(x)\exp\left(\frac{vx}{2D} - \frac{v^2t}{4D}\right), \tag{1.22}$$

which yields a differential equation of the same shape as (1.1b):

$$\frac{\partial M}{\partial t} = \frac{\partial}{\partial x}\left(D\frac{\partial M}{\partial x}\right), \tag{1.23}$$

and consequently can be solved in the same way as described above. For the general case of a semi-infinite medium without sorption, hence:

$$\begin{aligned} C(x,t) = \frac{1}{2\sqrt{\pi Dt}}\int_0^{\infty} \mathrm{d}x' C(x',0)\Bigg\{ &\exp\left[\frac{-(x'-x_0+vt)^2}{4Dt}\right] \\ &\pm \exp\left[-\frac{(x'+x_0+vt)^2 - 4xvt}{4Dt}\right]\Bigg\}, \end{aligned} \tag{1.24}$$

where the + sign in the second term signifies diffusion of nonvolatile penetrants without sorption, and the − sign signifies either diffusion of volatile penetrants, or diffusion of nonvolatile penetrants with the surface being a sink. In the more general case, one has to add to the drift term in (1.21) terms for the gradients of stress, pressure, gravity, and in the case of penetrant flow through pores the interaction of the electric field with fixed charges on the pore walls [44].

1.2.7 Electrolytes in Polymers

This case is closely related with that of embedded salts that form complexes with polymers, thus giving rise to strong ionic conductivity – comparable in magnitude to those of semiconductors. The diffusion coefficient D for a mobile species can be related here to the conductivity σ via the Nernst–Einstein equation,

$$\frac{\sigma}{D} = \frac{nq^2}{kT}, \tag{1.25}$$

with q = charge per carrier, n = number of charge carriers, T = absolute temperature, and k = Boltzmann's constant. The conductivity follows an Arrhenius relation of the form

$$\sigma = \frac{A}{T}\exp\left(-\frac{E_\mathrm{a}}{kT}\right). \tag{1.26}$$

The mobility u_i of an ion is given by: $u_\mathrm{i} = \sigma/(nq)$ [$\mathrm{cm^2\,V^{-1}\,s^{-1}}$]. The diffusion coefficient can also be related to the viscosity η of the polymeric medium through which the ions pass via the Stokes–Einstein relation:

$$D = \frac{kT}{6\pi\eta r}, \tag{1.27}$$

where r is the radius of the diffusion path (assumed to be cylindrical). As usual, various models – especially free-volume models and excess-entropy models – are used to describe D.

In polymer electrolytes, current transport is usually initiated by creating a potential gradient via electrical (i.e., voltage applied) or chemical (i.e., concentration gradient applied) means, thus giving rise to a force that leads to the drift. An important parameter in this respect is the so-called transference number t_i of a given charge specie i of totally m species. It is defined as the ratio of the mobility of a carrier i, divided by the sum of mobilities. Hence

$$t_i = \frac{\sigma_i}{\sum_{j=1}^{m} \sigma_i} . \tag{1.28}$$

One can also define a transference number t_+ for the whole polymeric system under consideration of $t_+ = D_+/(D_+ + D_-)$, where D_+ and D_- are the diffusion coefficients of the cations and the anions, respectively. Thus, for balanced-out charge transport $D_+ \approx D_-$, hence $t_+ = 0.5$. In fact, the majority of all polymer electrolytes show transference numbers in the range 0.2 to 0.6. However, there also exist systems with t_+ as high as 1 (e.g., $PESc_nLiClO_4$) ($n = 8$–50), poly(β-propiolacetone)$_n$$LiClO_4$, and PEG_4Napolystyrenesulfonate), or t_+ smaller than 0.02 (cross-linked triblock-copolymer PEO/$PP_{10}LiClO_4$, and $PEO_{16}MgCl_2$) [19]. The latter cases are explained by the lowering of the cation mobility due to heavy solvation in these cases.

Surprisingly, though polymer electrolytes have a wide-ranging potential for commercial applications in batteries, photoelectrochemical cells, sensors for moisture and hydrogen, electrochromic displays and even electrochemical transistors [19], to our knowledge there have not yet been performed any systematic studies on the transport of the above-mentioned electrolytes in ion-irradiated polymers.

1.2.8 Case I Diffusion with Sorption

Chemically active sites such as, e.g., radicals may impede the regular penetrant mobility, and hence act as traps. From the energetic point of view, these traps are places of especially deep potential depressions. There is a certain probability A to encounter such traps, and another probability B to escape from these traps again. Several different types of traps may be present in a sample simultaneously. For $B = 0$ the trapping sites bond the penetrant permanently, e.g., via irreversible chemical reactions. The case $0 < B < 1$ usually describes a reversible first-order trapping process (transient trapping). Traps can be *saturable*, i.e., bond only a characteristic number (usually 1) of mobile impurities, and thereafter vanish. It may also occur that traps are *unsaturable*, which means that the accumulation of an infinite number of impurities is possible, with unchanged capture cross-section. Finally, one might encounter the case that the capture cross-section of unsaturable traps increases with the number of trapped penetrants, which means that the continuous action of these traps leads to rapid penetrant precipitation, i.e., to the emergence of a new phase. Such traps were denoted as *"nucleation traps"* by Chadderton [46]. For simplicity, in the subsequent considerations these nucleation traps are included in the category of unsaturable ones. To account for the existence of traps, the diffusion equation (1.1a) is expanded by inclusion of both a source and a sink term [7]:

$$\frac{\partial C_{\mathrm{f}}}{\partial t} = \frac{\partial}{\partial x}\left(D\frac{\partial C_{\mathrm{f}}}{\partial x}\right) - \sum_i \frac{\partial C_{\mathrm{t}i}}{\partial t} + \sum_i \frac{\partial C_{\mathrm{r}i}}{\partial t}. \tag{1.29a}$$

In this equation, C_{f} is the concentration of mobile impuritites that did not encounter traps, $C_{\mathrm{t}i}$ is the concentration of impurities immobilized by trapping centers of type i, and $C_{\mathrm{r}i}$ is the concentration of impurities released from trapping centers of type i. The totally mobile fraction is hence $C_{\mathrm{mob}} = C_{\mathrm{f}} + \sum_i C_{\mathrm{r}i}$, and the total penetrant concentration C_{total} is equal to $C_{\mathrm{mob}}+\sum_i C_{\mathrm{t}i}$.

Equation (1.29a) can be rewritten in the form of a system of rate equations to describe the overall dynamics of migration, trapping, and detrapping [47–50]:

$$\frac{\partial C_{\mathrm{f}}}{\partial t} = \frac{\partial}{\partial x}\left(D\frac{\partial C_{\mathrm{f}}}{\partial x}\right) - A_i C_{\mathrm{f}} + B_i C_{\mathrm{t}i}, \tag{1.30a}$$

$$\frac{\partial C_{\mathrm{t}i}}{\partial t} = A_i C_{\mathrm{f}} - B_i C_{\mathrm{t}i}, \tag{1.30b}$$

where A_i and B_i are the rates of trapping and detrapping, respectively. These coefficients can be interpreted as the rates in the following reactions:

$$F + X_i \xrightarrow{A_i} T_i \quad \text{and} \quad T_i \xrightarrow{B_i} F + X_i, \tag{1.31}$$

where F represents a free penetrant, X_i a nonoccupied i-type trapping center and T_i the formed penetrant–trap complex. According to Ham [51], assuming a distribution of immobile spherical trapping centers of radius r_i, one has:

$$A_i = 4\pi\, r_i\, D\, C_{xi} \tag{1.32}$$

with C_{xi} being the concentration of active trapping centers. For an *unsaturable* trapping process special assumptions regarding C_{xi} are not necessary. In particular, this is valid if the concentration of these centers is very large compared with the solute concentration. However, if these two concentrations are of the same order of magnitude the variations in C_{xi} due to the formation of solute–trap complexes must be taken into account. In this case a *saturable* trapping process takes place that can be described, to a first order, by:

$$\frac{\partial C_{x_i}}{\partial t} = -\frac{\partial C_{ti}}{\partial t}. \tag{1.33}$$

Finally, assuming that the decomposition of the solute–trap complexes of the i-th type occurs by a thermally activated mechanism, the detrapping rate can be calculated by:

$$B_i = B_{0i} \exp\left(\frac{-E_i}{kT}\right), \tag{1.34}$$

where B_{0i} is a pre-exponential factor and E_i is the binding energy to the trap.

Though a general analytic solution of the system of rate equations (1.30a) and (1.30b) is not possible, one can solve analytically at least a number of specific cases. The reader is referred, e.g., to [16,48–50]. Let us mention here only a few peculiarities:

(1) Consider a capped but otherwise undisturbed polymer with $A(x = 0) = 1$, $A(x > 0) = 0$, and $B(x) = 0$. This means that whatever migrating penetrant touches the surface is immediately trapped there. Mathematically seen, the depth distributions in this case are identical with those after degassing of volatile matter from polymers, where the penetrant is lost forever once it has passed the surface. Therefore the same solution as in (1.11a) holds here too, however, with the difference that at $x = 0$ the trapped impurities pile up:

$$C(x = 0, t) = 1 - \int_{x=\partial x;\partial x\to 0}^{\infty} C(x, t)\,\mathrm{d}t = 1 - \mathrm{erfc}[x_0/(2\sigma^2 t)]. \tag{1.35}$$

(2) In the case of a homogeneous population of unsaturable traps, the depth-profile shapes of penetrants entering through the surface becomes exponential, and with $C(x, t) \sim \exp(-At)$, and A being a constant [16].

(3) Given a system where Case I diffusion with shallow saturable traps holds, and where the diffusion coefficient distribution $D(x)$ may be reciprocal to the penetrant's original concentration distribution $C(x, t_0)$, the trapping efficiency distribution $A(x)$ may be proportional to $C(x, t_0)$, the detrapping probability $B(x)$ may be regarded to be nonzero and depth independent, no unsaturable traps may be present, and where no penetrant loss through the surface occurs. Then it can be shown [52] that $C(x, t) = C(x, t_0)$, i.e., that the profile is stationary for whatever initial concentration distribution and internal mobility. Such a case is given for polymeric ionic conductors where the charge carriers (i.e., the ions) are loosely bonded to graft centers along the polymeric chains. Even in the case of inhomogeneous original charge carrier distributions $C(x, t_0)$ – reflecting inhomogeneous graft-center distributions – the ionic distributions do not change with time, as the graft centers are identical with the shallow traps, and as the ionic mobility is reciprocal to the average distance of the graft centers from each other. This implies that the smaller the graft center concentration, the higher the ionic mobility. Consequently, the more readily the ions will leave this graft-center-depleted region again, so that it remains permanently depleted of ions.

a

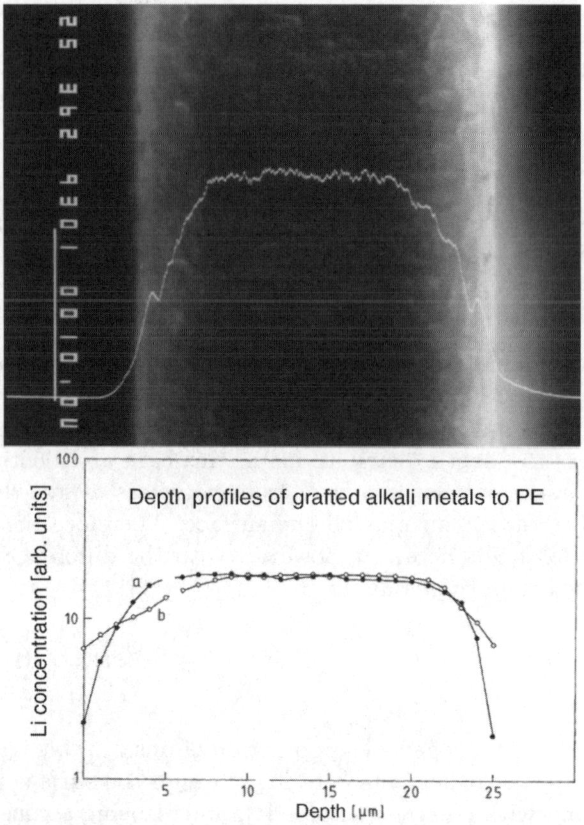

b

Fig. 1.1. (**a**) Cross-section through a 25-µm thick ionically conducting methacrylic acid-grafted and K^+-doped PE foil, SEM image. Also shown: K^+ depth distribution, as determined by AES. (**b**) *Open dots:* Li^+ distribution in the same foil, after replacing the K^+ by Li^+. NDP measurement. *Filled dots:* The K^+ distribution, for the sake of comparison. Remeasurements after arbitrary times yield the same results. It is seen that the stationary depth profiles are nonuniform, due to the production peculiarities of these foils [52]

As a realistic example of cases (1) and (3), see the ionically conducting polymers that emerge when grafting PE with methacrylic acid and subsequently doping them with some alkali ions (Li^+, Na^+, K^+) [52]. Figure 1.1 – here, nonuniform depth distribution of the conducting ions exist due to the production process of these foils. Though the alkali ions are highly mobile (as verified by the sample's conductivity), the alkali-ion depth profile is stationary and does not change with time. (A precondition for this stationarity are well-sealing polymeric surfaces. Once they are destroyed – e.g., by ion irradiation – corrosion effects set on, see Chap. 2.)

1.2.9 Chemical Reaction Kinetics

In any chemical reaction new matter is created at the expense of pre-existing ones. The dynamics of the corresponding decomposition reaction (and hence the nucleation of the new material) can be analyzed by the time dependence of the fractional decomposition $\alpha(t)$. Under isothermal conditions the rate constant k can be described by a function $g(\alpha)$ that depends on the reaction-controlling mechanism, the size and shape of the reacting particles [53]: $kt = g(\alpha)$. A plot of $g(\alpha)$ vs. t gives a straight line if the correct form of the function $g(\alpha)$ is found. Some possible functions are listed in Table 1.1. One can distinguish between diffusion-controlled reactions (symbolized by D_i), solid-state reactions that follow the first-order kinetics (F_1 function), phase-boundary-controlled reactions (symbolized by R_i), and random nucleation reactions (symbolized by A_i).

Table 1.1. Equations relating α and t, of possible chemical decomposition reactions. Diffusion-controlled reactions: D_1: function for a one-dimensional diffusion process governed by a parabolic law in a diffusion-controlled decomposition reaction; D_2: function for a two-dimensional diffusion-controlled process into a cylinder; D_3: Jander's equation for a diffusion-controlled reaction in a sphere; D_4: function for a diffusion-controlled reaction starting on the exterior of a spherical particle. Phase-boundary-controlled reactions: R_2: function for a circular disc, reacting from the edge inward; R_3: function for a sphere, reacting from the surface inward. A_1 to A_3: functions for various random nucleation processes

Mechanism that governs the reaction kinetics	Scaling of $g(\alpha)$	Fct.
One-dimensional diffusion	α^3	D_1
Two-dimensional diffusion	$(1-\alpha)\ln(1-\alpha)+\alpha$	D_2
Jander equation, three-dimensional diffusion	$[1-(1-\alpha)^{1/3}]^2$	D_3
Ginstling–Brounshtein equation, 3-dim. diffusion	$(1-2\alpha/3)-(1-\alpha)^{2/3}$	D_4
Two-dimensional phase-boundary reaction	$1-(1-\alpha)^{1/2}$	R_1
Three-dimensional phase-boundary reaction	$1-\ln(1-\alpha)^{1/3}$	R_2
First-order kinetic	$\ln(1-\alpha)$	F_1
Prout–Thompkins branching nuclei	$\ln[\alpha/(1-\alpha)]$	A_1
Random nucleation: Avrami equation	$-[\ln(1-\alpha)]^{1/2}$	A_2
Random nucleation: Erofe'ev equation	$-[\ln(1-\alpha)]^{1/3}$	A_3

If the solid-state reaction follows first-order kinetics, then the rate-determining step is the nucleation process, and there is an equal probability of nucleation at each active site. In phase-boundary-controlled reactions, the reaction is controlled by the movement of an interface at constant velocity and nucleation occurs virtually instantaneously. Usually it is assumed that the nucleation step occurs instantaneously, so that the surface of each particle is covered with a layer of the product. The nucleation of the reactant, however, may be a random process, and not followed by rapid surface growth. With the nuclei growing larger and larger, they will touch each other at a certain time so that their growth ceases. This process has been considered by Avrami and Erofe'ev [54], who have introduced the functions A_2 and A_3, respectively, for this purpose.

A general disadvantage of such a *"dynamic thermal analysis"* approach is that frequently more than one function $g(\alpha)$ fits the experimental results, and accordingly different kinetic data can be obtained. In these cases the selection of the proper model can be facilitated by combining isothermal and nonisothermal methods [55].

Many studies have been undertaken on the decomposition kinetics of various materials (for a recent one, see, e.g., [55]), however, until now no such study was ever undertaken for ion-irradiated polmyers. This is regarded as a deficiency, and it is recommended to fill this gap as the more detailed knowledge of such decomposition reactions would give deeper insight into the kinetics of the material's structural changes.

1.2.10 Numerical Solution of Diffusive Problems

For the practical application of the rate (1.30a) and (1.30b) in the general case, it is most convenient to use the approach of the finite-difference technique [56]. Here the equations are transformed into a system of algebraic equations:

$$\begin{aligned} C_{\mathrm{f}}(x,t+\delta t) = C_{\mathrm{f}}(x,t) &+ \frac{\delta t \delta x^2}{4}[D(x+\delta x) - D(x-\delta x)] \\ &+ \frac{D(x)\delta t}{\delta x^2}[C_{\mathrm{f}}(x+\delta x,t) - C_{\mathrm{f}}(x-\delta x,t)] \\ &- \sum_i A_i C_{\mathrm{f}}(x,t)\delta t + \sum_i B_i C_{\mathrm{t}i}(x,t)\delta t \end{aligned} \tag{1.36a}$$

$$C_{\mathrm{t}i}(x,t+\delta t) = C_{\mathrm{t}i}(x,t) + \sum_i A_i C_{\mathrm{f}}(x,t)\delta t - \sum_i B_i C_{\mathrm{t}i}(x,t)\delta t, \tag{1.36b}$$

where δt and δx are the time and the depth increment, respectively. Therefore, knowing the initial profiles at $t = t_0$, it is possible to calculate the concentration profiles at any later time $t = t_0 + N\delta t$ (N being an integer) by means of (1.28), where the coefficients D, A_i and B_i are chosen to get the best fit to the experimental results. In this way one can obtain the time evolution of the concentration profiles C_{f}, $C_{\mathrm{t}i}$, C_{xi} and C_{total}, using a very simple iterative numerical algorithm, which has been implemented in a computer code described elsewhere [57].

This code "DIFF", adapted to the diffusional problems in polymers, has been applied successfully many times, see, e.g., in [58]. It enables one to model the changes of a given penetrant distributions $C(x)$ within a polymer foil of arbitrary thickness in an iterative way, in dependence of arbitrarily selected depth distributions of the diffusion coefficient $D(x)$ and the absorption and desorption probabilities $A_{\mathrm{sat}}(x)$, $A_{\mathrm{unsat}}(x)$ and $B_{\mathrm{sat}}(x)$, $B_{\mathrm{unsat}}(x)$ in saturable and unsaturable traps, respectively. The penetrant can be chosen to be volatile or nonvolatile, and the two foil surfaces can either be open or act as sinks with predetermined sink strengths.

In the calculation, $D(x)$ is separated into a depth-independent term D_{c} and a depth-dependent function $f(x) : D(x) = D_{\mathrm{c}} f(x)$. Due to the nature of the underlying Fickian diffusion, a diffusional-profile broadening by, let us say, a factor n requires n^2 iterative steps in the calculation. This code is applied for the given tasks stepwise by trial-and-error until the optimum diffusion coefficient distribution is reached. Reasonable input conditions are chosen, and the resulting diffusional depth profiles are calculated. The comparison of these results with the true experimental findings leads to the decision of how to best modify the input conditions for the next simulation step. This trial-and-error procedure is usually finished when the overall agreement between measurement and calculation has become better than about 5%. From the number of iteration cycles n to obtain the best fit, and the channel width W of the given spectra, one can derive the diffusion coefficient D via: $Dt = nqW^2$, with $q = 1/6$.

An inherent danger of such complex simulation programs is that they are not a priori unambiguous. That is, different suitable combinations of input assumptions and parameters might describe the same experimental result. In order to decide with certainty which of the possible sets of input data describes reality, one has to perform a number of simulations for experiments subject to different corresponding critical parameters (e.g., to different temperatures, times, distributions of diffusion coefficients, trapping probabilities, etc.). Only this method allows one to discard inadequate models. For an example, see [59].

1.2.11 Clustering Phenomena

Suppose penetrants (atoms, ions, molecules) are abundant in a polymer at elevated concentration, then the penetrant–penetrant interaction is no longer negligible if $E_{\mathrm{mm}} > E_{\mathrm{pm}}$, i.e., if the penetrant is only poorly (or not at all) wetting, and if the metal–polymer chemical interaction is weak. (The wetting may be temperature dependent. For example, it has been observed that, e.g., for Cu in PI, unwetting sets in spontaneously at 170° C [60].) Here, E_{mm} is the cohesive energy of the penetrants to each other, and E_{pm} is the polymer–penetrant interface energy. As a result, clustering of the penetrant particles may occur. In fact, this is frequently observed for metals in polymers, such as, e.g., Ni, Cu, or Ag in PI [60].

Clustering of penetrants signifies the emergence of a new phase. The mathematical formalism of the growth and kinetics of nuclei for penetrant preipitation can be directly taken from the classical work of Smoluchowski [61] who established in 1918 a system of coupled rate equations to understand the coagulation of colloids in a liquid. Here each of the equations described the mobility of colloids of a certain size, and all equations were correlated with each other by the coagulation process. In our case we have now to replace colloids by penetrants, the coagulation process by the penetrant precipitation, and the final coagulation products by our penetrant clusters. Clearly, as the cluster diffusion coefficients correlate reciprocally with the cluster sizes, the larger clusters become largely immobilized. This explains why in the few experiments available [62] the originally quite fast metal diffusion into the polymers decreases rapidly with time.

The theory of nucleation of a new phase inside another one is quite well developed. Suppose we deal here with nonfractal precipitates with well-defined surfaces. Then the new phase is stable if the formation of new interface area requires an energy $\Delta E = \gamma \Delta A > 0$, with γ being the interface energy per unit area and ΔA being the increase in interface area. For a curved interface, e.g., of a spherical precipitate with radius r, the condition: $\Delta p = 2\gamma/r$ holds, with Δp being the pressure difference between the interior and the exterior of the precipitate [63,64]. Equilibrium between evaporation of matter from the precipitate into the polymer (i.e., redissolution), and condensation of penetrant from the polymer towards the precipitate (i.e., cluster growth) is given if

$$p(r) = p_0 \exp\left(\frac{2\,\gamma\,v_{\mathrm{m}}}{r\,k\,T}\right), \tag{1.37}$$

with v_{m} being the volume per penetrant particle and $p_0 = A\exp(-E_0/kT)$ being the penetrant's saturation pressure in the polymer (E_0 = potential barrier height to be overcome by the penetrant). This implies that penetrants above a critical radius $r_{\mathrm{c}}(T)$:

$$r_{\mathrm{c}} = \frac{2\,\gamma\,v_m}{k\,T}\,\frac{1}{\ln(p/p_0)} \tag{1.38}$$

and below some critical temperature T_{c} increase due to dominant condensation of penetrants, at the expense of precipitates below r_{c} that shrink due to dominant evaporation (Ostwald ripening). {Recently, it has been shown by Heinig et al. [65] that the inverse case also exists in which small clusters grow at the expense of large ones until size equilibrium is obtained. (This is one possible way to obtain self-organizing patterns.) This process can be described by the Gibbs–Thomson relation, by admitting negative surface tensions. This inverse Ostwald ripening holds, however, only for a well-defined window of the condensation/evaporation ratio; excessive evaporation would not leave the number of clusters constant but create new ones. Corresponding experiments for clusters in polymers have not yet been done but should also be feasible.}

If metal atoms diffuse from an evaporated layer into a polymer, the metal concentration distribution has its maximum at the interface, with the metal in greater depths being highly diluted. Consequently, the interface zone is the region of strongest cluster growth and of largest cluster sizes. Due to the above-derived stability criterion, the smaller clusters should shrink and finally vanish, whereas the larger near-surface clusters should show predominant growth at the expense of the smaller ones. In fact, clusters up to 150 nm diameter have been observed for, e.g., Cu in PI [60]. Of course, this cluster distribution is superimposed by the low-concentration one of dispersed, deeply diffusing metal atoms. The development of the evolution of metallic cluster distributions in polymers hence might lead to a net recession of the metal distribution towards the surface, i.e., it should eventually give the impression of negative effective diffusion coefficients, if the corresponding experiments were evaluated in terms of simple Fickian diffusion kinetics. In fact, the first RBS test experiments on the penetration of noble metals evaporated onto common polymers supported this expectation,[1] however, a theoretical elaboration of the above-sketched mechanism and systematic experiments are still missing. However, there have been performed at least the first Monte-Carlo simulations in the field of metal diffusion and metal agglomeration to clusters [66, 67].

An attempt has also been made to diffuse metal atoms (Cu) in polymers (PI) that had been implanted beforehand by atomic knock-on processes from a sputter-deposited layer [60]. In this case it was found that the final discontinuous metal-cluster distribution had its peak concentration at depths larger than the range of the knock-on metal atoms, which clearly indicates that the metal atoms had undergone some long-range diffusion before becoming immobilized by clustering. Again, a theory accompanying these observations is missing.

It should be noted at this point that there exist cases where metallic clusters produced on polymeric surfaces by evaporation eventually sink completely into the matrix. This occurs when the surface energy of the metal exceeds the sum of polymer surface energy and metal/polymer interfacial energy, thus establishing a driving force for the cluster to enter the bulk [68]. Eventually, this process also influences metal diffusion in polymers insofar as the filling of voids with metals would reduce the surface energy of the voids.

Bi- and multilayer structures comprising polymers and metals are used in various devices, e.g., as interconnections in electronic packages. Therefore, the interfacial characteristics, especially the diffusion of metals into polymers are of importance for the conservation of the properties of such structures under adverse conditions. On the other hand, the processes of diffusion and clustering of metals on polymers are of fundamental interest since they may lead to the creation of new materials. Up to now reliable data on metal diffusion in polymers are scarce and the diffusion mechanism is not yet well understood. Anyhow, the diffusion of metallic clusters in polymers does not follow the expected Stoke–Einstein relation, due to bridging interaction between the particles. This means that the transient network formed by polymeric

[1] Kaschny J, Fink D, Behar M, unpublished work (1998)

bridges slows down the particle mobility dramatically (in the example of Au in poly(t-butyl acrylate), PTBA, by a factor 10^3), and also the mobility of the polymer chains themselves is reduced by the metallic particles due to (in the case of Au) weak interaction [69].

1.2.12 Three-Dimensional Case I Diffusion

There are cases in polymer irradiation where the theory of one-dimensional diffusion is not sufficient to describe the observations. This is given whenever one deals with 3-dimensionally structured polymers, e.g., after producing an irradiation pattern through a mask, where lateral diffusional processes might also be of interest. In principle the expansion of (1.29a) to three dimensions is straightforward [26]:

$$\begin{aligned}\frac{\partial C_{\mathrm{f}}}{\partial t} &= \frac{\partial}{\partial x}\left(D\frac{\partial C_{\mathrm{f}}}{\partial x}\right) + \frac{\partial}{\partial y}\left(D\frac{\partial C_{\mathrm{f}}}{\partial y}\right) + \frac{\partial}{\partial z}\left(D\frac{\partial C_{\mathrm{f}}}{\partial z}\right) \\ &\quad - \sum_i \frac{\partial C_{\mathrm{t}i}(x,y,z)}{\partial t} + \sum_i \frac{\partial C_{\mathrm{r}i}(x,y,z)}{\partial t}.\end{aligned} \tag{1.29b}$$

For the solution please consult the corresponding books on diffusion. To our knowledge, such a case was not yet treated either for pristine or for irradiated polymers. Available codes that deal with the diffusive spreading of implantation profiles of keV implants in semiconductors in three dimensions [70] could, in principle, be readily applied to polymers.

Also, problems related with radial transport of matter in ion tracks have to be mentioned here. For example, one might think of using tracks as linearly extended defects to trap, e.g., diffusing metal atoms for the production of nanowires. In this and similar cases cylindrical diffusion problems have to be solved, with the basic equation being:

$$\frac{\partial C_{\mathrm{f}}}{\partial t} = \frac{1}{r}\frac{\partial}{\partial r}\left(D\frac{\partial C_{\mathrm{f}}}{\partial r}\right) - \sum_i \frac{\partial C_{\mathrm{t}i}(r)}{\partial t} + \sum_i \frac{\partial C_{\mathrm{r}i}(r)}{\partial t}. \tag{1.39}$$

Here, r is the radial distance from the track center. For the solution of this equation, please consult [6].

Technical polymers frequently incorporate various additives that, of course, influence the penetrant diffusion. These cases have also been treated in great length. For details, please consult [10].

1.3 Case II and Anomalous Diffusion

These cases frequently deal with systems where an external liquid penetrates into a solid. For aqueous penetration this signifies a restriction to hydrophilic solids. Our present knowledge in this area is descriptive to a large extent [3].

Theories that have been developed to interpret non-Fickian anomalities are still incomplete. The first step in all these cases is the transport of the penetrant through the liquid/solid interface. This problem, which involves the passage through an adsorbed liquid layer will be described in detail in the subsequent chapters of this volume, especially in connection with the penetration of liquids into latent ion tracks.

1.3.1 Swelling

Swelling is an effect that is important specifically for cross-linked matter and therefore for many ion-irradiated polymers in solvents. The swelling ability of a polymer is related to its water permeability and uptake, i.e., to its hydrophilicity. When a cross-linked polymer is placed in a solvent it swells, the extent of which depends on the cross-link density. The solvent swelling with time t leads to macroscopic polymer's dimensional changes $\Delta T/T_0 = T_{\mathrm{rel}}$ of a film with original thickness T_0 and final thickness T_∞, which are related to the solvent's diffusion coefficient D via [71]):

$$D = \pi \left(\frac{T_0}{2} \frac{k}{4 T_\infty} \right)^2 , \tag{1.40}$$

and

$$k = 100 \frac{\mathrm{d}\, T_{\mathrm{rel}}}{\mathrm{d}\sqrt{t}} . \tag{1.41}$$

A hygroelasticity coefficient μ can be defined via: $\mu = T_{\mathrm{rel}}\, W_{\mathrm{rel}}$ with $W_{\mathrm{rel}} = \Delta W/W_0$ being the relative weight gain.

Swelling equilibrium is attained when the dilution force acompanying the swelling equals the contraction force of the rubber-like elasticity. The swelling is believed to be a phenomenon similar to osmosis, and it can therefore be expressed as a function of osmotic pressure P_{os} and the volume fraction C_{s} of the solvent: $\mathrm{d}C/\mathrm{d}t = f(P_{\mathrm{os}}, C_{\mathrm{s}})$. Frequently, the correlation is more complicated than a simple exponential that Thomas and Windle [72, 73] and Hui et al. [74] proposed earlier.

The swelling of a polymer network can be described by a balance between penetrant/polmyer mixing and elastic free energy: $\Delta G_{\mathrm{swel}} = \Delta G_{\mathrm{mix}} + \Delta G_{\mathrm{el}}$ where the ΔG_i are Gibb's free energies of the individual processes. Its main driving force is the entropy of mixing. The most important parameters influencing the swelling appear to be [3]: a) the concentration C_x of the polymer in the solution, b) the molar mass M_{c} between the cross-links, c) the molar mass M of the polymer, and d) the volume fraction Φ of the polymer in the swollen state. However, up to the present, in spite of many models, no unified approach exists that accounts for all the swelling phenomena.

1.3.2 Case II Diffusion Models

Case II diffusion describes systems where phase transitions proceed from the surface of a polymer to its inside. Specifically, phase transitions from the glassy to the rubbery state are triggered by solvents with strong interaction with the polymer. The shape of Case II diffusing penetrant profiles is characterized by a sharp phase-transition front moving into the glassy polymer that is preceded by some Fickian-controlled diffusion zone (which advances with $\sqrt{t}$), and behind which the penetrant concentration is constant inside a rubbery polymeric phase, Fig. 1.2. The front velocity is zero until a critical concentration C_c is reached, and it then increases according to some power of the additional concentration [75]. It uses to assume constant speed rapidly.

The rate-controlling step at the advancing penetrant front is the speed of scaling-up of the available free volume for the penetrant, i.e., of the dynamics of the glassy polymer to turn rubbery. It may either be determined by the formation of crazes in response to the solvent osmotic stress [76], or it may be the time-dependent mechanical viscous resistance of the polmyer in response to the increase in volume and change of shape, i.e., to thermodynamic swelling (osmotic) stress [72, 73]. The first explanation appears to be limited to some appropriate systems only. Based on the latter assumption, Thomas and Windle calculated in their fundamental work [72] the osmotic pressure and the response of a glassy polymeric network thereupon, which is the fractional swelling.

It turns out that the polymer first takes up penetrant only very slowly but after a certain time the penetrant's molecular volume is rapidly accelerating towards equilibrium sorption. Parallel to this, the osmotic pressure rapidly vanishes in the swollen regime. For a reference frame moving with the Case II front, the conditions ahead of the front approach steady state. For steady-

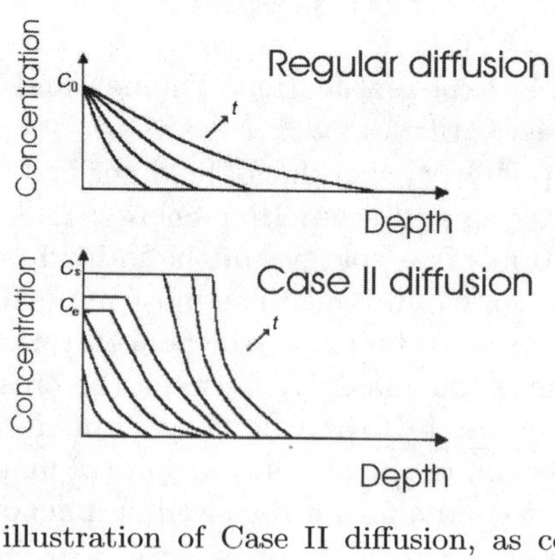

Fig. 1.2. Schematic illustration of Case II diffusion, as compared with regular (Case I) diffusion

state diffusion, described by a diffusion coefficient ahead of the boundary moving with constant velocity v, Fick's second law can be rewritten [77]:

$$\frac{\partial C}{\partial t} = \frac{\partial}{\partial x}\left(D\frac{\partial C}{\partial x} + v\,C\right) \tag{1.42}$$

(for a more general expression and discussion, see [78]). In this case, x describes now the distance ahead of the moving front. The solution for $x > 0$, if D is not a function of the solvent concentration, is $C = \Phi_\mathrm{c} \exp(-cx/D)$, with Φ_c being a proportionality factor.

Inserting the correlation between the osmotic pressure P_os at the front with the actual penetrant molecular volume Ω and its equilibrium volume fraction C_eq, one obtains an equation for the speed v of the penetration front [79]:

$$v = \left[\left(\frac{D}{C_\mathrm{c}}\right)\frac{\mathrm{d}C}{\mathrm{d}t}\bigg|_{C_\mathrm{c}}\right]^{1/2}, \tag{1.43}$$

with C_c being the critical penetrant concentration for Case II diffusion to set in.

The Case II diffusion equation (1.42) can be solved numerically via an iterative mechanism using the finite difference method [56] as decribed in Sect. 1.2.10. For this, the simulation code "DIFF" described in that paragraph was also expanded. This has been done by allowing for a stepwise increase of the maximum possible near-surface concentration at each iteration cycle with the speed v_sorpt, starting from an initial maximum possible penetrant uptake concentration C_glas characteristic of the glassy phase, and ending at another maximum possible penetrant uptake concentration C_rub, which stands for the final swollen rubbery phase. In the simplest case, the sorption speed v_sorpt is taken to be constant. It is known from experiment, however, that this is frequently not the case, and therefore an increase of v_sorpt with penetrant concentration can be optionally included. Furthermore, the glassy/rubbery phase front is allowed to migrate into the bulk with another preset velocity v_front, once a characteristic minimum penetrant concentration C_gel has been reached. Though in many cases C_rub and C_gel are identical, this is not required in the general case.

The observed sharp discontinuities in the penetrant concentration profiles can occur below as well as above T_g. The model of Case II diffusion of Thomas and Windle allows one to predict quantitatively the kinetics of Case II diffusion, including the magnitude of the velocity v of the sharp diffusion front [77]. Gall et al. [77] showed that Case II diffusion processes may also occur when polymers are uptaking organic solvent vapors. Both the velocity v of the penetrant front and the diffusion coefficient D of the penetrant ahead of the front decrease exponentially with the penetrant molecule's size d: $D = D_0 \exp(-\delta d)$, with D_0 and δ being proportionality factors. For linear molecules, δ is about 20% smaller than for spherical ones due to geometrically higher penetration capability, which results in three orders of magnitude higher D values [77].

Case II diffusion has been treated also for cases with other than the simple one-dimensional geometry. For example, in [80] it is solved for cylindrical annular films.

1.4 Penetration along Pores

1.4.1 Diffusional Penetration along Latent Tracks

Let us describe for simplicity latent ion tracks in a polymer as cylindrical regions in which penetrants diffuse preferentially along the track axis, then the penetration depth is expected to be independent of the track radius R and track direction θ (to the surface normal), for whatever diffusional case. The total penetrant uptake should increase with the intersectional area of tracks with the surface (for an individual track: $\pi R^2/\cos\theta$) – hence should scale with R^2, $1/\cos\theta$, and also with $\sqrt{t}$ in the case of regular diffusion, or with $\sim \exp(-At)$ in the case of presence of unsaturable traps – provided that the penetrant diffusion in the neighboring pristine matter is negligible.

1.4.2 Penetration along Etched Tracks

For the mathematical treatment of the diffusion coefficient D of idealized molecules with mass m that do not interact with each other, within a micro-capillary of inner radius R_{tot}, Grivtsova et al. [81] found:

$$D = \sqrt{kT/m}(R_{\text{tot}} - b), \tag{1.44}$$

with b being the range of the surface potential that determines the adsorption on the walls. To the accuracy of a numerical factor of 1.06 this expression coincides with the Knudsen diffusion coefficient. For the description of the geometrical diffusion paths along tracks, pipe-diffusion models are also adequate [82]. It is well known that the Rayleigh number is a good parameter to describe the liquid motion through pipes.

The case of diffusive motion of a penetrant into a pore that absorbs the penetrant on its walls is of special importance for catalytic processes, and it is also applicable for the chemical deposition of matter in etched tracks (see Sect. 2.2). The normalized depth distribution of the penetrant in a pore with radius r_{p} and length L that has still escaped absorption is given by

$$\eta = C(x)/C_{\text{s}} = \cosh[\Xi(1 - x/L)]/\cosh(\Xi), \tag{1.45}$$

where C_{s} is the initial penetrant concentration in the solution, $C(x)$ is the concentration of the mobile penetrant fraction along the pore at axial position x, η is a so-called effectiveness factor, $\Xi = L[2r_{\text{s}}/(r_{\text{p}}DC_{\text{s}})]^{1/2}$ is the so-called Thiele modulus [2], D is the penetrant diffusion coefficient that is assumed to be constant, and r_s is the absorption speed per unit pore surface [mole/area time]. This absorption may be correlated with some reaction, e.g., a catalytic process or a heterogeneous precipitation. For reactions of m-th order, $r_s = kC^m$, with k being the rate constant. By comparison of (1.45) with experimentally determined depth distributions of matter deposited in tracks, one can derive the penetrant's diffusion coefficient within the tracks.

1.4.3 Capillaric Penetration along Tracks

When the radius of a pore (e.g., a track) is sufficiently large, capillaric-transport mechanisms will dominate the penetration of liquids. For liquid motion in capillaries, the Rayleigh number becomes very small due to its proportionality with the fourth power of the radius, which signifies that convection does not play any role in this case.

As the penetration of a liquid through tracks sensitively depends on the track's internal structure one has to distinguish between several cases [83]:

- The *"Open-Cylinder" Model*: tracks are thought to be empty open cylinders with length $\approx$ ion range (for thick foils) or length $=$ foil thickness (for thin foils). This definitely holds for etched swift heavy-ion tracks, and also for latent tracks in cellulose nitrate or in very thin polyimide foils, according to recent TEM examinations [84].
- The *"Sponge" Model*: the track region is thought to be filled with sponge-like debris of radiochemical reaction products, carbonaceous clusters and voids, through which the penetrant can pass along the full track length. This is thought to be the model that best describes latent tracks in many polymers.
- The *"Open-Pore" Model*: the track is thought to be a porous medium, the near-surface part of which being directly accessible via open pores, and the remaining part of it being enriched with closed pores that are accessible only via additional diffusional steps.

The Open-Cylinder Model. For a theoretical treatment, the assumptions are made that the ion tracks are cylinders with radius R_{tot}, tilted by an angle Θ against the surface normal, and being embedded in polymer foils of thickness d. Furthermore it is assumed that the penetrating liquid is wetting (with σ being the surface tension), that the liquid penetrates only from one side of the foil into the track so that the embedded gas can readily escape, and that the total height of liquid penetration (in the case it proceeds upward) is small enough that gravity does not yet play any role.

Two components contribute to the viscous flow of a liquid into a capillary, the bulk flow and the surface flow. In the first case, the penetrant moves parallel to the pore wall at a radial distance $0 \le r_{\text{penetr}} < R_{\text{eff}}$, and in the second case, at a distance $R_{\text{eff}} \le r_{\text{penetr}} \le R_{\text{tot}}$ from the track center. Here, the correlation between an "effective" track radius R_{eff} and the true track radius $R_{\rm tot}$ is given by the relation $R_{\rm eff} = R_{\rm tot} - b$, with b describing the distance of interaction between the liquid's molecules and the track walls. This distance is estimated to be of the order of 1 to 10 nm, depending on the system under consideration. Usually the penetrant–wall interaction leads to the transient or permanent penetrant's sorption, so that for larger pores the surface flow can be neglected. The latter dominates, however, in the case of very narrow pores. In the following considerations we shall neglect this effect, which is thereafter treated in a separate chapter.

The driving force K for liquid penetration is $K = 2\sigma\pi R_{\rm tot}/\cos(\Theta)$, as $2\pi R_{\rm tot}/\cos(\Theta)$ is the length of the track zone intersecting the polymer surface. This force is also equal to $\Delta p\pi(R_{\rm tot}/\cos(\Theta))^2$, with Δp the pressure difference between filled and empty capillaric regions. Hence:

$$\Delta p = 2\,\sigma\cos(\Theta)/R_{\rm tot}. \tag{1.46}$$

When the solution penetrates the capillary, Hagen–Poiseuille's law is applied:

$$V = \pi\,R_{\rm eff}^4 \Delta\,p\,t/(8\,\eta\,x). \tag{1.47}$$

Inserting now (1.47) into (1.46) and rewriting the total penetrating volume per unit time V/t as:

$$V/t = \pi\,R_{\rm tot}^2\,u(t)\,/\cos(\Theta) \tag{1.48}$$

(with $u(t) = {\rm d}x/{\rm d}t$ being the liquid's penetration speed through the capillary), one obtains a homogeneous differential equation with the solution

$$x(t) = \sqrt{[\sigma\,R_{\rm eff}^4 \cos^2(\Theta)\,t/(2\,\eta\,R_{\rm tot}^3)]}. \tag{1.49}$$

The open-cylinder model then yields for the liquid-volume uptake proportionality with $x(t)$, which signifies that the larger the track diameter the more liquid penetrates the deeper, in contrast to the common experience that small capillaries promote liquid penetration more than large ones. The reason for this finding is the additional presence of viscosity in the calculations, which leads to scaling with R^2, whereas capillarity scales only reciprocally with $\sqrt{R}$. It also signifies that the larger the tilt angle Θ the shorter the liquid's penetration depth per time, and the less is the overall liquid uptake. Finally, it signifies that high surface tension and low viscosity promote the penetration.

A comparison with diffusional penetration along the tracks (see Sect. 1.4.1 of this volume) shows that the penetration depth and the total uptake scale in the same way with time (i.e., $\sim \sqrt{t}$), but differently with R and Θ. Consequently, the observation of the time dependence does not give any clue to decide between the diffusional and capillaric models, but the observation of the dependence on track radius and tilt angle does. This possibility has indeed been applied recently to characterize different irradiated polymers concerning their penetrability. For example, it was found that hydrated ionic penetrants such as Li^+ ions diffuse into latent tracks in polyimide, but migrate into irradiated silicone rubber via capillarity.

If, for the open-cylinder model, enclosed gas along the tracks cannot escape, the compression of the gas signifies an increase of its pressure p_{gas}, thus reducing the liquid's penetration speed and depth. For equilibrium, the maximum liquid penetration depth is

$$x = d/2 - [kT/(8\pi\sigma\cos\Theta)][(R_{\mathrm{tot}} - b)/R_{\mathrm{tot}}^2]. \tag{1.50}$$

Due to the extremely high capillaric pressure in narrow tracks, the second term of (1.50) becomes very small, i.e., that the remaining gas bubbles are of negligible size (some $10^{-7}\%$ of the original size for track radii in the nm range) so that the penetrant fills up the track nearly perfectly. Gas under such a high pressure is expected to dissolve in the liquid completely, so that one may conclude that the presence of gas in porous latent, or small etched ion tracks should not play any role in the penetration of liquids.

Penetration of an aqueous liquid into hydrophilic material signifies water uptake, with the material's volume expansion (i.e., swelling) being the main consequence. Swelling of a bulk polymer volume element enclosed by rigid walls can only lead to reduction of the available free intrinsic volume – including the one along the ion tracks – with increasing exposure time. As the penetrants start their migration at the surface, the latter will be affected most by swelling. Taking into account that the flow through a track scales with R_{tot}^3, one understands that even a minor reduction of the track radius by swelling leads to a major reduction of the flow-through volume. In other words, the onset of swelling will rapidly seal the track, which has two important consequences: (a) The total penetrant volume uptake will come to saturation. (b) Due to the proceeding compression of the near-surface track region towards greater depths with increasing exposure time, the penetrant in the near-surface track region will rapidly be squeezed towards greater depths, thus extending x strongly. Due to missing delivery of more liquid, x will thereafter come to saturation – frequently filling up already the whole track at that moment. For a simple estimation of the order of magnitude of this effect, neglecting the depth dependence of swelling along the track, one gets as an upper limit

$$R_{\mathrm{tot}} = R_{\mathrm{tot}}[(1-\xi) + \xi\exp(-St)], \tag{1.51}$$

with S being the speed of track swelling and ξ being the maximum possible reduction of the track radius by swelling. For small R_{tot} (as given in the case of latent tracks), $\xi = 1$, and for large R_{tot} (e.g., for etched tracks), $\xi < 1$. Hence swelling is less influential the larger the track radius, or the larger the track's tilt angle. Using this estimation for whatever swelling speed, one obtains a dramatic growth of penetrant depths in the moment when swelling becomes marked, stemming from the rapid squeezing of the near-surface penetrant towards greater depths.

The Sponge Model. There are experimental indications that the bulk zones of latent tracks in polymers might best be described as zones of intermittently distributed excess free volume and clusters of carbon-enriched radiochemical reaction products that had been identified on a number of occasions [85]. In other words, these tracks might have a sponge-like fine structure. To describe this case in a simple macroscopic way, one can replace the penetrant's real viscosity by an effective viscosity η_{eff}, which contains implicitly the effect of all the obstacles (radiochemical debris, clusters) in the penetrant's pathway. Using this approach, one can use all the formulae derived for the *open-cylinder model*, and one immediately sees that the same parametric dependence holds, though the penetration will, of course, now be considerably retarded. Swelling will be more efficient in the sponge model than in the cylinder model, as the liquid's penetration will be hindered now not only by the recession of the track walls, but also by the expanding obstacles in the penetrant's pathway. As both track formation and the liquid's penetration are statistical processes, the overall penetrant distribution will be smeared out due to statistically varying boundary conditions of capillaric penetration. This contrasts with the diffusion model, where the distribution results from the (statistical) penetration process itself.

The Open-Pore Model. The experimentally observed finite liquid's penetration depths in irradiated polymers could also signify that latent tracks act as open (i.e., accessible) pores just up to a certain depth and not beyond. In this case, the measured penetrant distributions in their initial stage would just reflect the statistical variation of the open-pore lengths. Further liquid penetration would then require the much slower diffusional processes. Though one cannot exclude this model a priori, we note that it would yield a different dependence on penetration parameters (such as the track radius and direction) than was experimentally observed: In such a case, both the total penetrant uptake C and penetration depth x should be rather independent of the track radius, x should scale with $\cos(\Theta)$ and C with $1/\cos(\Theta)$, and the penetration should come to an abrupt end once the whole accessible pores have been filled out. This *"open-pore model"* was never verified by observations for latent tracks.

Conical Pores

In the above consideration, both latent and etched tracks were thought to be cylindrical in geometry. This is, however, given only in those few cases where the track etching speed exceeds the bulk etching speed by far. In most other cases, ion-track etching leads to conical etched-track shapes. Therefore it is of importance to treat the flow of matter through tracks of this shape, too. This has in fact been done by Beriozkin et al. [86], by integrating the Hagen–Poiseuille law over the whole length of the cones. The result of their calculation was an equation that correlates a track diameter D_{HD} that is characteristic of hydrodynamic flow with the penetrant's viscosity η, the penetrant flow Q, the number n of pores per cm^2, the whole membrane area S_m, and the applied transmembrane pressure P:

$$D_{HD} = (128\,\eta\,l\,Q/\pi\,n\,S_m\,P)^{1/4}. \tag{1.52}$$

1.5 Interfaces

Polymer/Polymer Interfaces

A key question is the degree to which the long polymer chains of two different adjacent polymers are tangled up with each other and bond to each other, as this determines the strength and toughness of the interface. Most pairs of homopolymers display a low entropy of mixing, and they retain abrupt interfaces, with little chemical attachment and consequently weak adhesion. If two polymers are miscible to some degree, a little thermal treatment will lead to interdiffusion and entanglements of the polymer chains across the interface. By this means, those chains constitute links to bind the interface, often quite strongly.

However, even two polymers that are completely soluble in each other – so that the bulk composition is absolutely uniform – will have a different composition at their interface. The reason is their difference in surface energies. For example, olefines have a lower surface energy than polymers containing carbonyl or other polar groups, and so the polymer with the lower surface energy will enrich on the interface to minimize the total free surface energy. It is the balance of three factors that determines both the amount of enrichment and the thickness of the enriched layer: (1) the intermolecular cohesive force (which is proportional to the surface energy), (2) the gain in entropy of mixing (which counteracts the surface enrichment), and (3) the tendency to maximum disorder, i.e., to a broad interface region (as this increases the system's entropy, hence lowers the surface energy) [87]. Interface (and also surface) enrichment is much more important for polymers than for any other material, as the great lengths of the polymer chains opposes intimate mixing. The enriched layer is always of the order of the random walk size of a polymer chain, i.e., around 1 to 10 nm. It is clear that this surface-segregation effect should be influenced by ion irradiation, essentially due to the change in points (1) and (2), and due to formation of radicals.

Both temperature and composition are parameters with which one can control the balance between mixing and phase separation of two polymer phases. In the phase-separation regime, domain growth can set in. Surfaces and interfaces have the effect of leading the domain growth towards a coherent composition wave perpendicular to the surface ("surface-directed spinodal decomposition") [88]. In this way, sharply defined polymer multilayer structures can be produced by a one-step coating process under nonequilibrium conditions. Again, ion irradiation could be used for the fine tuning of these structures. Such a study has, however, not yet been performed.

For bonding immiscible polymers A and B, it is sometimes possible to create an appropriate diblock A–C of copolymer to be used as a joining layer, such that the new constituent C is miscible with both A and B, resulting in the final structure {A/C/B}. As one might expect, the adhesion energy for such a bond will increase with the surface density of copolymer linkages. Large molecular weight molecules occupy too much interface per area per link. A low molecular weight of polymers is clearly preferable, down to the point when the chain length starts to become too small to provide strong entanglement in the host [89].

The main parameters controlling the kinetics of {polymer A/polymer B} interdiffusion is the *mutual diffusion coefficient* D_m that describes the transfer of monomers A to the polymer B, and of momers B to the polymer A. In the simplest case, both diffusion coefficients D_m are equal. On the other hand, if one monitors one labelled molecule A migrating within the polymer A, what is seen is the self-diffusion. For unirradiated polymers, the self diffusion can usually be assumed to be Fickian, and it is described by a self-diffusion coefficient D_s. This case has been described above. For more details of the basic physics of interdiffusion of pristine polymer/polymer layers, see [89]. Examinations on the behavior of interdiffusion of ion-irradiated polymer/polymer systems are still largely missing. It is known, however, that stacks of polymers readily weld with each other during ion irradiation so that they cannot be separated from each other after the irradiation without destroying them. There is some similarity with the better-examined metal–polymer interdiffusion.

Metal/Polymer Interfaces

A thin metal film with thickness d deposited on a rigid substrate (e.g., a polymer held at low temperature) will, in the general case, be subject to tensile stress S, resulting in the storage of elastic energy U [88]:

$$U = \frac{S^2 t(1-v)}{E}, \tag{1.53}$$

where E is Young's modulus and ν the Poisson number. Above a critical energy, the film will lose its stability and detach from the substrate spontaneously, if

$$S^2 d = \frac{W_{\mathrm{ad}} E}{2(1-v^2)}, \tag{1.54}$$

with W_{ad} being the work of adhesion. Typically, a refractory metal film with $S \sim 10^{9-10}$ dyne/cm^2 may reach this threshold at a thickness of a few μm [45]. For polymers held at ambient or elevated temperatures the maximum tolerable film thickness will be even larger, due to their strongly reduced rigidity at that temperature.

Ion irradiation has long since proven to be a good tool for the improvement of the adhesion of any two layers to each other. In the case of low-energy ion beams, the improvement of adhesion is essentially due to the ballistic intermixing of the interface, as a consequence of nuclear energy transfer from the projectile ions to the target atoms. Eventually, this intimate intermixing may give rise to chemical reactions between the two bonding partners.

The surfaces of polymers prepared from the solution generally present a "terminated" structure of tangled chains, with saturated bondings. Chemical metal–polymer bonding, therefore, usually demands the disruption of polymer chains to expose active sites. The chemical bonding is always accompanied by the formation of metal–carbon complex bonds. Interface bonding can also often be enhanced by the addition of chemically active species, e.g., the use of O_2 in conjunction with IBAD (ion-beam-assisted deposition), or implantation of Cr at the interfaces [88].

Both ion-induced chain scissions and cross-linking make irreversible changes in the polymer and its elastic properties. Scission ultimately degrades the polymer itself and the irradiated layer fails; alternatively, if cross-linking dominates, adhesion failure occurs between the strong cross-linked polymer regions and the underlying unirradiated polymer.

Presputtering of polymers such as PTFE, PES, etc., by low-energy ions may lead to deep surface roughness that may assist metal layer adhesion. The thus-achieved increase of effective interface area enables one to obtain a higher density of surface activation without problems of bulk damage [88].

References

1. Perlmutter DD, Scrosati B, Diffusion processes in the doping of polymer films. Solid State Ion **27**, 421–123 (1988)
2. Cunningham RE, Williams RJJ, Diffusion in Gases and Porous Media. Plenum Press, New York (1980)
3. Aminabhavi TM, Aithal US, An overview of the theoretical models used to predict transport of small molecules through polymer membranes. J Macromol Sci Rev Macromol Chem Phys **C28**, 421–474 (1988), and numerous references therein
4. Dalnoki-Veress K, Forrest JA, de Gennes PG, Glass transition reductions in thin freely-standing polymer films. J de Phys IV (Proc) **10**, 221–226 (2000)
5. Faupel F, Willecke R, Thran A, Diffusion of metals in polymers. Mater Sci Eng **R22**, 1–55 (1998)
6. Crank J, Mathematics of Diffusion, 2nd edn. Clarendon Press, Oxford (1975)
7. Frisch HL, Sorption and transport in glassy polymers – a review. Polym Eng Sci **20**, 2–13 (1980)
8. Shewmon PG, Diffusion in Solids. McGraw Hill, New York (1963)
9. Boltaks BI, Diffusion in Semiconductors. HJ Goldsmith (ed.), Infosearch, London (1963)
10. Frisch HC, Stern SA, CRC Cri Rev Solid State Mater Sci **11**, 123–187 (1983), and numerous references therein
11. For a description of the molecular diffusion model, see e.g.: Pace RJ, Datyner A, Model of sorption of simple molecules in polymers. J Polym Sci Polym Phys **18**, 1103–1124 (1980)

12. DiBenedetto AT, Molecular properties of amorphous high polymers. I A cell theory for amorphous high polymers, and II. An interpretation of gaseous diffusion though polymers. J Polym Sci **A1**, 3459–3487 (1963)
13. Calcagno L, Percolla R, Foti G, Interdiffusion in polystyrene crosslinked by ion irradiation. Nucl Instrum Methods **B91**, 426–430 (1994)
14. Vrentas JS, Duda JL, Ling HC, Free-volume theories for self-diffusion in polymer-solvent systems. I. Conceptual differences in theories. J Appl Polym Sci Polym Phys **23**, 275–288 (1985)
15. Russel TP, Concerning voids in polyimide. Polym Eng Sci **24**, 345–349 (1984)
16. Kie T, Tiefenverteilung und Ausheizverhalten von implantierten 6Li-Atomen in verschiedenen Metallen. (in German). PhD. Thesis, Free University Berlin (1985)
17. George SC, Thomas S, Transport phenomena through polymeric systems. Prog Polym Sci **26**, 985–1017 (2001)
18. Kalinin B, Suetin PE, Mutual diffusion coefficients of high-temperature gases – Heat-transfer. Soviet Research **7**, 146–150 (1975), Suetin PE, Loiko AE, Kalinin BA, Measuring the mutual gas diffusion coefficient at low temperatures. Inzhenerno-Fizicheskii Zh **19**, 933–935 (1970) (in Russian), and: Kalinin B, Suetin PE, Trudy Uralstrogo Industrialnogo Politech Inst Acta **189**, 96–98 (1971) (in Russian)
19. Tonge JS, Shriver DF, Polymer Electrolytes. Chap. 5 of: "Polymers for Electronic Applications", JH Lai (ed.), CRC Press, Inc. Boca Raton (Florida), USA, p. 157–210 (1989)
20. Thran A, Kiene M, Zaporojtchenko V, Faupel F, Condensation Coefficient of Ag on Polymers. Phys Rev Lett **82**, 1903–1906 (1999)
21. Strunskus T, Keine M, Willecke R, Thran A, v. Bechtolsheim C, Faupel F, Chemistry, diffusion, and cluster formation at metal-polymer interfaces. Mat & Corros **49**, 180–188 (1998), and: Thran A, Strunskus T, Zaporojtchenko V, Faupel F, Evidence of noble metal diffusion in polymers at room temperature. Appl Phys Lett, submitted (2002)
22. Faupel F, Diffusion and interface formation at polymer-metal interface. Proc 1st Intl Conf on Polymer-Solid Interfaces, Pireaux JJ, Bertrand P, Bredas JL (eds.), IOP Publ. Bristol, UK (1992), and: Faupel F, Proceedings of the Conference on "Polymer-Solid Interfaces", Namur, Belgium, 2–6 Sept. 1991, IOP Publ. Ltd., UK (1992), pp. 171–188 (1991), and many references therein
23. Farrel G, Carter G, Diffusive processes in a solid during tempering. Vacuum **17**, 15–19 (1967)
24. See, e.g., Fink D, Chadderton LT, Cruz SA, Fahrner WR, te Kaat EH, Melnikov AA, Varichenko VS, Zaitsev AM, Ion track doping. Radiat Eff Defects Solids **132**, 81–90 (1994)
25. Kowalczyk SP, Kim YH, Walker GF, Kim J, Polyimide on copper: the role of solvent in the formation of copper precipitates. Appl Phys Lett **52**, 375–376 (1988)
26. Carslaw HS, Jaeger JC, Conduction of Heat in Solids, 2nd edn. Clarendon Press, Oxford (1989)
27. Kelly R, The diffusion of "attached" inert-gas activity. Can J Chem **39**, 2411–2422 (1961)
28. Fink D, Die Verwendung von (n,p)- und (n,α)-Reaktionen für die Messung von Reichweite- und Diffusionsprofilen und Channeling- und Blocking-Effekten. PhD. Thesis, Free University Berlin (1974) (in German)

29. Fink D, Chung WH, Klett R, Döbeli M, Synal HA, Chadderton LT, Wang L On the dyeing of ion tracks in polymers. Nucl Instrum Methods **B108**, 377–384 (1996)
30. Eßer M, Diplomarbeit TU Clausthal (1993) (in German)
31. Eßer M, Lösungsmittel-induzierte Delegation molekularer Sonden in latenten Kernspuren und ihre photophysikalische Analyse. PhD. Thesis. University Clausthal-Zellerfeld, Germany (1996) (in German)
32. Park GS, Transport principle. Solution, diffusion, and permeation in polymer membranes. In: Synthetic membranes: science, engineering, and applications. Bungay PM, Lonsdale MK, de Pinho MN (eds.), Reidel Publ., Dordrecht, (1986), pp. 57–107
33. Stern SA, The separation of gases by selective permeation. In: Membrane Separation Processes, Meares P (ed.), Elsevier, Amsterdam, (1976), pp. 195–326
34. Remmert G, Transporteigenschaften und Geometrie von Schwerionenspuren in Polymerfolien. PhD. Thesis. Johann Wolfgang Goethe Universität, Frankfurt am Main (1994) (in German)
35. Turowski T, Schockwellenmodell zur Beschreibung des ionendichteabhängigen Diffusionsverhaltens bestrahlter Polymerfolien. PhD. Thesis. Christian-Albrechts-Universität Kiel (2001) (in German)
36. Vieth WR, Diffusion in and through Polymers, Hanser Verlag, München (1991)
37. Danesi PR, Horwitz EP, Rickert PG, Rate and mechanism of facilitated Americium (III) transport through a supported liquid membrane containing a bifunctional organophos-phorous mobile carrier. J Phys Chem **87**, 4708–4715 (1983)
38. Jirage K, Hulteen JC, Martin CR, Effect of thiol chemisorption on the transport properties of gold nanotubule membranes. Anal Chem **71**, 4913–4918 (1999)
39. Tomasella E, Effect of UV and Gamma irradiation on Rn222 permeation through polyvinyl chloride (PVC). Application to radium-containing source packing for the purpose of storage. France: Franche-Comté University. DEA-Report No. 1700 (1994)
40. Escoubes T, Dolveck JY, Davenas J, Xu XL, Boiteux G, Ion beam modification of polyimide membranes for gas permeation. Nucl Instrum Methods **B105**, 130–133 (1995)
41. Martin CR, Nishizawa M, Jirage K, Kang M, Lee SB, Controlling transport selectivity in gold nanotubule membranes. Adv Mater **13**, 1351–1362 (2001)
42. Remmert G, Eyal Y, Fischer BE, Spohr R, Gas permeability and cross section of latent ion tracks in polymers. Nucl Instrum Methods **B105**, 197–199 (1994)
43. Smoluchowski M. Ann Phys **48**, 1103 (1915) (in German)
44. Pintauro PN, Verbrugge MW, The electric-potential profile in ion-exchange membrane pores. J Membr Sci **44**, 197–212 (1989)
45. Brochard-Wyart F, Kinetics of polymer–polymer interdiffusion. In: Fundamentals of Adhesion. Lee L-H (ed.), Plenum Press, New York and London, (1991), pp. 181–206
46. Chadderton LT, Nucleation of damage centres during ion implantation of silicon. Radiat Eff **8**, 77–86 (1971)
47. Hurst DG, Diffusion of fission gas. Calculated diffusion from a sphere taking into account trapping and return from traps. AECL-1550, Nov. 1962
48. Gaus H, Über den Einfluß von Fallen auf die Diffusion (in German) Z Naturforsch 20a, 1298–1308 (1965)
49. Biersack JP, Depth distribution of implanted atoms under the influence of surface erosion and diffusion. Radiat Eff **19**, 249–256 (1973)

50. Sizmann R, The effect of radiation upon diffusion in metals. J Nucl Mater **6970**, 386–412 (1978)
51. Ham FS, Theory of diffusion-limited precipitation. J Phys Chem Solids **6**, 335–351 (1958)
52. Fink D, Klett R, Müller M, Omichi H, Hosoi F, Vacík J, Hnatowicz V, Chadderton LT, Influence of energetic ions on grafting to polyethylene. Appl Phys **A63**, 441–446 (1996)
53. Sharp JH, Bindley WG, Achar BNN, Numerical data for some commonly used solid state reaction equations. J Am Ceram Soc **49**, 379–382 (1966)
54. Coats AW, Redfern JP, Kinetic parametcers from thermogravimetric data. Nature **201**, 68–69 (1964)
55. Mahfouz RM, Monshi MAS, Al-Owais A, Alshehri S, Al-Osaimi M, Abd El-Salam NM, γ-irradiation effects on kinetics and mechanism of the thermal decomposition of zinc acetate. Radiat Eff Defects Solids, submitted (2002)
56. Smith GD, Numerical Solution of Partial Differential Equations: Finite Difference Methods, 2nd edn. Clarendon Press, Oxford (1978)
57. Press WH, Teukolsky SA, Vettering WT, Flannery BP, Numerical Recipes in FORTRAN, 2nd edn. Cambridge University Press, Cambridge (1992/93)
58. Fink D, Klett R, Behar M, Sánchez G, Kaschny JR, Hertlein WG, Changes in the photoresist inhibitor distribution after ion irradiation and thermal treatment. Nucl Instrum Methods **B132**, 660–670 (1997)
59. Fink D, Behar M, Kaschny JR, Klett R, Chadderton LT, Hnatowicz V, Vacík J, Wang L, On the redistribution of $^{6}Li^{+}$ ions implanted into poylpropylene foils. Appl Phys **A62**, 359–367 (1996)
60. Marin N, Serruys Y, Metallic Impurities diffusion in a polymer film (kapton) under and out of irradiation, Nucl Instrum Methods **B131**, 109–120 (1997)
61. v. Smoluchowski M. Phys Chem **92**, 129–168 (1918)
62. Soares MRF, Kaschny JRA, dos Santos JHR, Amaral L, Behar M, Fink D, Diffusion and solubility of Au implanted into the AZ1350 photoresist. Nucl Instrum Methods **B166/167**, 615–620 (2000), and [2–7] therein
63. Walton AJ, Three Phases of Matter, Clarendon Press, Oxford (1984), pp. 1–182
64. Spohr R, Ion Tracks and Microtechnology: Principles and Applications. Vieweg, Braunschweig (1990)
65. Heinig KH, Schmidt B, Strobel M, Bernas H, Inverse Ostwald ripening and self-organization of nanoclusters due to ion irradiation. Mater Res Soc Symp Proc **647**, pp. 014.6.1–12 (2001), and: Heinig KH, Schmidt B, Müller T, Bernas H, oral presentation at the Intl. Conf. on Radiation in Insulators, Lisbon, Sept. 3–7, 2001; unpublished
66. LeGoues FK, Silverman BD, Ho PS, The microstructure of metal–polyimide interfaces. J Vac Sci Technol **A6**, 2200–2204 (1988)
67. Silverman BD, Single particle diffusion into a disordered matrix: Simulation of a metal–polymer interface. Macromolecules **24**, 2467–2474 (1991)
68. Kovacs GJ, Vincent PS, Formation and thermodynamic stability of a novel class of useful materials: close-packed monolayers of submicron microdisperse spheres just below a polymer surface. J Colloid Interface Sci **90**, 335–351 (1982)
69. Cole DH, Shull KR, Rehn LE, Baldo P, Metal–polymer interactions in a polymer/metal nanocomposite. Phys Rev Lett **78**, 5006–5009 (1997)

70. Fink D, Müller M, Nakagawa ST, Hirata K, Shi B, Wang KM, 3-dimensional implantation profiles: The spatial distribution of Nd in Si. Proc Intl. Conf. on "Ion Implantation Technology", Kyoto, Japan, 22–26. 6. 1998, Matono J, Takaoka G, Yamada I (eds.), publ. by: The Inst. of Electr. & Electronics Engineers, Piscataway, USA 1999, pp. 708–711
71. Marom G, Broutman LJ, Measuring diffusion coefficients by hygroelasticity. J Appl Polym Sci **126**, 1493–1496 (1981)
72. Thomas NL, Windle AH Discontinuous shape changes associated with Case II transport of methanol in thin sheets of PMMA. Polymer **18**, 1195 (1977)
73. Thomas NL, Windle AH, A theory of Case II diffusion. Polymer **23**, 529–540 (1982)
74. Hui CY, Wu KC, Lasky RC, Kramer EJ, Case II diffusion in polymers. I. Transient swelling. J Appl Phys **61**, 5129–5136 (1987)
75. Astarita G, Sarti GC, A class of mathematical models for sorption of swelling solvents in glassy polymers. Polym Eng Sci **18**, 388–395 (1978)
76. Sarti GC, Solvent osmotic stresses and the prediction of Case II transport kinetics. Polymer **20**, 827–832 (1979)
77. Gall TP, Lasky RC, Kramer EJ, Case II diffusion: Effect of solvent molecule size. Polymer **31**, 1491–1499 (1990)
78. Edwards DA, Non-Fickian diffusion in thin polymer films. J Polym Sci **34**, 981–997 (1996)
79. Mills PJ, Palmstrom CJ, Kramer EJ, Concentration profiles of non-Fickian diffusants in glassy polymers by Rutherford backscattering spectrometry. J Mater Sci **21**, 1479–1486 (1986)
80. Cunningham RE, Williams RJJ, Diffusion in Gases and Porous Media. Plenum Press, New York (1980)
81. Griftsova LA, Churaev NV, Calculation of the gas diffusion coefficient in microcapillaries using molecular data. Inzhenerno Fisicheskie Zhurnal **47**, 806–809 (1984)
82. See, e.g.: Le Claire M, Rabinovich J, A mathematical analysis of diffusion in dislocations. III Diffusion in a dislocation array with diffusion zone overlap. J Phys **C16**, 2087–2104 (1983), and earlier papers cited therein
83. Fink D, Müller M, Capillaric penetration of etchant solution into swift heavy ion-irradiated silicone rubber. Nucl Instrum Methods **B170**, 134–144 (2000)
84. Eyal Y, Gassan K, Observation of latent heavy-ion tracks in polyimide by means of transmission electron microscopy. Nucl Instrum Methods **B156**, 183–190 (1999)
85. For a recent overview, please consult: Fink D, Klett R, Chadderton LT, Cardoso J, Montiel R, Vazquez H, Karanovich AA, Carbonaceous clusters in irradiated polymers as revealed by small angle X-ray scattering and ESR. Nucl Instrum Methods **B111**, 303–314 (1996)
86. Beriozkin VV, Zasgorsky DL, Nechaev AN, Tsiganova TV, Mitrofanova NV, Apel PYu, Mchedlishvili BV, The track membrane porous structure and selective properties investigation. Radiat Meas **34**, 593–595 (2001)
87. Jones R, A superficial look at polymers. Phys World, March 1995, pp. 47–51
88. Jones RAL, Norton LJ, Kramer EJ, Bates FS, Wiltzius P, Surface-directed spinodal decomposition. Phys Rev Lett **66**, 1326–1329 (1991)

89. Baglin JEE, Interface structure, adhesion, and ion beam processing. In: Materials and Processes for Surface and Interface Engineering, Paleau Y (ed.), Nato Advanced Science Institute Series E: Applied Sciences, Vol. **290**. Kluwer Academic Publ., Dordrecht/Boston/London, (1995), pp. 111–149

2 Transport Processes in Low-Energy Ion-Irradiated Polymers

D. Fink and V. Hnatowicz

2.1 Transport of Implanted Penetrants; General Remarks

In contrast to ion implantation into metals, semiconductors or ceramic materials, where the implanted ions distribute nicely according to the theoretically predicted ballistic distributions, for polymeric targets one has frequently observed strong deviations from the expected range profiles. It is accepted that these profiles result from the diffusional redistribution of the as-implanted particles, their mobility being influenced by the defects in the polymer that act as traps for the mobile species. In fact, not only the mobility of implanted ions, but also that of other penetrants (e.g., gases, liquids, etc.) is affected by these defects.

Ion irradiation is known to destroy the polymers via mechanisms that are mediated by *electronic and collisional ("nuclear") energy* transfer (also called: "energy loss"). Collisional defects – i.e., production of knock-on atoms, voids and microcavities – preferentially show up as deep potential depressions with limited radius in the surrounding medium, whereas electronic defects primarily are excited atoms or ions (so-called radicals), usually with no extra free volume associated with them. The depth distributions of both these processes are usually strongly different from each other, so that they can be distinguished easily. It appears that electronic defects often have a limited lifetime due to recombination processes (i.e., short-range electronic and ionic transport processes), whereas collisional defects appear to exhibit long-term stability [1]. As both these defects act as traps they can be probed via suitable tracer transport experiments. In pristine and in ion-irradiated polymers the transport of matter is nearly always a diffusive one, which is influenced by the above-mentioned traps.

The question whether a penetrant in an ion-irradiated solid is predominantly captured by electronic or collisional ("nuclear") defects – i.e., radiation-induced free volume and/or radicals, respectively – apparently depends on both the type of the penetrant and the sample age, i.e., on the time in between the sample irradiation and the penetrant-uptake procedure. For example, noble gases were always found to be bound only to nuclear defects [2], whereas positrons were found to probe preferentially the electronic defects [3,4]. Whereas light reactive ions such as Li^+ implanted into polymers always preferentially bind to their self-created electronic defects [5], the same ions in aqueous solutions sometimes decorate their electronic, but sometimes also their nuclear defects [6], even in three dimensions. The reason for the latter behavior is that in aqueous solution, the ions are surrounded by a hydrate shell and hence fit only to the larger collision-induced nuclear defects in a dense polymer matrix. If the polymer matrix is less closely packed, the ions have access to both nuclear and electronic defects, so that in this case the electronic traps occasionally outnumber the nuclear ones by far, due to the higher abundance of electronic defects.

If, nevertheless, preferential decoration of nuclear defects is observed, the reason is that electronic defects are often metastable and therefore tend to vanish (as part of the "aging" process of irradiated polymers) due to recombination processes, whereas nuclear defects survive more often. This means that the ratio between nuclear and electronic defects changes with increasing sample age in favor of the nuclear component. This can be followed, e.g., by EPR examinations that are sensitive to electronic distortions of the considered samples, hence preferentially probe defects of electronic origin. For example, it was observed that the EPR signal of ion-irradiated nylon decreased linearly by a factor 4 within 4 days, after which it no longer changed [7]. One could conclude that the half-lifetime of most electronic defects in nylon is about two days, and that the ratio of originally EPR-active electronic to nuclear defects was 4:1.

Another point that has to be considered when studying ion irradiation of polymers is their strong fluence dependence. With increasing ion fluence, hence increasing polymer destruction, the material loses its insulating properties so that electronic defects play a decreasing role as traps, and consequently the importance of nuclear defects to trap migrating penetrants increases.

Chain scission results in the formation of small polymeric fragments, part of which are volatile and can escape from the sample by fast diffusion processes. Thus excess free volume V_{rad} is created, additionally to the intrinsically pre-existing microcavities V_{FH} and the free volume due to thermal expansion V_{FS}. The amount of V_{rad} depends strongly on the examined system – i.e., on the transferred energy density and on the type of the irradiated polymer. It can range from negligible fractions (such as given in, e.g., polyimide irradiated by low-energy ions [8]) up to nearly 100%, i.e., up to nearly complete vanishing of the polymer along the ion tracks (such as given in "self-developing resists" like cellulose nitrate [9,10]).

Essentially at elevated temperatures the *self-diffusion* of the polymeric chains tends to annihilate the radiation-induced excess free volume V_{rad}. Eventually existing strain fields between the pristine polymer and the ion-irradiated zone may act as an additional driving force for this relaxation process. If excessive free-volume production renders the remaining polymeric structure instable, it may even collapse to a carbonaceous material of higher density (*densification, compaction*).

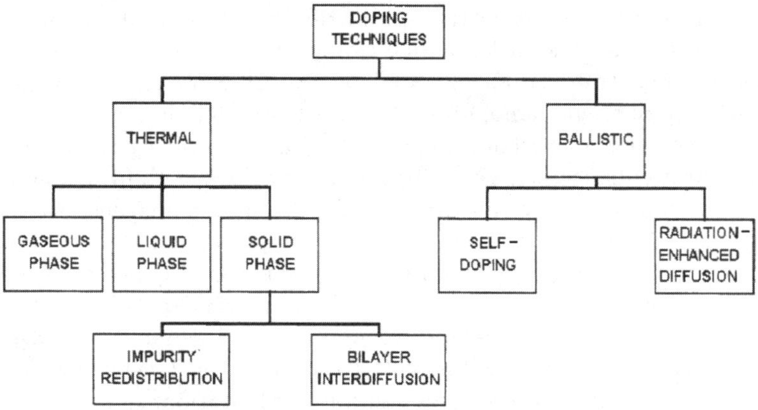

Fig. 2.1. Genealogy of the different ways of doping a polymer with impurities

The above-mentioned tendency for the decrease in free volume of irradiated polymers is also influenced by the *cross-linking* effect. As a typical example, one should mention the results of low-energy ion irradiation of PEEK, which were obtained by positron annihilation [11]. Cross-linking tends to bring neighboring chains much closer to each other, and therefore leads to a decrease in free volume. On the other hand, cross-linking retards the change in free volume, as the radiation-induced interconnections of neighboring polymer chains lead to a dramatic reduction of their mobility in the ion-irradiated zone. As a result of these geometrical obstacles, the irradiated zones may – depending on the specific case – eventually be regarded as rather rigid and hard entities embedded within a plastic medium, with only little relaxation of the neighboring pristine matter into the ion-track region. This is what gives the ion-irradiated zones in polymers some long-term stability.

The concentration distributions of penetrants incorporated in polymers do not necessarily always have exactly the same shape as the energy-loss distribution and hence the distribution of free volume. For example, compare Figs. 2.19 and 2.20, where the electrolyte distribution is found to be somewhat broader than the calculated nuclear energy transfer distribution. (This cannot be explained by the experimental depth resolution and/or the error of the tomographic reconstruction procedure only.) An exact identity of the penetrant distributions with the corresponding energy transfer curves can only be expected if the polymer's penetrant uptake has come to saturation, i.e., for very high $D\,t$ values (D = diffusion coefficient of the penetrant, t = diffusion time), and when stationary penetrant distributions are reached, so that the whole available free volume can be regarded as being filled by the penetrant. As long as this state has not yet been obtained, the dynamic process of penetrant incorporation – i.e., both the penetrant depth distributions and their diffusion coefficients – are governed by the distribution of free volume, the presence of eventual traps, and the distribution of the penetrant's diffusion coefficient. The latter is determined by both the total free volume available for penetrant transport, and by the microscopic structure of the transient or permanent nanopores that constitute this free volume. Thus, a change in the pore microstructure, e.g., by interconnecting them, may lead to a dramatic enhancement in diffusional penetrant mobility, without any major alteration in the overall available free volume being necessary. There are hints that this is the case, e.g., in ion-irradiated PEEK [12]. On the other hand, there exist cases [13] where radiation-induced cross-linking appears to create geometrical obstacles and retards the penetrant mobility.

When studying diffusion processes, one has to distinguish between thermal and radiation-enhanced diffusion (RED) processes. The latter takes place either during, or immediately after the ion implantation. It usually exceeds the thermal mobility at ambient temperature by many orders of magnitude. Whereas, in the first case, it is the intrinsic and thermally created free volume that promotes the atomic transport, it is the radiation-induced excess free volume that promotes the transport in the latter case. This means that radiation-enhanced diffusion is similar to that in, e.g., semiconductors or ceramics, where the energy transfer leads to the formation of mobile excess vacancies that enable the nonthermal transport. The polymeric excess free volume is created, e.g., by the loss of volatile components. It is expected not to collapse immediately but to undergo some thermal migration during which it might enable the transport of embedded impurities. This excess free volume can either be formed during the implantation of the same ion specie that thereafter undergoes RED ("self-radiation-enhanced diffusion"), or it can be formed by ions B different from the diffusing specie A. Below we shall first treat the different cases of self-radiation enhanced transport processes in irradiated polymers in more detail, before we treat RED more generally.

Finally, when dealing with trapping of mobile ions, one has to distinguish between intrinsic and radiation-induced traps. Both can be either saturable and unsaturable traps. It has turned out that, in general, traps of nuclear origin behave as unsaturable traps, whereas traps that emerge from electronic polymer destruction rather are saturable ones. This points to a different nature of the two trap species. Whereas the electronic defects appear to be radicals, i.e., excited or charged polymer segments that are prone to undergo chemical reactions, the nuclear traps are rather related with the production of excess free volume, due to the knock-off of polymer atoms from their previous locations.

As electronic defects are no longer chemically active after bonding with an impurity, and as no excess volume is created on that occasion, there is no tendency to absorb more than just one atom, i.e., these defects act as saturable traps [14]. On the other hand, the collision-based excess free volume, signifying a deep potential depression, might attract more than only one atom, which may be chemically reactive or inert such as a noble gas atom. Especially the latter ones are well known to cluster into bubbles in practically any material, e.g., in metals or semiconductors. In these earlier studies it had been found that several arriving noble gas atoms can share one vacancy V (e.g., around 5 He atoms per vacancy in a tungsten lattice: He_5V), until the combined strain of the noble gas impurities forces a neighboring lattice atom to leave, thus creating a second vacancy (He_6V_2) that is subsequently occupied by more arriving gas atoms. It is supposed that a similar mechanism of precipitate growth to clusters (in the case of gaseous penetrants: bubbles) also acts in polymers.

2.2 Thermal Diffusion of Implanted Ions

From the absence of any detectable He depth profiles in polymers one can conclude that the mobility of He therein always exceeds $10^{-14}\,\mathrm{cm^2\,s^{-1}}$. The thermal long-term mobility of B in PET at ambient temperature does not exceed $5 \times 10^{-15}\,\mathrm{cm^2\,s^{-1}}$, [15]. Heavy ions that show a Gaussian profile shape immediately after their implantation in polymers, redistribute thermally during annealing by letting their concentration maximum collapse and developing a tail towards the bulk. The measured profiles of, e.g., 100-keV Bi in AZ111 can be simulated well by assuming a bulk diffusion coefficient of $D_\mathrm{b} = 1.2 \times 10^{-14}\,\mathrm{cm^2\,s^{-1}}$ and a near-surface diffusion coefficient of $D_\mathrm{s} = 1.2 \times 10^{-15}\,\mathrm{cm^2\,s^{-1}}$, [16]. The reason for the decrease in near-surface mobility is given by the retardation due to radiation-induced-trapping–detrapping processes, by which an effective diffusion coefficient D_eff emerges that can be approximated (provided that $A(x)$ and $B(x) \neq 0$) by:

$$D_\mathrm{eff}(x) = D_0(x)/[1 + A(x)/B(x)]. \tag{2.1}$$

2.3 Radiation-Enhanced Diffusion

2.3.1 Self-Radiation Enhanced Transport of Ions Implanted into Polymers after Low Fluence Irradiation

Heavy noble gas ions inplanted into polymers show disagreements between measured ranges and range stragglings with theory, see Vol. I, Table 4.4 and Fig. 2.1. In general, the as-measured distributions are shorter than theoretically predicted, which might either be attributed to a compaction effect of the irradiated polymer layer [17], or – more probable – to a diffusional transition to traps that follow the nuclear damage distributions. It is difficult to distinguish both cases unambiguously.

One observes in these cases an additional tail towards the bulk (Fig. 2.1) indicating that part of the implanted material remained mobile after the implantation. As subsequently performed thermal annealing experiments revealed that the thermal diffusion coefficients at ambient temperature are by

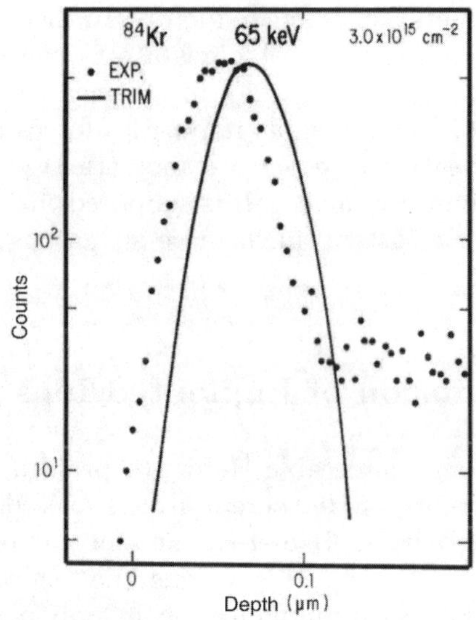

Fig. 2.2. Redistribution of a noble gas in a polymer. Example: 65-keV Kr in photoresist AZ111, as measured by RBS. The ion distribution does not correspond to the predicted range distribution (as shown here). The implanted particles exhibit considerable mobility after the implantation as can be seen from the pronounced backward tail. In the near-surface region, nuclear damage traps the mobile implants. The Kr distribution resembles the predicted nuclear damage distribution (not shown here) closely [2, 18, 19]

far inferior to the strong mobility observed during the implantation, one has to attribute the latter to radiation-enhanced diffusion (RED). Similarly as in metals or semiconductors, where RED is attributed to irradiation-produced excess vacancies that promote the migration far beyond the irradiated region, we assume that here also RED results from the radiation-induced excess free volume due to degassing of volatile products, that enables a faster impurity diffusion than the intrinsic and thermally activated free volume only.

A consequence of the self-radiation-enhanced mobility of implanted noble gases in polymers is that one can never detect the whole amount of implanted ions in the sample. The fraction of retained gas increases with the ion fluence, due to the increasing probability of trapping at self-created damage sites, Table 2.1 [20].

Light ions implanted into polymers at room temperature often redistribute partly, Fig. 2.3a. There appears to exist a threshold energy below which the profiles follow the theoretical prediction for range profiles, and above which redistribution sets in, see, e.g., Fig. 2.3b. This threshold is different for each system. For example, for ^{10}B implanted into the AZ111 photoresist this threshold lies at around 30 keV, for ^{19}F in AZ111 at 70 keV, and for ^{6}Li in AZ111 at 90 keV [5]. However, for a given target one finds reasonable scaling

Table 2.1. Fraction of Xe atoms retained in the implanted region as a function of the nominal implantation dose for implantation into photoresist [20]

Nominal implantation fluence [Xe/cm^2]	Ratio of the measured to the nominal implanted fluence
1×10^{14}	0.20
5×10^{14}	0.50
8×10^{14}	0.70
1×10^{15}	0.72

with the electronic stopping power, Fig. 2.4. One recognizes that the redistributed fraction f is zero below a lower energy-transfer threshold (about 170 eV/nm in the given case), and unity above an upper energy transfer threshold (estimated to be about 280 ± 60 eV/nm for AZ111) [21]. Such relations are strongly target dependent, and appear to reflect the material's resistance against ion-beam-induced free-volume formation. For example, when studying the redistribution of Xe implanted into diamond, an S_e value as high as 36 keV/nm was found for a redistributed fraction $f = 0.64$ [21].

In general, the redistributed profiles can be characterized by a superposition of a regular profile and another one that has an additional tail directed towards the surface (Fig. 2.3); in some cases one also observes a gradual transition between these two curve shapes, Fig. 2.5.

The nonregular fraction increases with ion energy. For example, for 100-keV Li and B implantation into AZ111, this component amounts to only 5 to 10% of all implanted ions (Fig. 2.3), but it increases already to about 15% at 150 keV. The importance of this nonregular fraction also depends on the polymer specie and the projectile ion; whereas for most polymers this fraction amounts to only 5 to 10%, there are also cases (e.g., F in AZ111) where nearly 100% of the implanted ions get redistributed (Fig. 2.5). The reason for the emergence of the nonregular component is a redistribution of some fraction of the implanted ions that can be explained by the diffusional transition from the original implantation profiles towards those governed by damage sites, at which the mobile ions are bound either transiently or permanently [5, 18, 19]. This mechanism appears to be a radiation-enhanced, rather than a thermal mobility of the implanted ions, because subsequent annealing hardly affects the ion distribution any longer. It still remains open whether this redistribution takes place during, or immediately after, the implantation.

The order of magnitude of this diffusion is given for the Dt values as, e.g., $5.6 \times 10^{-13}\,\mathrm{cm}^2$ for 100-keV B in PA [22], $1.2 \times 10^{-8}\,\mathrm{cm}^2$ for 5-MeV B in PMMA [23], and $2.9 \times 10^{-13}\,\mathrm{cm}^2$ for 100-keV B in S1813 [24]. For the diffusion coefficients, one can deduce $3 \times 10^{-12}\,\mathrm{cm}^2\,\mathrm{s}^{-1}$ for 5-MeV Li in PMMA [23], and 3.6×10^{-12} and $1.8 \times 10^{-12}\,\mathrm{cm}^2\,\mathrm{s}^{-1}$ for 100-keV B in the photoresist S1813 in axial and transversal directions, respectively, of the 3D

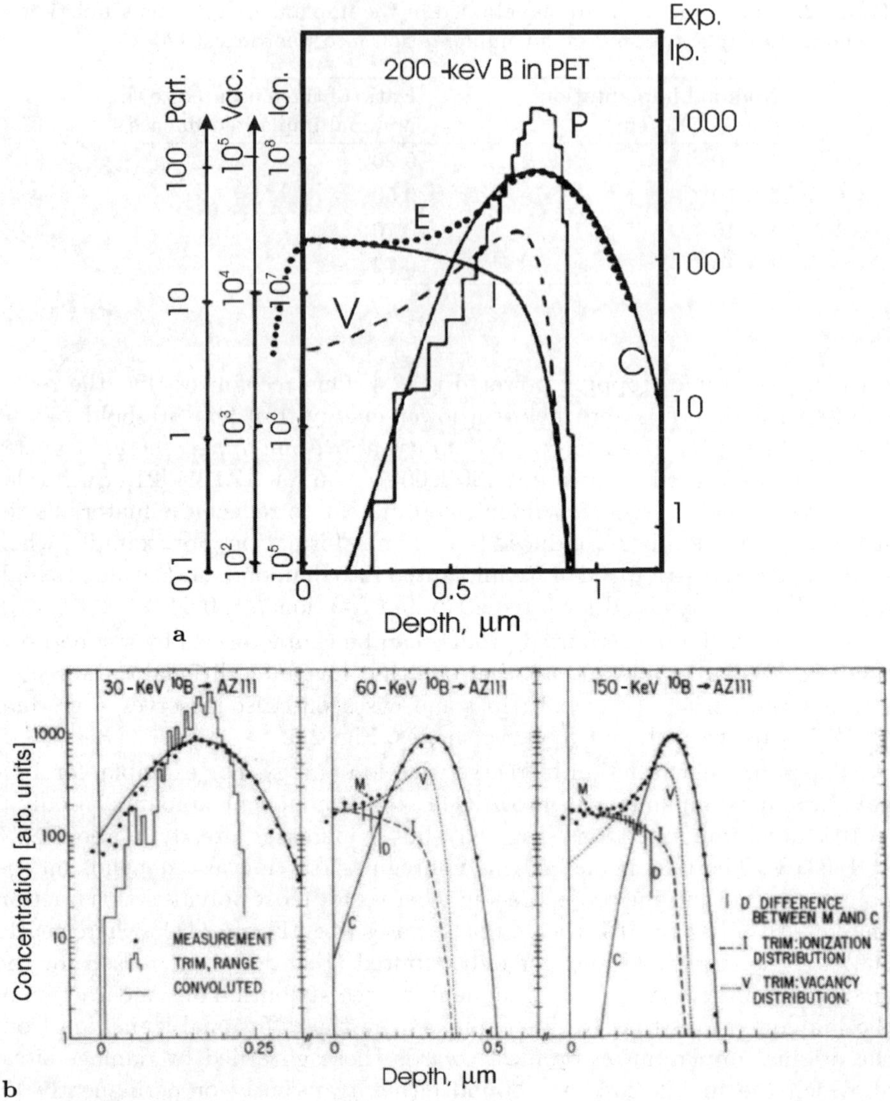

Fig. 2.3. Typical implantation profiles of a light ion implanted into a polymer. (**a**) 200-keV B in PET. It is seen that the measured profile (E) exhibits a bimodal shape, being superimposed by an implantation profile (P,C) and a profile that follows the distribution of transferred electronic excitation energy (I). In order to allow a comparison of the regulary implanted fraction with theoretical prediction (P), the latter has to be convoluted (C) with the depth-resolution function. Apparently, nuclear damage (V) does not influence the implanted particles' redistribution in this case. (**b**) Threshold behavior of the redistribution of implanted light ions in polymers. Shown here for the example of B in AZ111 photoresist [21]

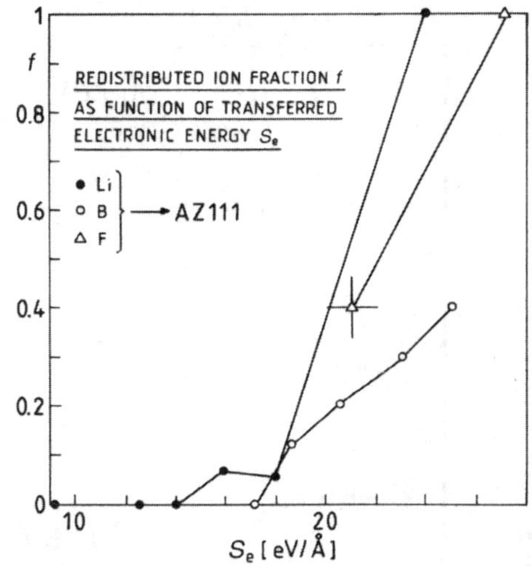

Fig. 2.4. The mobile fraction f of light ions after implantation into the photoresist AZ111, as a function of the transferred electronic energy per pathlength S_e [21]

redistribution profile [24]. The mobility is higher along the beam direction than in the transversal direction. This makes sense insofar as free volume is formed preferentially along the track center due to high transferred energy density but not in the outer track regions. The compacted matter there (due to shock waves, cross-linking, etc.) does not contribute markedly to the transverse RED.

In connection with these nonregular profiles, it is also worth noting that the change in shape occurs at energies where the electronic stopping power is much larger than the nuclear one (e.g., $S_e = 5 \times S_n$ for B in AZ111, $S_e = 6 \times S_n$ for B in PET, and $S_e \approx 30 \times S_n$ for Li in AZ111). Secondly, the transitions occur for different projectiles (B, Li, F in AZ111) at approximately the same value of the electronic stopping power, $S_e \approx 18\,\mathrm{eV/Å}$. Thirdly, the redistributed ion fraction depends on the type of the implanted ion.

In general, the as-implanted ions are only trapped by defects of electronic origin; there are, however, also cases where both electronic and nuclear defects contribute. For example, in the photoresist S1813 (Shipley Europe Ltd.) approximately a 1:1 ratio is found between particles trapped at nuclear and at electronic defects, respectively. Taking into account the 10 times lower abundance of nuclear defects, this signifies a ten times higher trapping efficiency of the nuclear defects than that of electronic damage sites. The redistribution can well be understood by diffusion simulations such as the one shown in Fig. 2.6 [24] (see also Fig. 5.10 in Vol. I).

Meanwhile, quite a number of such studies have been performed for several systems. It is interesting to mention in this connection [14] that implanted Li^+ ions were subject to redistribution not only at the self-created damage,

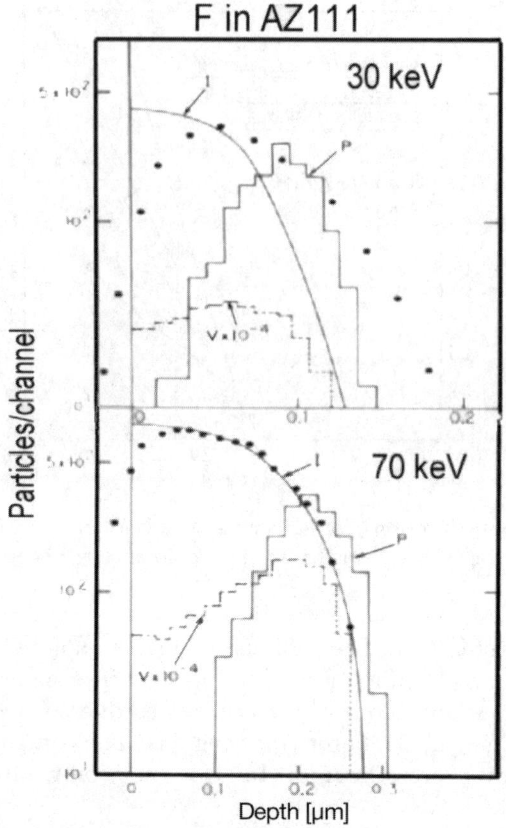

Fig. 2.5. Transition of the shape of the distribution of fluorine ions implanted into the photoresist AZ111, as a function of energy. Whereas at 70 keV all detected fluorine clearly follows the distribution of electronic damage, the shape of the depth profile as measured after 30-keV implantation appears to be a transition between the implantation and the electronic damage profiles. Experimental limitations did not allow measurements at lower energies where the fluorine profile should follow even closer the regular implantation profile. NRA measurement, exploiting the ^{19}F $(p, \alpha\gamma)$ ^{16}O nuclear reaction [2]

but simultaneously also to (spatially separated) damage by other (Kr) ions. The fact that Li precipitation also occurred there proved that a fraction of the implanted ions had left their region of deposition, to migrate freely within the undisturbed polymeric bulk until being trapped by the Kr damage sites. In this study it was also shown that diffusional simulations, though containing a great number of free parameters, can nevertheless point unambiguosly to a specific diffusional mechanism responsible to fit the measured distributions, by performing adequate tests. In the specific case cited there (150-keV Li in PP) radiation-enhanced diffusion in the presence of saturable traps of collisional origin was found to be the responsible mechanism.

Finally, it may be of interest that in the case of light-ion irradiation of polymer / semiconductor bilayer structures (e.g., photoresist on silicon) the implanted ions redistribute only in the polymer but follow the regular range profiles inside the silicon, with a sharp transition between both distributions at the interface [18].

Hydrogen and helium ions have never been identified in polymers after their implantation, in spite of considerable attempts. For ^{3}He, the instrumental threshold of the NDP measuring technique in coincidence, exploiting the ^{3}He(n,p)^{3}H reaction, is less than 2×10^{19} atoms/cm^3, corresponding to about 0.02% atomic [25]. The reason for the absence of these implants appears to be a very high thermal mobility, at ambient temperature, of at least $1 \times 10^{-14}\,\mathrm{cm^2\,s^{-1}}$ of both gases in polymers, combined with a very small trapping probability in intrinsic or radiation-induced defects.

Other Effects of Radiation-Enhanced Mobility

The transient enhanced mobility of the molecular chain ends can initiate cross-linking [26] (see also Sect. 2.9).

2.3.2 Radiation-Enhanced Transport of Ions in Polymers

In order to understand the most general case of the RED mechanism, sequential implantation experiments were performed, first by a specie A that represents the mobile ions to be probed, followed by a specie B that induces the RED. The depth distributions of A were measured and simulated with the diffusion code DIFF to extract from the best fit decisive parameters such as the radiation-enhanced diffusion coefficients D_{RED} and the trapping and detrapping probabilities A and B, respectively.

Specifically, both A and B were selected to be noble gases to avoid any possible interfering chemical effects. For example, A = 80-keV Xe at implantation doses of 10^{14} to $10^{15}\,\mathrm{cm^{-2}}$, and B = 670 to 800-keV Ne, Ar, Kr, and Xe at an irradiation fluence of 5×10^{13} to $2 \times 10^{15}\,\mathrm{cm^{-2}}$, with the target material being the photoresist AZ1350 [27]. The values obtained for the product of RED and diffusion time: $D_{\mathrm{RED}}t$ of A are in the order of some $10^{-10}\,\mathrm{cm^2}$, from which diffusion coefficients around $D_{\mathrm{RED}} \approx 1.7 \times 10^{-13}\,\mathrm{cm^2\,s^{-1}}$ follow. This has to be compared with thermal diffusivities of the order of $\approx 1.3 \times 10^{-14}\,\mathrm{cm^2\,s^{-1}}$, i.e., the RED process signifies a moderate enhancement. The derived D_{RED} value corresponds to the thermal diffusion coefficient that would be obtained after thermal annealing at around 330°C (if the photoresist were not already affected by pyrolysis at that temperature). Typical trapping probabilities are slightly below $10^{-3}\,\mathrm{s^{-1}}$, and the detrapping probabilities are of the order of a few $10^{-5}\,\mathrm{s^{-1}}$.

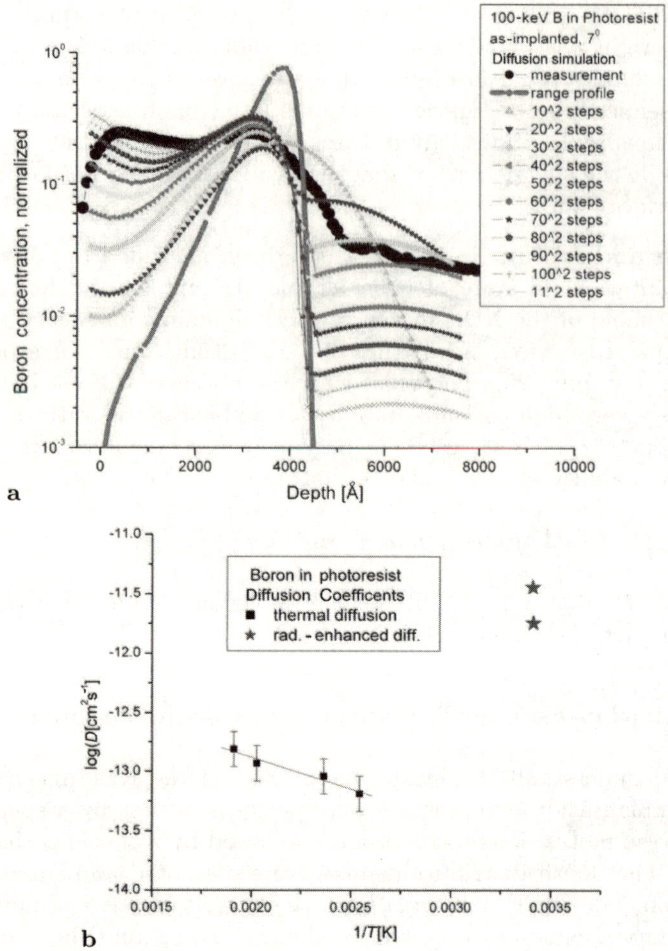

Fig. 2.6. (**a**) Simulation of the redistribution of boron, implanted at 100 keV into the photoresist S1813 at room temperature. The ballistic range profile is seen to change by radiation-enhanced diffusion towards a profile that is governed by electronic and nuclear damage in the near-surface side, and by a long diffusional tail in the bulk side. From the number of iteration cycles, one can deduce the diffusion coefficient of the redistributing boron. The still visible difference between the as-simulated and as-measured profiles vanishes nearly completely, if the as-simulated profiles are convoluted with the experimental depth resolution (not shown here). (**b**) Comparison between the radiation-enhanced and the thermal diffusion of B in S1813, as derived from diffusion simulations for the as-implanted and as-annealed profiles

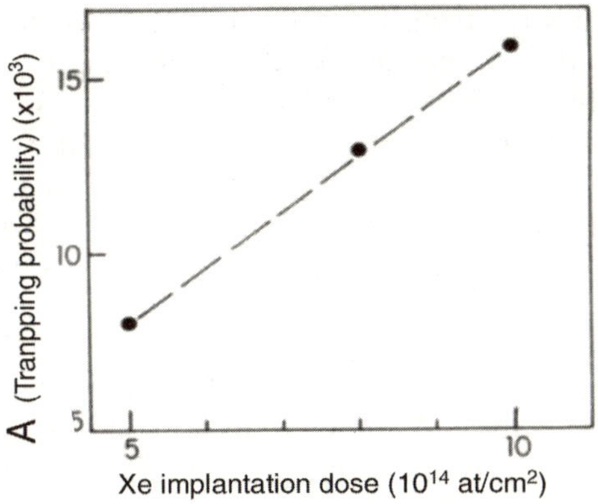

a

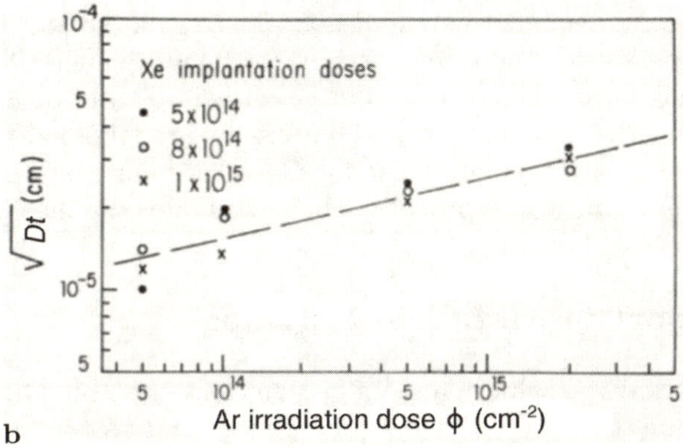

b

Fig. 2.7. Parametric dependences of RED of noble gases in polymers. (**a**) Trapping probability as a function of the implantation dose of the probing ion. (**b**) Mean free radiation-enhanced diffusion path as a function of the irradiation fluence of the postbombardment projectile (in this case: Ar), for different implantation doses

It was found that the product of RED and diffusion time: $D_{\mathrm{RED}}t$ of A depends on the fluence of B. The trapping probability A depends only on the implantation dose of A, and the detrapping probability B is independent of both the implantation dose of A and the irradiation dose of B. The trapping probability A of A is independent of the ion specie B. On the other hand, the detrapping probability B of A increases with increasing mass of the ions B. More precisely, B scales with the transferred nuclear energy density of B. Some of these correlations are depicted in Fig. 2.7. The transferred electronic energy seems to play a lower role for the RED of A.

Furthermore, though the diffusion lengths $(D_{\mathrm{RED}}t)^{1/2}$ are independent of the implantation fluence of A, they follow a Φ^{α} relationship, with Φ = fluence of B and $\alpha = 0.22$ for A = Ar. Typically $10^{17}\,\mathrm{cm}^{-3}$ trapping centers are formed in the target material, whereas the ballistic theory (TRIM) predicts as many as 10^{22} nuclear defects/cm^3. This means that the trapping efficiency within nuclear defects is relatively low, of the order of 10^{-5} only, or that many nuclear defects have annealed. From the detrapping probability B, one can estimate an activation energy for detrapping of the order of 250 meV, whereas the thermal activation energy is of the order of only 80 meV.

Comparing a noble gas, Xe, and a chemically highly reactive implant, Cs, one finds that the radiation-enhanced mobility in AZ1350 after Ar postbombardment is much higher for the noble gas than for the alkali ion. If one compares the fraction of the ions (implanted at 5×10^{14} cm^{-2}) that do not undergo radiation-enhanced diffusion after postbombardment, one finds {0.73 to 0.40} for implanted Xe, and {0.61 to 0.48} for implanted Cs, after postbombardment with Ar ions at fluences ranging from 5×10^{13} to 2×10^{15} Ar ions/cm^2, respectively. For an implanted ion fluence of 1×10^{15} cm^{-2}, these retention ratios shift to about {0.93 to 0.67} for Xe and {0.84 to 0.64} for Cs, respectively, for the same postbombardment fluences. Whereas these retention ratios have a linear dependence with implantation fluence for Xe (indicating that the number of trapped atoms is proportional to the damage produced), this relation is nonlinear for Cs. This points to a more complex trapping mechanism, where probably chemical bonds are also involved.

2.4 Transport of Metals in Polymers

There are two ways to achieve a metal–polymer interaction. One is the direct deposition of metals onto polymers (or vice versa), and the other is the implantation of metal ions into a polymeric target. Though the first case is not linked with ion irradiation, it will be briefly summarized here as well, as its findings are also important for the case of metal implantation into polymers.

2.4.1 Transport of Metals through Metal/Polymer Interfaces

The behavior of the interface depends on the metal/polymer bonding. Two principle cases can be distinguished, weak and strong bonding, with typical metals such as Cu and Cr. In the Cr/PI(PMDA-ODA) system, several delocalized metal/polymer bonding complexes are found. The most stable complexes formed are those where the Cr atom sits above the six-membered ring of the PMDA or ODA subunits, i.e., at the position of the highest coordination. Also, the position above the five-membered ring of the PMDA subunit was found to be occupied by Cr. Thus the metal–polymer bonding is different from the localized bonding observed for, e.g., metal oxides or carbides. The case of Cu/PI is similar to the case of Cr, except that the chemical interaction at the interface is much weaker. This indicates the necessity to activate the polymer surface either chemically or by ion irradiation, to obtain good Cu/PI bonding. Since Cu is more electron rich than Cr, it is not expected to occupy high coordination sites [28].

Though there has been invested much work in the study of the transport processes after metal deposition onto polymer surfaces [29,30], there is comparatively little work available on the migration processes in metal/polymer interfaces after ion irradiation. Most of these papers deal with the possibilities of how to improve adhesion, the latter being better the greater the amount of interface intermixing.

2.4.2 Transport of Ion-Implanted Metals in Polymers

Whereas reactive metals in polymers do not have long-range mobility but act as efficient diffusion barriers, isolated atoms of less-reactive metals diffuse deeply into polymers. However, the very pronounced aggregation tendency of metal atoms frequently inhibits the atomic diffusion and gives rise to cluster diffusion instead. Therefore, in order to study the pure atomic metal diffusion, metals (Ag, Au, Bi) have been implanted at very low fluence ($10^{12}\,\mathrm{cm}^{-2}$) and low energy (20 keV) into a photoresist [31].

Thus the maximum implanted concentration is kept below 10^{-2} at.%, i.e., below the solubility limit (0.3 at.% at 250°C [32]), and the radiation damage is minimized ($10\,\mathrm{eV/nm}^3$) so that the metal's diffusive properties are influenced only marginally. After annealing above T_g, the implantation profiles broadened regulary. Diffusion coefficients of the order of 10^{-15}–$10^{-17}\,\mathrm{cm}^{-2}$ could be derived, and from the Arrhenius plots activation energies around 600 to 650 meV were determined. Frequently the mobility in the irradiated area is slightly lower than that in the unirradiated polymer, due to hindrance by trapping–detrapping processes (Fig. 2.8).

Whereas at very low fluences (for 30-keV Ag in AZ1350: below $5 \times 10^{12}\,\mathrm{cm}^{-2}$), variations of the implantation fluence do not affect the metal

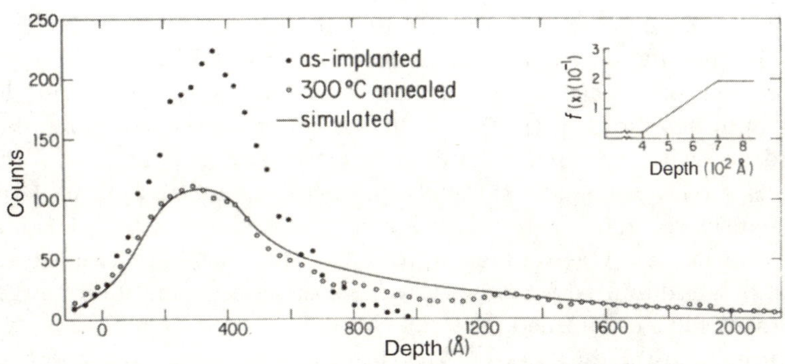

Fig. 2.8. Thermal motion of Bi implanted into photoresist AZ111 at very low fluence. The as-implanted profile broadens due to regular diffusion, which is slightly reduced in the irradiated zone due to trapping-detrapping mobility. The insert shows the depth-dependent function $f(x)$ that describes the change in the diffusion coefficient according to: $D(x) = f(x) * D_0$

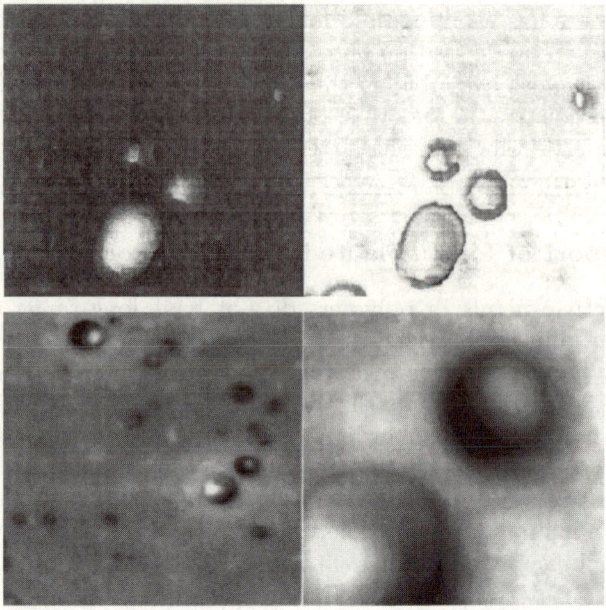

Fig. 2.9. (**a**) Metal precipitates, obtained after implantation of 100-keV Au up to 10^{15} cm^{-2} into PDHS, and subsequent annealing up to 400°C for 6 h. Image size: 130 × 150 nm; *left:* height profiles, *right:* friction profiles of AFM micrographs. (**b**) AFM image of Au precipitation in PMPySi. The polymer film was irradiated with 50-keV Au ions at 10^{14} cm^{-2}, and then annealed at 400°C for 1 h. Pronounced Au precipitates of 100 to 150 nm diameter show up in the nonwetting polymer surface. Image sizes: *left:* 1.5 μm × 1.5 μm, *right:* 300 nm × 300 nm. Unpublished. Courtesy of L. Farenzena and R.M. Papaleo

diffusion coefficients, the latter are found to decrease slightly when the fluence is further increased. For example, when going from 5×10^{12} cm^{-2} in the aforementioned case to 5×10^{13} cm^{-2}, the diffusion coefficients decrease by nearly a factor of 2 [31]. This indicates an enhanced trapping efficiency, concomitant with our understanding of polymeric ion damage. When the fluence is further increased the metal atoms start interacting preferentially with themselves, and cluster diffusion is reported upon sample annealing, with activation energies ranging from 0.4 to 2 eV [31]. Metal clustering appears to be an unlimited process without saturation, so that quite large metal precipitates can be obtained, Fig. 2.9.

High-fluence ion implantation was used, e.g., in [33] to study diffusion and clustering of different metals in polyimide (PI). The metal behavior in the polymer surface layer was studied by the RBS and ERDA techniques and by cross-sectional TEM. It was found that the diffusion and clustering depend strongly on the chemical reactivity of the metal [29, 30, 33]. In the Cu/PI, Pd/PI and Ag/PI systems clustering of metal atoms on the polymer surface or in the polymer near-surface layer is observed to lead to creation of metallic nanoparticles. Subsequent irradiation of the system comprising Cu deposited onto PI with 300-keV Kr^+ ions leads to inward diffusion of Cu atoms and creation of Cu nanoparticles, about 15 nm in diameter, inside of the modified PI [29].

No metal mobility was observed in the same system with the much more reactive Ti coating. Reactive metal layers in contact with polymer surfaces, e.g., Ni or Co on PI, strongly interact with each other, thus often forming diffusion barriers that prevent further metal migration in the polymer. Diffusion barriers can also be formed by preirradiation of, e.g., PET or PMMA with 2-MeV He^+ ions before metal deposition. In [33] PI samples were implanted with Ag^+, Cu^+ and Pd^+ ions to fluences from 10^{15} to $5 \times 10^{17}\,cm^{-2}$, and metal nanoparticles with a mean diameter of about 5 nm were observed in the highly carbonized PI surface layer. Electron diffraction patterns suggest that the particles have a crystalline structure.

Similar nanoparticles were also created by high-fluence implantation of Ag into PMMA [34] and silicone resin [35], and of Au in PMPySi, see Fig. 2.9b.

For unirradiated PI, Cu and Ag in polymers, the diffusion coefficients are in the $10^{-15}\,cm^2\,s^{-1}$ range when heated in vacuum. In N_2, O_2, moisture or air atmosphere, Ag and Cu exhibit higher diffusion at temperatures in excess of about 548 K. The reason is oxidative degradation of PI even at small amounts ($\leq 1\%$) of oxygen. Ag diffusion values show a marked transition at 540 to 560 K, while Cu diffusion coefficients vary gradually when increasing the temperature. At 613 K, Cu diffusion values have reached the $10^{-14}\,cm^2\,s^{-1}$ range, while Ag diffusion coefficients have become as large as $9 \times 10^{-13}\,cm^2\,s^{-1}$.

Please take into account that all experimental results in the polymer field are subject to many ambiguities. Even small parametric changes may influence the diffusivities by many orders of magnitude. These may be changed in the polymeric structure, chemical composition, additives, or polymer corrosion. As an example of the latter case, it has been found that the diffusion coefficients of silver in polyimide are increased by about two orders of magnitude if the polyimide foil uptakes less than 1% oxygen only [36].

2.5 Transport Processes of High-Fluence Implanted Ions

Fluences exceeding the $10^{15}\,cm^{-2}$ region lead to even more dramatic deviations from low-fluence predictions. Here, sputtering, the material's progressive carbonization, density changes, radiation-enhanced diffusion and crack formation start playing a non-negligible role. In general, the as-implanted profiles roughly resemble the theoretically expected distributions of nuclear damage, rather than those of the theoretical ion range or electronic damage distributions [22]. They are always shorter and mostly narrower than the nuclear-damage distributions predicted by theory. This appears to reflect both the neglection of cross-terms between nuclear and electronic stopping [37] and the

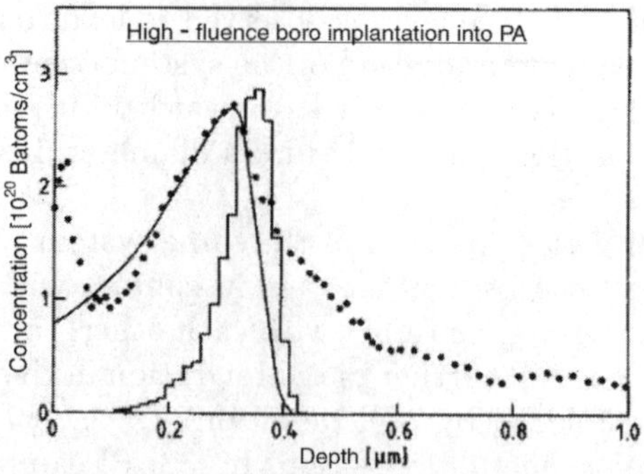

Fig. 2.10. Typical depth profile of high-fluence implanted boron in polyamide (PA); comparison of measurement (*points*) with low-fluence theory (histogram: range profile, and *solid line:* nuclear damage distribution according to TRIM [38]). Theory adjusted to the experimental depth scale. The detector resolution does not influence the measured profile markedly. After [22]

material's compaction, which sets in at low fluences. Additionally, the high-fluence implantations lead to the emergence of tails extending to the near-surface and far sample regions, which points to a considerable mobility of the implants, once their solubility limit is exceeded. Often the sample surface acts as a sink for the diffusing matter, which leads to a surface precipitation of the implants, Fig. 2.10. Such pronounced surface peaks have never been found after low-fluence ion implantation, and hence are considered as just another indication for the enhanced mobility of implants above their solubility limit.

In order to understand the redistribution in more detail, a diffusional simulation was undertaken. This work (Fig. 2.11) reveals that the subsurface profile shape can be completely understood in terms of a depth-independent radiation-enhanced diffusion in the order of $Dt = 10^{-10}$–10^{-11} cm^2, and with trapping in unsaturable traps of collisional origin, plus trapping in the near-surface regime. The emergence of unsaturable traps is a new feature in high-fluence ion implantation, as the ions' redistribution after low-fluence implantation is always only affected by trapping in saturable traps. This indicates the onset of formation of precipitates of the implanted matter at high fluences.

For the example of 100-keV boron in various polymers it was shown that isochronal annealing of the high-fluence implanted samples always reveals two stages, as summarized in the thermal desorption spectra (TDS) of Fig. 2.12. First, at slightly elevated temperatures (typically $\sim 50°$C) everpresent intrinsic impurities are released into the whole polymer (i.e., even beyond the implanted zone), and are trapped in the near-surface highly radiation damaged zone. As a result, this leads to a continuous increase of the total amount of near-surface boron, eventually even beyond the originally implanted amount. For most polymers the gettering efficiency of both the implanted and the intrinsic boron is highest at around 50°C.

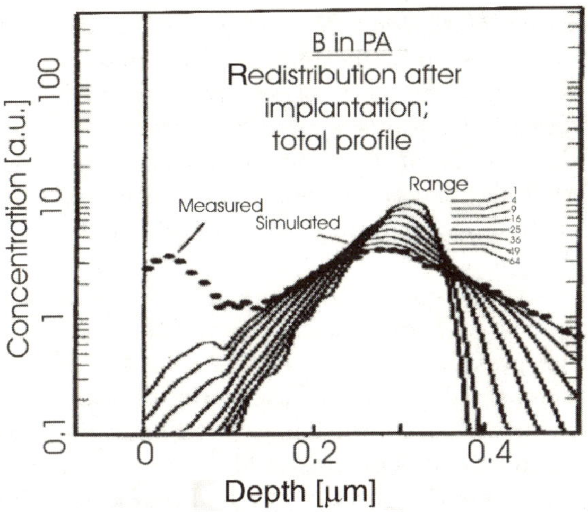

Fig. 2.11. Redistribution of 100-keV B implanted into PA by radiation-enhanced diffusion. Computer simulation in steps, each step corresponding to $Dt = 5.6 \times 10^{-13}\,\mathrm{cm}^2$. The number of steps is marked near each curve. The initial curve is the theoretically predicted range profile, corrected in the depth scale to fit to the experiments. The points mark the measured boron distribution. After [22]

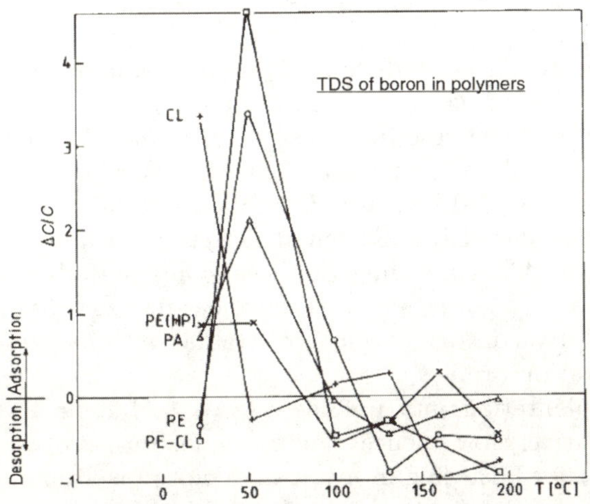

Fig. 2.12. Thermal-desorption spectrum of 100-keV boron implanted into various polymers at high fluences. PA (polyamide), CL (cellulose), PE (natural and high-pressure-treated polyethylene), and mixtures of these polymers. Two stages can be distinguished, first ion-induced defects getter diffusing matter (where $\Delta C/C > 1$), and then a stage continuous desorption (i.e., $\Delta C/C < 1$). After [22]

The second stage of thermal desorption sets in at $\sim 100°C$, when one observes a smooth decrease in the bulk boron content with annealing temperature. This is attributed to boron desorption from the radiation-induced defects. This desorption is quite similar for different polymers, in spite of their different glass temperatures. The reason for this is found in the relatively high destruction yield that renders the concept of glass temperatures obsolete for these modified materials. As the TDS spectrum reveals a continuous release, one must conclude that the (collisionally based) traps exhibit a broad spectrum of different bonding energies. For some polymers (PE and CL) part of the released mobile boron precipitates at the surface; in other systems (PA, high-pressure-treated PE) the already existing surface boron precipitations decrease. The shape of the boron distributions (reflecting the nuclear trap distribution) hardly changes with annealing; only at the highest examined annealing temperatures (190° C) do the distributions broaden that can be explained by the onset of thermal boron mobility with diffusion coefficients of the order of 10^{-13}–$10^{-12}\,cm^2\,s^{-1}$.

Ion irradiation of polymers at the highest fluences have revealed another new phenomenon: occasionally also deep-lying double-peaked structures are found, extending far beyond the range, which cannot yet be properly explained [39,40]. (Examples: 40-keV Fe^+ and Co^+ irradiation of PI at 2.5×10^{16} to $1.25 \times 10^{17}\,cm^{-2}$ [40].)

A great deal of the implanted material is usually found at the surface where it precipitates and – if reactive – reacts with oxygen to form oxide (e.g., K_2O [39]).

In this connection it is worthwhile discussing ion implantation into carbon that might be regarded as the utmost decay product of irradiated polymers. In contrast to the above-described observations, *light* ion implantation into, e.g., pyrographite yields exponential profile shapes, which can be understood as range profiles, modified by (radiation-enhanced) diffusion with subsequent trapping at homogeneously distributed defects with a defect density of (5 to 10) $\times 10^{15}$ cm [43]. Annealing of these samples leads to profile shapes that are additionally governed by the nuclear-damage distributions of the implanted ions. By contrast, *heavy*-ion irradiation into pyrographite yields regular implantation profiles.

Gas-ion implantation into polymers leads to bubble formation and to blistering at relatively low fluences. For example, Shrinet et al. [42] found that the critical dose for blistering in mylar and kapton is around $7 \times 10^{15}\,cm^{-2}$. The blisters observed are very large (10 to 150 µm), as compared with the particle range. The emergence of gas bubbles and blisters indicates that when the gas ions are implanted at sufficiently high flux, they – as well as the newly formed gaseous reaction products – cannot escape rapidly enough from the polymer so that they become enriched to form bubbles that finally deform the polymer surface.

2.6 Transport Processes of Liquid Penetrants within Ion-Irradiated Polymers

Instead of using mobile as-implanted ions to probe their self-created damage distributions, one also can take any other suitable marker for this purpose. This marker can be gaseous, liquid, or solid, or one can implant positrons instead of ions.

2.6.1 Liquid Penetrants in Ion-Irradiated Polymers

Most of the examinations reported below have been performed with 3 mole/l or 5 mole/l aqueous LiCl solutions, the Li component having been used as the tracer to probe the corresponding damage distributions via NDP. In order to obtain an idea about the dynamic processes of liquid penetration into ion-irradiated polymers, diffusional simulations were performed. For PI samples irradiated at fluences below 1×10^{13} to $3\times10^{13}\,\mathrm{cm}^{-2}$ *all* penetrant depth profiles are similar to those observed for pristine PI under the same conditions – monotonously decreasing from a surface maximum to depths far exceeding 10 µm. For samples irradiated at higher fluences, however, the profiles show bimodal distributions, with near-surface regions of high penetrant concentration, ranging deeper than the projected range by $\sim 10\%$ for He and by $\sim 100\%$ for heavier ions. This component is superimposed by a decreasing tail, similar to that in the unirradiated PI, but with a concentration up to one order of magnitude higher. Figure 2.13 shows an attempt to simulate the electrolyte penetration in such a case. It turned out that in this case only those simulations that can reproduce the measured electrolyte distributions well are these that assume that the nuclear damage centers act as saturable traps, i.e., that each damage site can host just one tracer atom.

For low-energy ion-irradiated PI (3-MeV C^+ at $\approx 10^{14}\,\mathrm{cm}^{-2}$ [43], 200-keV Ne^+ at $\approx 10^{12-13}\,\mathrm{cm}^{-2}$ [6], and 100-keV B^+ at $5\times10^{16}\,\mathrm{cm}^{-2}$ [44], the overall depth distributions of an aqueous LiCl tracer solution (as determined by NDP) is constituted of two components, one that describes the penetration into the pristine material and is independent of the ion-track direction, and another that is due to penetration along the excess free volume of the ion-irradiated zone. The latter component follows more or less the distributions of the ion's transferred nuclear energy, though the ratio of electronic to nuclear stopping powers is in all these cases much higher than unity, ranging from about 5 to 1200. Possibly, electronic defects are too small to host the penetrating Li^+ ions, as they are surrounded by a relatively large hydrate shell in their aqueous environment.

In the overwhelming number of examined systems (aqueous LiCl, methylene blue, styrene, Co-carborane, or Pb-acetate solutions, styrene solution, butyl-Li solution in THF, and fullerene solution in toluene, penetrating into PVDF, PI, PE, or PP irradiated with 300-keV H to 5.35-GeV Xe ions at fluences between 10^9 and $10^{15}\,\mathrm{cm}^{-2}$) liquid penetrants decorate the electronic

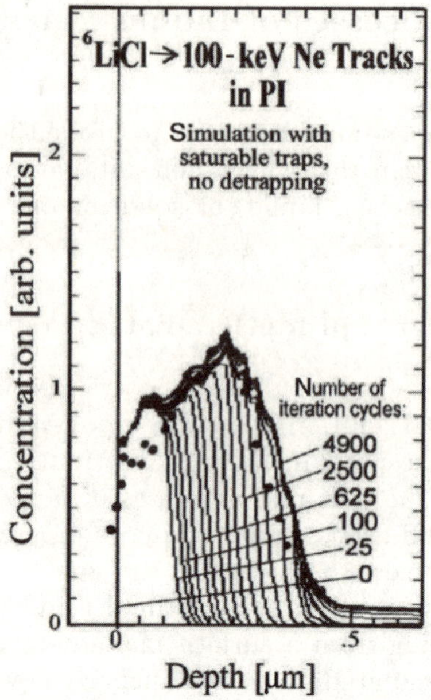

Fig. 2.13. Diffusion simulation of the electrolyte penetration from the outside through the sample surface into 100-keV Ne predamaged PI. Simulation shown for the case of regular diffusion with $D = 2 \times 10^{-13}\,\mathrm{cm^2\,s^{-1}}$, with saturable traps distributing according to the nuclear energy transfer [6] (TRIM scale adjusted to experiment). The numerical simulation program proceeds in iterative steps, the numbers of which being given. The points indicate the measured electrolyte spectrum. Apart from decoration of the nuclear-damage distribution, the electrolyte is seen to exhibit a long tail penetrating deep into the polymer's bulk

defects [44], however occasionally also collisional defects may dominate the trapping process (e.g., in the case of penetration of a LiCl solution into 100 or 200-keV Ne irradiated PI [6]). Positrons as electrically charged probes are always found to decorate the electronic damage sites in polymers, see, e.g., Fig. 2.14.

Deviations from the pure electronic or nuclear energy transfer distributions were found after aging. PI irradiated with 100-keV He, Ne, Ar, Kr, and Xe at fluences between 10^{12} and $10^{16}\,\mathrm{cm^{-2}}$ were marked with 5 mole/l LiCl solution at ambient temperature about 10 years after their irradiation and storage under ambient conditions [45], and the Li depth distributions were then probed with NDP. In this case, saturation with the penetrating marker was achieved for Ar^+-, Kr^+-, and Xe^+-irradiated PI after 1 h exposure at ambient conditions, for any irradiation fluence. The same was found for He^+ irradiated PI at fluences below $1 \times 10^{15}\,\mathrm{cm^{-2}}$, however, increasing the He^+ fluence by a factor of 10 resulted in an increase of the penetrant uptake at prolonged times by a factor of 4. In the case of Ne^+-irradiated PI, the ex-

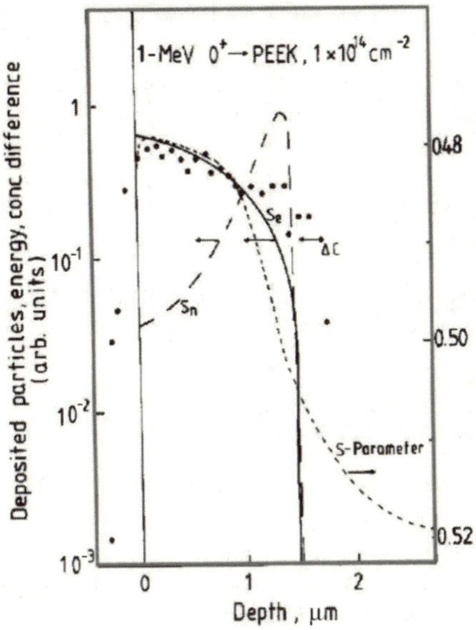

Fig. 2.14. The as-measured positron annihilation depth distribution (S-parameter; *thin dashed line, right scale*) and the electrolyte distribution (*points, left scale*; as determined by NDP) in ion-irradiated PEEK follow both the electronic energy transfer distribution (*solid line*), but not the nuclear damage distribution (*thick dashed line*) [3,4]

tension of the exposure time from 1 to 25 h led to an increase of the dopant content by a factor of 2.6 for fluences below 1×10^{13} and above $1 \times 10^{15}\,\mathrm{cm}^{-2}$, and by a factor of 2.0 between 3×10^{13} and $3 \times 10^{14}\,\mathrm{cm}^{-2}$. The different time dependences of the penetrant uptake for different implanted projectiles at different fluences are due to the different thicknesses of the irradiated layer and to different degrees of compaction and carbonization. In conclusion, one may say that most ion-irradiation effects in polymers exhibit long-term stability, with only a few changes occurring during aging.

One-dimensional diffusion simulations indicate that one gets good agreement between simulation and the measured marker distribution only when assuming a hindered diffusion towards deep saturable traps. The term "deep" signifies that the marker cannot desorb again from the traps, and the term "saturable" means that the defect can bond only one Li^+ ion, after which it loses its trapping property for the next Li^+ ion passing by. One can deduce a lower limit of some $2 \times 10^{-13}\,\mathrm{cm}^{-2}\,\mathrm{s}^{-1}$ for the marker solution's diffusion coefficient into low-energy ion-irradiated PI.

Both the anionic and cationic components of an ionic penetrant in solution can have quite different diffusion coefficients. Thus, it was found that the effective diffusion coefficients of Cs^+ cations and $[(C_2B_9H_{11})Co]^-$ anions of an aqueous metallocarborane solution, penetrating into PE that was irradiated

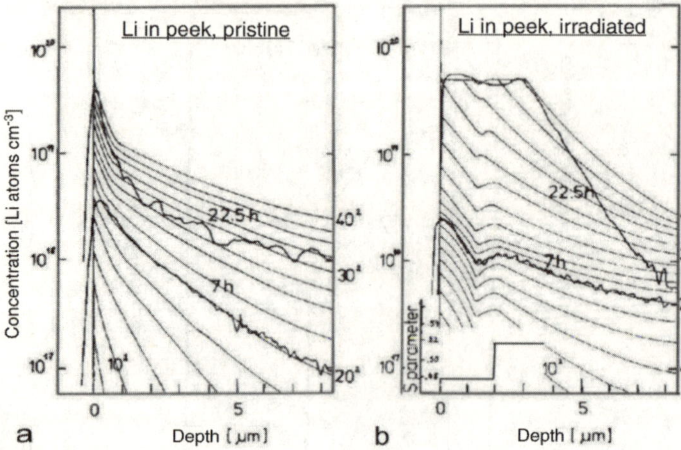

Fig. 2.15. Distribution of electrolyte penetrating into (**a**) pristine, (**b**) irradiated PEEK. *Thin lines:* diffusion simulation with depth-dependent diffusion and trapping coefficients; detrapping probability = 0. The insert of (*b*) shows the as-measured S-parameter distribution

by 150-keV F^+ ions up to fluences of 5×10^{13} to 1×10^{15} cm^{-2}, differed by up to a factor of 10^4 from each other [46]. As a result, dramatic differences in the stoichiometric ratio emerged, ranging to deviations between a factor 5 and 40. The lower uptake and mobility of the larger metallocarborane component is ascribed to the much larger dimensions of that anion, as compared with the small cation. In this case, both the boron and cesium uptake increased with temperature (determined at 24, 50, and 85°C), and also the diffusion coefficients followed a more or less Arrhenius-like behavior with an activation energy of around 1 eV. The uptake decreases with increasing ion damage. The depth distributions of both components follow more or less the electronic energy loss profiles for low-fluence implanted PE, but tend to follow the classical Fickian diffusion in high-fluence irradiated PE. However, already in low-fluence irradiated PE, an additional diffusional tail emerges that extends far beyond the implanted region.

In the case of cluster-ion impact (e.g., 1 MeV/atom C^{n+}, $n = 1$–8 [43]), the penetrant profiles were found to consist of two components, one that follows the nuclear-energy transfer distribution, and a near-surface one with a tail towards the bulk. With increasing cluster projectile size, the near-surface component increases dramatically, whereas the relative importance of the bulk component gradually decreases. This is understood by the near-surface disintegration of the clusters into individual components, with a much enhanced near-surface energy transfer as compared with single-atomic ions. In the extreme case of C_{60}^+ impact, the penetrant distribution follows, however, an exponential shape ranging far beyond the projectile's range. This is inter-

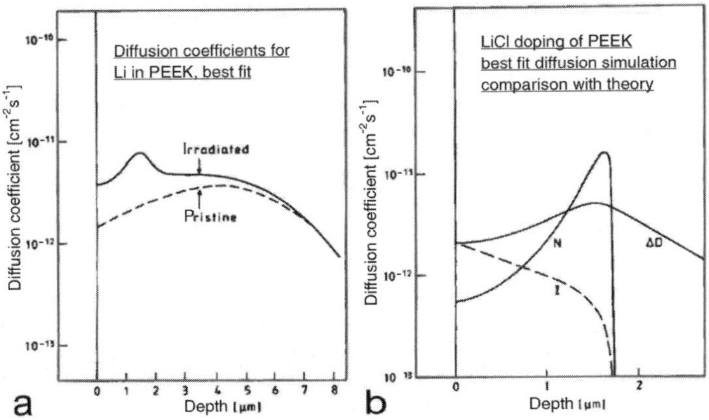

Fig. 2.16. (**a**) Depth-dependent diffusion coefficient, obtained from an optimum fit of the diffusion simulation to the experimental results of Fig. 2.15. (**b**) Comparison of the difference in diffusion coefficients ΔD for irradiated and pristine PEEK with nuclear (N) and electronic (I) energy transfer curves

preted as the buildup of long cracks along the projectile trajectory, due to the extremely large energy transfer per pathlength that exceeds that of swift heavy ions considerably [47].

PEEK has some peculiarities, insofar as it behaves differently from PI, PET, PSiO and others. It exhibits a clear Case II diffusional behavior for marker penetration. This is seen most pronouncedly for the irradiated matter, where the penetrant mobility is faster so that a concentration plateau is formed (Fig. 2.15). The experimental spectra can in this case only be simulated if the diffusion coefficient D is assumed to be pronouncedly depth dependent (Fig. 2.16a). The depth dependence of D in the case of irradiated PEEK is found to be influenced by nuclear, rather than by electronic damage in the case of long exposure (Fig. 2.17b) and by electronic damage in the case of short exposure (Fig. 2.14); [48].

Another system that shows clear Case II diffusion is the diffusion of a fluorinated hydrocarbon, 3,5-bis (trifluoromethyl) aniline (TFMA), in 30-keV H^+-irradiated (10^{13} to $10^{15}\,cm^{-2}$) poly(styrene). It is the swelling of PS by sorption of TFMA that determines its diffusion. With increasing degree of ion-induced cross-linking, the diffusion coefficient decreases by about two orders of magnitude [13].

Leaching of ion-irradiated and LiCl-solution-soaked samples led to different results. Whereas rinsing in distilled water for 1 min did not change the Li distribution in Ne^+-irradiated PI (for whatever irradiation fluence and penetration time), Xe^+-irradiated samples showed considerable leaching of the Li marker by factors of 2 and 5 for samples exposed for 1 and 25 h, respectively (independent of ion fluence). The high stability of the penetrant

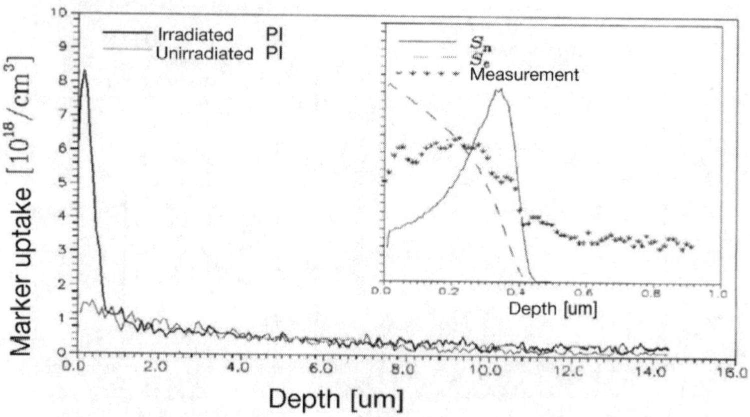

Fig. 2.17. Electrolyte transport in a high-fluence irradiated polymer [44]. Insert: detail as compared with theory

in the Ne^+-irradiated samples may perhaps be explained by the much deeper marker penetration in this case [49].

In the case of very high fluence irradiation of polymers by light ions (e.g., 100-keV B^+ at $5 \times 10^{16}\,\mathrm{cm}^{-2}$), a long tail in the Li^+-marker distribution, extending up to at least 15 µm (i.e., far beyond the ion range), develops additionally to the ion redistribution along the nuclear damage profile [44], which one could speculatively attribute to long-ranging crack formation due to the strong density differences between the bulk polymer and the ion-irradiated carbonized surface, Fig. 2.17.

2.6.2 Transport of Penetrants in Three Dimensions

By means of the modified tomographic approach MOTOR (see Vol. I, Chap. 3), one can reconstruct damage distributions of ion impact into polymers even in three dimensions. Figures 2.18 and 2.19 illustrate such an example. Here, the same case was treated in 3 dimensions that had been simulated above in Fig. 2.13 in one dimension [50]. The LiCl-penetrant distributions were fitted by 3D split Gaussians:

$$G(x, r) = A \exp(-\{1/2[(z - R_\mathrm{p})/\sigma]\}^2) \exp(-\{1/2[r/\sigma_\mathrm{trans}]\}^2)\,, \qquad (2.2)$$

with $\sigma = \sigma_1$ for $z < R_\mathrm{p}$ and $\sigma = \sigma_2$ for $z > R_\mathrm{p}$ being the stragglings in the longitudinal (beam) direction, σ_trans being the straggling in the transversal (radial) direction, R_p being the mean projected range, and A being a normalization factor. In the case of the highest fluence ($10^{14}\,\mathrm{cm}^{-2}$), these Gaussian distributions had to be truncated at the concentration of saturation to obtain a realistic distribution. Comparison with theory according to TRIM [38]

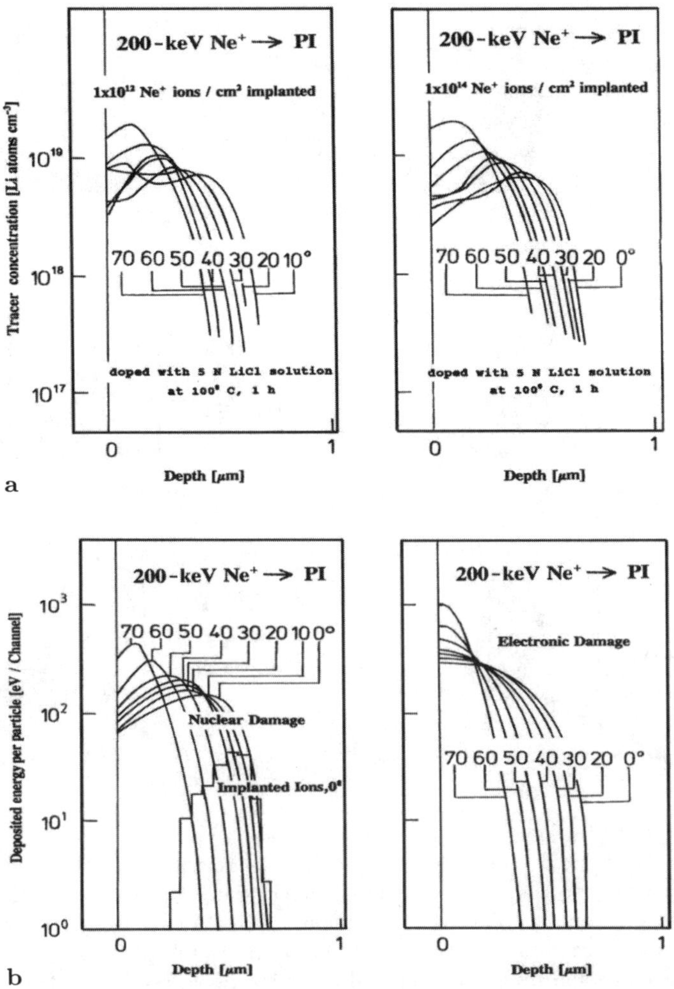

Fig. 2.18. Distribution of an electrolyte penetrating into a polymer that was previously irradiated by low-energy ions. Example: 200-keV $^{10}Ne^{+}$ preirradiation of PI under various angles and fluences; subsequent penetration of 5 mole/l LiCl solution at ambient temperature. Measurement of the electrolyte distribution via Li as tracer element by NDP. Shown are (**a**) the electrolyte distributions for two different fluences (1×10^{12} and $1 \times 10^{14}\,cm^{-2}$), and (**b**) theoretical predictions for the deposited particles: nuclear and electronic damage. Comparison of (**a**) and (**b**) shows that the electrolyte appears to distribute according to the nuclear-damage distribution of the implanted ions [6]. The curves for 0° and 10° in (**a**) are practically identical within the experimental accuracy

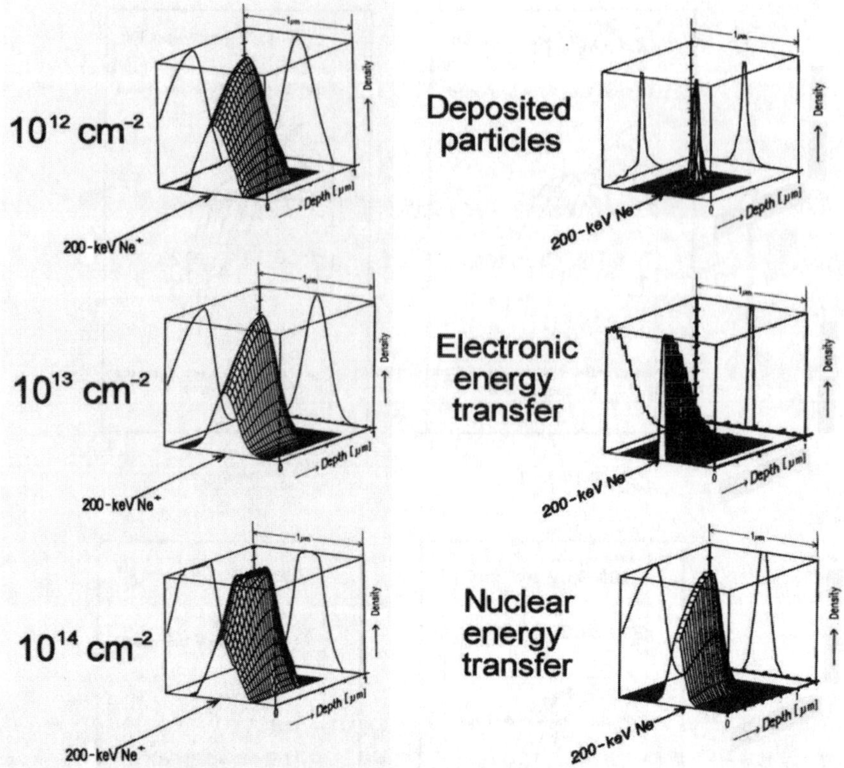

Fig. 2.19. *Left side:* The tomographically reconstructed 3D distributions of the electrolyte according to Fig. 2.18, and *right side:* The corresponding theoretical predictions. It is clearly seen that the electrolyte distributes in the ion-irradiated polymer according to the nuclear-energy transfer distribution even in three dimensions [6]. Such a tomographic reconstruction is possible as the energy density was chosen to be the same for all directions, and as the total penetrant concentrations of all one-dimensional input depth profiles are identical, see Fig. 2.18

makes it clear that the Li^+ marker ions redistribute according to the three-dimensional distribution of the nuclear, rather than of the electronic damage or of the deposited projectile ions.

Please note that the concentrations of marker uptake are nearly of the same order of magnitude for all cases, independent of the fluence of the Ne^+ ions ($10^{12-14}\,cm^{-2}$). This finding has been obtained also on other occasions. It may be understood by assuming that each impinging ion is capable of compressing part of the track neighborhood to some extent so that the efficiency of an individual track to trap mobile penetrants decreases reciprocally with fluence. (In fact, e.g., irradiation of porous glasses shows the same effect of decreasing porosity with increasing ion fluence.) Above a critical fluence of about $10^{13}\,\mathrm{cm}^{-2}$ the marker distributions broaden in the longitudinal direction but roughly maintain their average depths and lateral widths. The broadening is thought to be the consequence of the maximum possible tracer uptake coming to saturation, which signifies the formation of a plateau.

The formation of radiation-induced free volume with pore sizes large enough to host the Li^+ ions in solution (i.e., with their hydrate shell of several Å radius) stems from nuclear collisions (and not from electronic bond breaking) with subsequent degassing of radiochemical fragments. This is remarkable insofar as the deposited electronic energy density, $0.36\,\mathrm{eV/nm^3}$, exceeds that of the nuclear energy density ($0.08\,\mathrm{eV/nm^3}$) by about a factor of 5 in the given case (200-keV Ne $\rightarrow$ PI at $1 \times 10^{12}\,\mathrm{cm}^{-2}$), but the electronic damage is inefficient for hosting the Li^+ ions.

This finding is also remarkable as Li^+ and B^+ projectile ions of comparable energy are known to redistribute after their implantation into PI and other polymers according to their *electronic*, rather than their nuclear energy transfer distribution [20, 21, 51]. This apparent disagreement clears up when taking into account that in the latter case one deals with "naked" Li^+ or B^+ ions without any hydrate shell, which are correspondingly smaller than those surrounded by water molecules. Apparently, electronic defects can just host the naked ions but not the hydrated ones.

Even for the highest fluence, this finding does, in principle, not change. From the ratio of lithium uptake to ion fluence one deduces that each Ne^+ ion is capable of hosting roughly 200 Li ions within its created free volume. Taking into account that each Ne^+ ion creates around 100 nuclear defects, one may conclude that nuclear defects are not very efficient in trapping hydrated Li^+ ions – or that a considerable fraction of them anneals out.

The marker uptake turned out to be limited to a maximum of around 1×10^{20} Li atoms/cm^3, which signifies that the maximum possible free volume in the irradiated matrix is around 0.1% atomic. This maximum free volume is correlated with a totally deposited nuclear energy density of about $8\,\mathrm{eV/nm^3}$. If more energy is deposited, the efficiency to produce excess free volume drops to zero. This behavior signifies that all possible free volume stems from individual ion impacts only, and that at the onset of ion-track overlapping no more additional free volume is formed. Similarly, the marker uptake as a function of the total deposited energy density in, e.g., low-energy ion irradiated PI or PET is constant above some $\sim 10\,\mathrm{eV/nm^3}$ and remains unaltered up to at least $10^3\,\mathrm{eV/nm^3}$.

2.7 Penetration of Gases into Ion-Irradiated Polymers

Gaseous penetrants behave quite similarly to liquid penetrants in irradiated polymers (see also Table 2.4). The experiments performed on the penetration of gases into low-energy ion-irradiated polymers were restricted essentially to the examination of iodine and mercury [8, 52, 53]. The ion irradiation of

Table 2.2. Penetrant content (in atoms cm^{-2}) in a 300-nm thick surface layer of pristine PET, and PET irradiated to a fluence of 5×10^{13} cm^{-2}. Penetration time in brackets [53]

Penetrant	Li (2 h)	I (3 h)	Hg (30 d)
Pristine PET	2×10^{16}	1.2×10^{16}	1×10^{15}
Irradiated PET	7×10^{16}	4×10^{16}	3×10^{15}

polymers such as PET or PE always leads to an increase of the content of incorporated penetrants by a factor of the order of 3 to 4, regardless of their chemical nature and penetrant conditions (see Table 2.2). The iodine uptake in pristine polymer is significantly higher for PET than for PE, but the effect of the ion irradiation on the iodine incorporation is much stronger for PE than for PET. The total uptake of iodine in PE first increases linearly with fluences until it saturates (at around 10^{14} cm^{-2} for 150-keV F^+ and As^+-irradiated samples).

The influence of the transferred energy loss type on the free volume has been demonstrated nicely by, e.g., Davenas et al. [54]. Whereas 2-MeV α-particle irradiation of PI leads to an increase of gas permeability, 170-keV N^+ impact onto PI induces an important reduction in the methane transport capability through PI foils. In general, atomic displacements appear to induce new free volume, resulting in an enhancement of penetrant diffusion. Radiolysis-induced high concentrations of dangling bonds are responsible for trapping of penetrants. High-fluence nuclear-energy transfer is efficient to create densification due to cross-linking and carbonization. This leads to reduced permeability and a lower concentration of potential trapping sites for penetrant atoms. In other words, polymer surfaces treated in such a way act as efficient diffusion barriers [54].

As is often the case of liquid penetrants, the depth profiles of both I and Hg penetrants in pristine polymers are featureless curves, gradually decreasing from their maximum value towards the bulk. In the case of low-fluence (10^{13-14} cm^{-2}) ion-irradiated PET and PE profiles are superimposed by an additional component that roughly follows the distribution of electronic-energy transfer. For high-fluence-irradiated PET and PE (10^{15} cm^{-2}) both these components vanish in favor of a third one that roughly follows the nuclear-damage distribution.

This can be explained by the coexistence of the material's structural changes in the course of the ion irradiation via the two well-known basic processes of chain scissioning and cross-linking, with the consequences of creation and annihilation of excess free volume, which leads to a diluted phase and a compacted phase, respectively. The concentration $N(x)$ of incorporated iodine atoms at a depth x is then supposed to be equal to the products of the local fraction of the diluted phase $P(x)$ and the concentration of potential iodine trapping centers $C(x)$ therein: $N(x) = C(x)\,P(x)$. Now $C(x)$ is

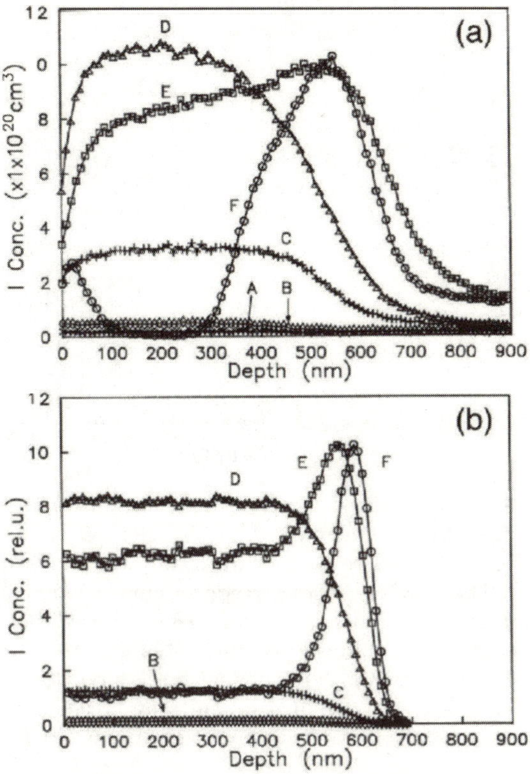

Fig. 2.20. Concentration profiles of iodine incorporated in 150-keV F^+-irradiated PE at fluences of A: 1×10^{11}, B: 1×10^{12}, C: 1×10^{13}, and D: 1×10^{14} cm^{-2}; (**a**) as-measured and (**b**) simulated according to (2.3)

assumed to be proportional to the product of ion fluence Φ and the total stopping power $S_{\text{tot}}(x)$ of the ions at depth x: $C(x) = \Phi S_{\text{tot}}(x)$, and $P(x)$ is obtained assuming that the rate $\mathrm{dP(x)/d\Phi}$ at which $P(x)$ disappears is proportional to $S_{\text{tot}}(x)$, to the remaining fraction of diluted volume $P(x)$, and to the chemical yield of cross-linking $G: \ \ P(x) = \exp(-\beta G S_{\text{tot}}(x)\Phi)$, with β being the effective volume occupied by one cross-link. Then:

$$N(x) = S_{\text{tot}}(x)\Phi \exp(-\beta G S_{\text{tot}}(x)\Phi). \tag{2.3}$$

The thus-obtained iodine concentration distributions are compared with experimental findings in Fig. 2.20. It is seen that both figures show, in principle, the same tendency concerning the evolution of the depth profiles, thus indicating that both electronic- and nuclear-energy transfer cooperate in this case. The differences between experiment and simulation that also show up quite clearly illustrate the present state-of-the art of understanding the transport processes occurring in irradiated polymers.

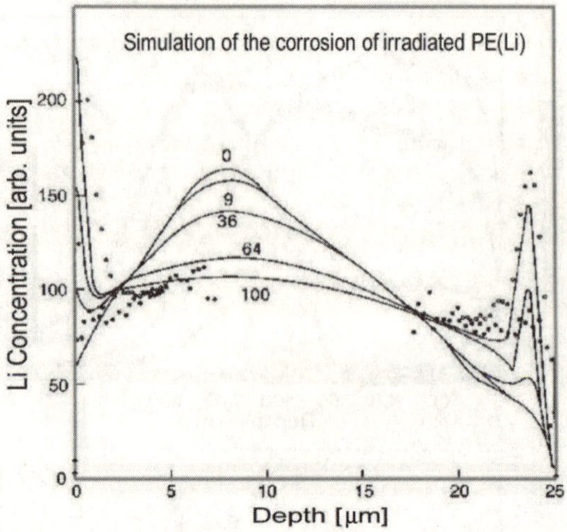

Fig. 2.21. Corrosion of Li-doped methacrylic acid-grafted PE. The original Li^+ distribution (described by the line marked by "0") changes towards the line marked by "100" after more than half a year exposure to ambient air – see the *open dots* (experimental points). The *smooth curves* (except for curve 0) are results of a computer simulation; the numbers denote the iteration cycles required. Simulation performed with a Li room-temperature diffusion coefficient of $2 \times 10^{-14}\,\mathrm{cm^2\,s^{-1}}$ and a surface trapping probability of 20%, and under the assumption that no permanent trapping would take place in the bulk PE. *Full dots:* Li distribution after removing the near-surface precipitation by dipping the sample briefly into water (NDP measurements)

2.8 The Influence of Surfaces on the Transport of Penetrants

Still another point deserves attention. It appears that frequently polymers are covered with a protective surface layer that largely prevents the rapid indiffusion of, e.g., ambient oxygen. In this way, even sensitive polymeric components (e.g., Li^+ ions in ionic-conducting grafted polyethylene [55]) are stabilized for quite a time. This sealing may be due to superficial polymer oxidation and surface compaction to a denser structure. Also, the precipitation of (intrinsic or external) impurities (i.e., "dirt") at the surface may help turn the latter into a diffusion barrier.

Anyhow, these protective layers can be disturbed or destroyed by low- or high-energy ion irradiation, either due to sputtering or ion-beam mixing. The consequences can be a disastrously rapid oxygen indiffusion with associated internal corrosion of the polymeric bulk. This effect has been compared by Hnatowicz with the "opening of a door, through which thereupon unwanted guests enter in great numbers".

As an example, see Fig. 2.21, [55]. Li^+ ions in methacrylic acid-grafted PE with a nonuniform but stationary depth distribution (see Chap. 1 of Vol. I) start redistributing, once the ionically conductive polymer is ion irradiated. The reason is found in the destruction of the sealing sample surface by the ion irradiation (10^{15} He^+ ions/cm^2), which enables the exposure of the chemically reactive Li to penetrating ambient air. The corrosive action of both air and water vapor leads to the formation of immobile Li_2O, LiOH, and Li_2CO_3 that precipitate at the sample surface within half a year after the irradiation. Simultaneous with this surface precipitation, a depletion of mobile Li^+ in the bulk is observed, see the transition from curve 0 (original Li distribution) to curve 100.

2.9 Transport of Intrinsic Polymer Components after Ion Irradiation

There are two counteracting topics with which one has to deal in this connection, with the increased and with the retarded mobility of radiochemical products. The increased mobility may lead to changes in the local density, composition, and the overall shape of the polymer sample. This effect is counteracted by the action of some of the radiation-induced defects that act as traps for the mobile products, and thus contribute to a radiation-induced enhancement of thermal polymer stability.

2.9.1 Mobility of Radiochemical Products

Quite a lot of work has been invested to study the composition of the radiochemical products (see Vol. I, Chap. 7) as this gives the decisive clue for the radiochemical processes going on in polymers during ion irradiation. These studies concentrate on those products that are small enough to escape from the host and therefore can be recorded easily in neighboring mass spectrometers. If such studies are performed in a time-resolved mode, one can derive additionally the migration speed of such volatile products in the host polymers. This has, in fact, been done at various times. From the multitude of results we select here as a representative example just the measurements on the diffusion coefficients of the molecular fragments H_2, C, C_2H, C_2H_2, C_3H_3, C_3H_5, C_4H_8, C_4H_9, C_5H_{10}, C_5H_{11}, and C_7H_8 in 400-keV He^+-irradiated PVT [56]. Their diffusion coefficients D, if depicted as a function of their mass M, yield a smoothly decaying curve $D(M)$, decreasing from $D(H) = 2.5 \times 10^{-8}\,cm^2\,s^{-1}$ according to the above-cited sequence down to $D(C_7\,H_8) \sim 9 \times 10^{-10}\,cm^2\,s^{-1}$.

An indirect consequence of this mobility appears to be that occasionally defect structures may cluster, which can be verified, e.g., by etchant attack. In this way, e.g., a PET sample irradiated with 200-keV B^+ at $1 \times 10^{13}\,cm^{-2}$

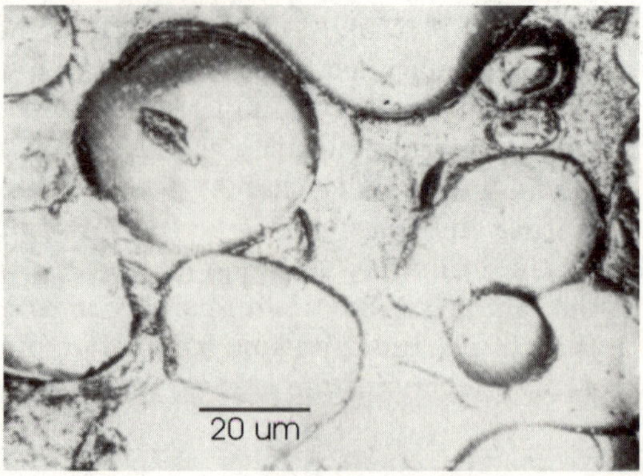

Fig. 2.22. Etchant attack of a 200-keV B^+-irradiated PET foil by KOH, half a year after the irradiation. Optical microscopy

showed after attack with 5 mole/l KOH half a year after the ion irradiation large irregular etch holes that exceeded the implantation range by nearly two orders of magnitude, Fig. 2.22. By contrast, etchant attack of the as-irradiated sample led to slow and homogeneous removal of the irradiated layer only.

2.9.2 Photoresist Inhibitor Mobility

In contrast to the positive photoresist Novolak of the AZ1350 series, the corresponding inhibitor (see Vol. I, Chap. 1) contains sulfur, so that the mobility of the inhibitor (or its radiochemical destruction products) within the photoresist can be probed by means of RBS, using sulfur as its marker [57]. The sulfur is bound to the inhibitor as SO_3. The comparatively small sulfur signal, corresponding to some 1–3 wt% in the dry material, can be well resolved in the RBS spectra when using a lin/log display of the spectra, Fig. 2.23.

It is found that the sulfur spectra change after ion implantation or thermal treatment insofar as a pronounced surface precipitation shows up. Similar effects are not found for the other observable components C, N, and O, which points to the stability of the photoresist itself. Systematic examinations with 20-keV Ne^+ and 15–20-keV F^+ irradiation at fluences of 6.4×10^{14} to $4 \times 10^{15}\,cm^{-2}$, isochronal furnace annealing up to 500°C, and subsequent simulations with a diffusion simulation code yielded the following results [57]:

(1) The sulfur enriches a near-surface layer of about 60 nm thickness, followed by a depletion of the sulfur concentration up to a depth of at least 200 nm, see Fig. 2.24, i.e., both the S-enriched and S-depleted layers extend far beyond the ion-irradiated region. The implanted ions themselves remain where they came to rest, and do not participate in the sulfur migration.
(2) For pristine samples and those implanted at low fluence these changes set in at 200°C. For samples implanted at higher fluences these changes

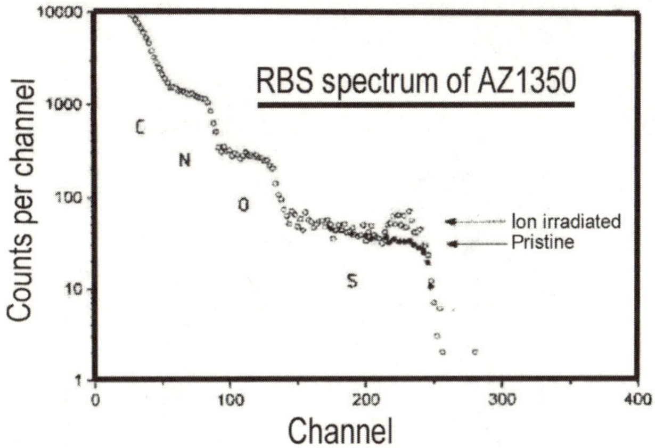

Fig. 2.23. Typical RBS spectrum of the AZ1350 photoresist, pristine (*full circles*) and F^+ ion irradiated (*open circles*). The C, N, O, and S signals can be clearly distinguished

can already be observed at ambient temperature. After annealing at 300 to 350°C, the sulfur precipitation vanishes in the low-fluence implanted samples. High-fluence implanted samples are not affected by thermal annealing but maintain their sulfur precipitation.

(3) No good fit between experimental results and diffusion simulations could be obtained for the Ne^+-irradiated samples when assuming a redistribution under the influence of irradiation-induced defects only. Good agreement could only be achieved when admitting additionally the oxidation of the sulfur-marked molecules by rapidly indiffusing oxygen, Fig. 2.24. Only about 1 at.% of freely mobile oxygen is required to achieve this effect. In this case, the simulation yields diffusion coefficients D of the order of $7 \times 10^{-17}\,\text{cm}^{-2}$ (at ambient temperature) to $1.0 \times 10^{-14}\,\text{cm}^{-2}$ (at 450 to 500°C), and maximum trapping rates A of 1×10^{-2}.

So it appears that the photoresist is usually covered with a protective surface layer that prevents its oxidation, but that the ion irradiation is capable of destroying this layer, to enable oxygen indiffusion with eventually fatal consequences. As a similar behavior was also observed on other occasions (see above) one has to regard this process as a more general phenomenon. Apparently it does not matter here whether the surface is destroyed by low-energy light ions (e.g., 100-keV He^+; [55]) or high-energy heavy ions (such as 135-MeV Ar^{5+}; [58]) – only a minimum energy density of about $3 \times 10^{-5}\,\text{eV/nm}^3$ has to be deposited in the surface region before marked oxygen indiffusion takes place, independent of the projectile species and energy applied. It is worth mentioning that this energy density is the same as that that had been determined as the threshold energy density for the onset of significant penetrant uptake in PI [44]. Consequently, also in the other cases mentioned

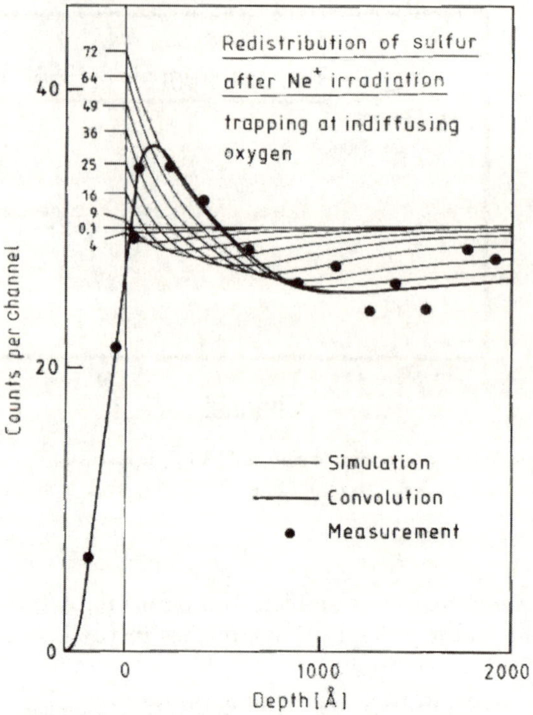

Fig. 2.24. Depth distributions of sulfur in the AZ1350 photoresist after 20-keV Ne^+ irradiation. Curve 0: idealized initial sulfur distribution. *Smooth lines:* Simulated concentration distributions after different diffusion steps (marked by the numbers; each step corresponds to $Dt = 2.3 \times 10^{-13}\ cm^{-2}$). *Thick curve:* Convolution of the best diffusional simulation with the detector resolution. *Points:* As-measured depth profile. The simulation was performed under the assumption that unlimited oxygen diffuses regularly into the polymer and bonds to the mobile inhibitor, i.e., that the oxygen acts as an unsaturable trap to immobilize the inhibitor motion. Superimposed on this is the trapping in radiation-induced defects that act as saturable traps

above, similar changes in depth distributions with near-surface precipitations and far depletions were found.

In contrast to Ne^+ irradiation, F^+ irradiation of the photoresist leads to an efficient expulsion of all sulfur-containing molecules from the ion-irradiated zone so that the depth profiles of sulfur redistribution shift into the bulk by a distance corresponding to the F^+ ion range. The reason is not yet known, but it is clear that it must be the chemical nature of the fluorine that promotes the replacement of the sulfur.

2.9.3 Thermal-Stability Enhancement of Photoresists

In an earlier study it was shown that polymers heavily bombarded with high-energy Ar ions become more resistant to thermal treatment [59], in other words, that the diffusional loss of the pyrolytic decay products decreased. These results have meanwhile been reconfirmed, and put on a sound scientific basis [60, 61]. Especially the photoresist AZ1350-J has been studied by these authors by irradiation of 20- to 1600-keV He, B, F, and Ne ions at fluences between 10^{13} and 10^{16} cm^{-2}. Two different groups of systems were selected, one where the electronic energy transfer dominated, and one with dominant nuclear energy transfer. Subsequently, the samples were annealed in vacuum up to 450°C and then analyzed by RBS. Their results can be summarized as follows:

The pristine photoresist slowly starts decomposing at 300°C, and reaches dramatic decomposition rates above 400°C. For example, at 450°C annealing for 30 min the mass losses are already as large as 30 and 60% for the C and O components, respectively. Irradiated samples show considerably less loss of carbon and oxygen: after irradiation with 800 keV Ne at 1×10^{15} cm^{-2} (i.e., in the regime of dominant electronic-energy transfer) the mass loss reduced to 5 and 15%, respectively, and at a Ne pretreatment up to 2×10^{15} cm^{-2}, virtually no more mass loss was recorded. Even after irradiation with the lighter He ions at a lower energy of only 400 keV at 2×10^{15} cm^{-2}, the mass loss was still much less than in the pristine case, with 20 and 10%, respectively. Only at annealing temperatures as high as 500 to 600°C do the material losses again become marked, with 20% and 50% C release, respectively. If pre-bombardment was done with, e.g., Bi ions instead, the photoresist AZ111 maintains some overall stability up to about 350°C [16].

The threshold fluence to obtain some thermal protection is 2×10^{15} cm^{-2} for both Ne and B, and it is around 8×10^{15} cm^{-2} for He ions. Plotting the same data vs. the amount of energy deposited by electronic-energy transfer during the irradiation, one obtains a universal curve, see Fig. 2.25. As both noble gases and chemical reactive projectiles follow the same trend, one has to conclude that it is the radiation damage rather than the radiochemistry that is responsible for this stability.

The energy density of 1×10^4 eV/nm^3 can be regarded as the threshold energy density transfer for which no more C and O losses are observed at 450°C annealing. The same observation is made when observing the total sample volume, instead of the C or O content. This effect can be attributed to the buildup of electronic damage centers that act as traps to hinder – or even completely prevent – the loss of volatile reaction products. That is, after the irradiation the polymer sample has certainly undergone dramatic chemical changes, but its net chemical formula and its total mass remained unaffected.

Also the thickness of irradiated photoresist is correlated with the transferred electronic-energy density due to the accumulation of excess free vol-

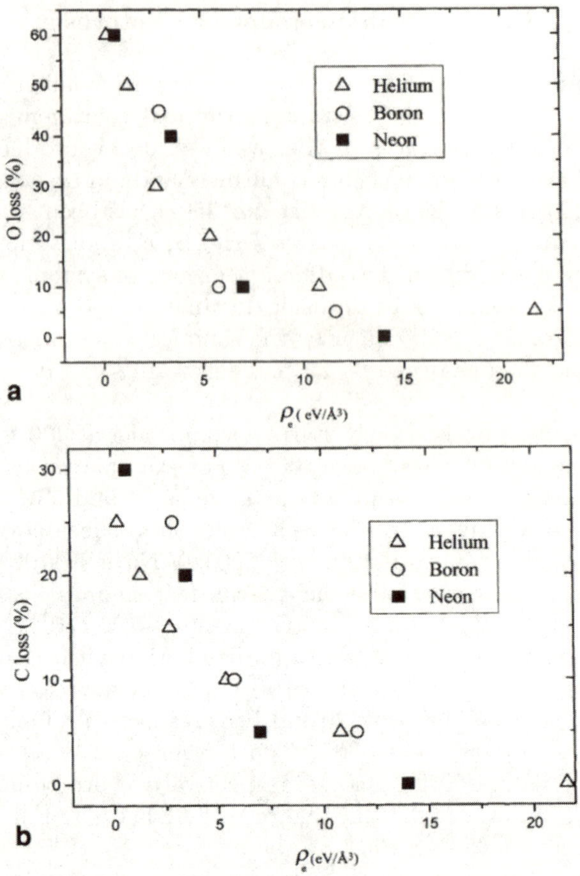

Fig. 2.25. (**a**) Oxygen, (**b**) carbon loss of photoresist AZ1350-J after thermal annealing at 450°C as a function of the electronic-energy density $\rho_e = \Phi S_e$ for He, B, and Ne irradiations; after [60,61]

ume. It increases rapidly from, let us say, 0.80 µm for the pristine film, to some 1.40 µm after depositing 3×10^3 eV/nm^3, and comes to saturation at about 1×10^4 eV/nm^3 at a thickness of 1.55 µm [60].

After irradiation in the regime of dominant nuclear energy transfer [62,63], chemical effects rather than radiation damage were found to be responsible for the change in thermal stability of the irradiated films. For example, 50-keV Bi, Sn, and Sb implantations improved the thermal stability, whereas the implantation of the noble metals Ag and Au did not affect the thermal decomposition behavior. In the first case, the threshold energy density for thermal stability is of the order of 7×10^3 eV/nm^3, i.e., comparable with the above case.

Table 2.3. Comparison of the mobility [$cm^2\,s^{-1}$] of UV-generated charge carriers and large molecules (here: ferrocene) in several polymers[a)]

Host polymer	Charge carrier mobility	Mobility of ferrocene
PS	–	1.1×10^{-13}
PMMA	7.0×10^{-12}	8.8×10^{-13}
PBM	2.2×10^{-10}	4.6×10^{-12}
PVBC	7.5×10^{-10}	6.8×10^{-12}

[a)] PBM – poly(benzyl methacrylate),
PVBC – poly(vinylbenzyl chloride) [64]

2.10 Transport of Charge Carriers in Irradiated Polymers

Energetic radiation impinging into polymers creates charge carriers. This is exploited, e.g., in photoconductive polymers. Here, donors that are photo-excited (usually with low efficiency, typically with quantum yields of 10^{-3} to 10^{-2}) transfer electrons to strong electron acceptors doped into the polymer matrix. Applied electric fields are helpful in separating the product ion pairs permanently. Table 2.3 illustrates that the charge carrier mobility is typically an order of magnitude higher than that of large molecules.[1] It is not yet known how radiation defects influence this charge-carrier mobility.

2.11 Summary and Concluding Remarks

It has become clear from the previous examination that the transport of penetrants in irradiated polymers is a highly complex matter. Therefore the above findings are summarized in Table 2.4. From the table it becomes clear that in nearly all cases the true implantation profiles as predicted by the ballistic theory play no role. Also, purely diffusional effects are nearly completely absent in the case of irradiated polymers. The majority of penetrants appear to favor the trapping at electronic defects in the case of low fluence, and at nuclear defects in the case of high-fluence irradiated polymers. However, one also notes that the dominant type of traps depends on the polymer. As a summary, the present data base does not yet allow an unambiguous prediction of the fate of implanted or indiffused penetrants, and hence experiments in this field will still have to continue for quite a while.

Some questions remain that concern occasionally observed unusually long transport processes in polymers. For example, Davenas et al. [65] reported

[1] This table refers to UV-generated charge carriers as corresponding values for ion-created charge carriers appear to be still missing.

Table 2.4. Overview of the transport behavior of different penetrants in different polymeric systems. Only the dominant mechanisms are given. The notation A:B signifies: solution of A in solvent B. The notation I(g) signifies that iodine is introduced as a gaseous penetrant. Occupation of two segments signifies that both mechanisms act simultaneously

<table>
<tr><th>Dominant:
System:</th><th>Regular diffusion</th><th>Trapping in nuclear defects</th><th>Trapping in electronic defects</th></tr>
<tr><td rowspan="5">Transport of penetrants in self-irradiated polymers</td><td colspan="2">Ar, Kr, Xe ions in polymers [2]</td><td></td></tr>
<tr><td></td><td></td><td>Li, B, F in many polymers (e.g., [5])[a)]</td></tr>
<tr><td></td><td colspan="2">B → S1813 [25]</td></tr>
<tr><td>Low-fluence implanted metals in polymers [31]</td><td></td><td></td></tr>
<tr><td></td><td>High-fluence implantation of B in polymers [22]</td><td></td></tr>
<tr><td rowspan="8">Transport of penetrants in preirradiated polymers</td><td></td><td>$LiCl:H_2O$ → PI [6, 43]</td><td>$Pbacetate:H_2O$ → PE [67]</td></tr>
<tr><td></td><td></td><td>$KCl:H_2O$ → PP [68]</td></tr>
<tr><td></td><td>I (g) → PI [8]</td><td>E^+ → polymers [3, 4]</td></tr>
<tr><td></td><td colspan="2">I(g) → PE [69]</td></tr>
<tr><td colspan="2">$LiCl:H_2O$ → PEEK [12, 48]</td><td>Methylene blue: H_2O → PI [70]</td></tr>
<tr><td>*$[(C_2B_9H_{11})Co]_2Cs$: H_2O → PE, high-fluence irradiated [46]</td><td></td><td>*$[(C_2B_9H_{11})Co]_2Cs$: H_2O → PE, low-fluence irradiated [46]</td></tr>
<tr><td></td><td>I, Hg → high-fluence irrad PET [52]</td><td>I, Hg → low-fluence irrad PET [52]</td></tr>
<tr><td></td><td>Loss of C- and O-containing radiochemical products [60, 61]</td><td></td></tr>
</table>

a) Only for this case regular profiles were also found occasionally hydrogen loss from $(CH)_x$ foils from zones far deeper than the range of the implanted atoms. Another phenomenon concerns sputtering: 130-keV $^3He^+$ implantation into {200 nm Au / 2 µm PET / 10 nm Al} foils resulted in some Al emission from the back side of the foil, though the He^+ ions should have come to rest inside the polymer layer [66]. A possible explanation is that the polymer had some unexpected microporosity in both cases that allowed the projectiles to travel much further into the material than expected. Similar unexpected observations were also made on other occasions.

References

1. Fink D, Klett R, Hu X, Müller M, Schiwietz G, Xiao G, Chadderton LT, Wang L, Mathis C, Hnatowicz V, Vacík J, Characterization of aged latent ion tracks in polyimide. Nucl Instrum Methods **B116**, 66–71 (1996)
2. Guimarães RB, Behar M, Livi RP, de Souza JP, Amaral L, Zawislak FC, Fink D, Biersack JP, Dose and energy dependence of implanted ion profiles ($9 \leqq Z1 \leqq 83$) in the AZ111 photoresist. Nucl Instrum Methods **B19/20**, 882–886 (1987)
3. Hirata K, Kobayashi Y, Hishita S, Ujihara Y, Damage depth-profiling of Au^+ and O^+-irradiated amorphous PEEK by monoenergetic positron beams. Appl Phys **A64**, 491–495 (1997)
4. Kobayashi Y, Kojima I, Hishita S, Suzuki T, Asari E, Kitajima M, Damage-depth profiling of an ion-irradiated polymer by monoenergetic positron beams. Phys Rev **B52**, 823–828 (1995)
5. Fink D, Biersack JP, Chen JT, Städele M, Tjan K, Behar M, Olivieri CA, Zawislak FC, Distributions of light ions and foil destruction after irradiation of organic polymers. J Appl Phys **58**, 668–676 (1985)
6. Fink D, Müller M, Petrov A, Klett R, Palmetshofer L, Hnatowicz V, Vacík J, Červená J, Chadderton LT, Aqueous marker penetration into ion irradiated polyimide. Nucl Instrum Methods **B191**, 662–668 (2002)
7. Balasubramanian V, Kelkar DS, Kurup MB, Electrical conduction in ion implanted Nylon-6 films. Nucl Instrum Methods **B113**, 257–260 (1996)
8. Davenas J, Xu XL, Diffusion of iodine into polyimide films modified by ion bombardment. Nucl Instrum Methods **B71**, 33–38 (1992)
9. Fink D, Müller M, Szimkowiak P, Klett R, Vacík J, Hnatowicz V, Chadderton LT, Rutherford backscattering of laterally heterogeneous structures: The determination of radial density distributions in ion tracks in collodium. Nucl Instrum Methods **B134**, 87–97 (1998)
10. Geis MW, Randall JN, Deutsh TF, Efremov NN, Donelly JP, Woodhouse JD, Nitrocellulose as a self-developing resist with sub-micrometer resolution and processing stability. J Vac Sci Technol **B1**, 1178–1181 (1983)
11. Kobayashi Y, Haraya K, Hattori S, Sasuga T, Effect of electron-irradiation on the free volume of PEEK studied by positron annihilation. Nucl Instrum Methods **B91**, 447–449 (1994);
    and: Fink D, Vacík J, Červená J, Hnatowicz V, Kobayashi Y, Hishita S, Ghosh S, Klett R, On the uptake of aqueous tracer solutions by pristine and ion-irradiated PEEK. Nucl Instrum Methods **B134**, 61–72 (1998)

12. Vacík J, Červená J, Hnatowicz V, Fink D, Strauss P, Study of LiOH etching of polyethylene terephthalate irradiated with 11.4 MeV/amu Pb ions by neutron depth profiling and a particle transmission. Nucl Instrum Methods **B146**, 509–512 (1998)
13. Umezawa K, Nakanishi S, Gibson WM, Frisch HL, Kitamura A, Study of diffusion and trapping of a fluorinated hydrocarbon in ion beam irradiated poly(styrene) by nuclear reaction resonance analysis. Phys Lett **65**, 1501–1503 (1994)
14. Fink D, Behar M, Kaschny JR, Klett R, Chadderton LT, Hnatowicz V, Vacík J, Wang L, On the redistribution of $6Li^+$ ions implanted into polypropylene films. Appl Phys **A62**, 359–367 (1996)
15. Fink D, Müller M, Stettner U, Behar M, Fichtner P, Zawislak FC, Koul S, Non-regular depth profiles of light ions implanted into organic polymer films. Nucl Instrum Methods **B2**, 150–154 (1988)
16. Guimarães RB, Amaral L, Behar M, Zawislak FC, Fink D, Range measurements and thermal stability study of AZ111 photoresist, J Appl Phys **63**, 2502–2506 (1988)
17. Bello I, Carter G, Knott KF, Waworth L, Stephens GA, Farrell G, Gas trapping induced by argon implantation and irradiation of polymer films. Rad Eff **89**, 189–204 (1985)
18. Guimarães RB, Amaral L, Behar M, Fichtner PFP, Zawislak FC, Fink D, Implanted boron depth profiles in the AZ111 photoresist. J Appl Phys **63**, 2083–2085 (1988)
19. Guimarães RB, Amaral L, Behar M, Fink D, Zawislak FC, Depth profiles of Li ions implanted in the photoresist AZ111. J Mater Res **3**, 1422–1426 (1988)
20. Behar M, Amaral L, Kaschny J, Zawislak F, Guimarães RB, Fink D, Radiation induced diffusion of Xe implanted into the AZ1350 polymer. Nucl Instrum Methods **B46**, 313–316 (1990)
21. Fink D, Müller M, Stettner U, Behar M, Fichtner P, Zawislak FC, Koul S, Non-regular depth profiles of light ions implanted into organic polymer films. Nucl Instrum Methods **B2**, 150–154 (1988)
22. Vacík J, Červená J, Fink D, Klett R, Hnatowicz V, Popok V, Odzhaev V, High fluence boron implantation into polymers. Rad Eff Defects Solids **143**, 139–156 (1997);
    and: Vacík J, Hnatowicz V, Červená J, Peřina V, Popok V, Odzhaev V, Fink D, High fluence boron implantation in polyimide. Nucl Instrum Methods **B148**, 1126–1130 (1999)
23. Fink D, Müller M, Ghosh S., Hnatowicz V Vacík J, Tomographic study of the three-dimensional distribution of a high-fluence implant in a polymer. Appl Phys **A68**, 429–434 (1999)
24. Soares MRF, Fink D, Müller M, Petrov A, Behar M, $^{10}B^+$ Ion implantation into the photoresist S1813. Submitted to Appl Phys A (2003)
25. Fink D, Müller M, Stettner U, Behar M, Fichtner P, Zawislak FC, Koul S, Non-regular depth profiles of light ions implanted into organic polymer films. Nucl Instrum Methods **B2**, 150–154 (1988)
26. Al-Najjar SAR, Bull R, Durrani SA, Track profile and range studies of heavy and light ions in CR-39: Effect of thermal pre-treatment on etching characteristics. Proc of the 11th Intl. Conf. on solid state nuclear track detectors, Bristol. pp. 167–170 (1981)

27. Kaschny JR, Amaral L, Fink D, Behar M, Radiation induced diffusion of Xe to a polymer film. Rad Eff Defects Solids **125**, 289–298 (1993)
28. Ho PS, Haight R, White RC, Silverman BD, Faupel F, Chemistry, microstructure and adhesion of metal-polymer interfaces. In: Lee L-H (ed.), Fundamentals of Adhesion, Plenum Press, New York and London, 1991, pp. 383–406
29. Marin N, Serruys Y, Metallic Impurities diffusion in a polymer film (kapton) under and out of irradiation. Nucl Instrum Methods **B131**, 109–120 (1997)
30. Marin N, Serruys Y, Diffusion of metals deposited on a polyimide film (Kapton) under and out of irradiation. Nucl Instrum Methods **B105**, 175–180 (1995)
31. Soarez MRF, Amaral L, Behar M, Fink D, Diffusion of Ag implanted into the AZ1350 photoresist. Nucl Instrum Methods **B191**, 690–694 (2002)
32. Soarez MRF, Kaschny JRA, dos Santos JHR, Amaral L, Behar M, Fink D, Diffusion and solubility of Au implanted into the AZ1350 photoresist. Nucl Instrum Methods **B166–167**, 615–620 (2000), and references [1–7], and [13] cited therein
33. Kobayashi T, Iwaki M, Doi Y, Iwaki M, Metallic nanoparticles formed in polyimide by ion implantation. Nucl Instrum Methods **B175**, 548–553 (2001)
34. Stepanov AL, Abdullin SN, Petukhov VYu, Osin N, Khaibullin RI, Khaibullin IB, Formation of metal-polymer composites by ion implantation. Philos Mag **B80**, 23–28 (1999)
35. Khaibullin RI, Osin YN, Stepanov AL, Khaibullin IB, Synthesis of metal/polymer composite films by implantation of Fe and Ag ions in viscous and solid state silicone substrates. Nucl Instrum Methods **B148**, 1023–1028 (2000)
36. Shanker K, MacDonald JR, Thermal stability of metal films on polyimide. J Vac Sci Technol **A5**, 2894–2897 (1987)
37. Grande PL, Zawislak FC, Fink D, Behar M, Range parameters study of medium-heavy ions implanted into light substrates. Nucl Instrum Methods **61**, 282–290 (1991)
38. Biersack JP, Haggmark LG, The transport and ranges in matter. Nucl Instrum Methods **174**, 257–269 (1980)
39. Nakao A, Kobayashi T, Iwaki M (2001), Surface modifications of polystyrene by K implantation. Nucl Instrum Methods **B191**, 685–689 (2002)
40. Popok VN, Khaibullin RI, Bazarov VV, Valeev VF, Hnatowicz V, Machkova A, Odzhaev VB, Anomalous depth distribution of Fe^+ and Co^+ ions in polyimide under high fluence implantation. Nucl Instrum Methods **B191**, 695–699 (2001)
41. Fink D, Müller M, Klett R, Vacík J, Hnatowicz V, Červená J, Three-dimensional implantation distribution of lithium implanted into pyrographite, as revealed by solid state tomography in combination with neutron depth profiling. Nucl Instrum Methods **B103**, 423–428 (1995)
42. Shrinet V, Chaturvedi UK, Agrawal SK, Nigam AG, A simple set-up for in-situ observation of the critical dose of blistering during ion implantation in polymers. Nucl Instrum Methods **B1**, 617–620 (1984)
43. Fink D, Klett R, Chung WH, Grünwald R, Döbeli M, Ames F, Chadderton LT, Vacík J, Hnatowicz V, Doping of C_n^+ ($n = 1, 3, 5, 8$) cluster ion tracks in polyimide. Rad Eff Defects Solids **140**, 3–20 (1996)
44. Klett R, Charakterisierung von hochenergetischen Schwerionenspuren in Polyimid, PhD. Thesis, Humboldt-University, Berlin, 1996 (in German)

45. Fink D, Klett R, Hnatowicz V, Vacík J, Mathis C, Omichi H, Hosoi F, Chadderton LT, Wang L, Bonding of dopants to irradiated polymers. Nucl Instrum Methods **B116**, 434–439 (1996), and references therein, and: Červená J, Vacík J, Hnatowicz V, Mackova A, Peřina V, Fink D, Decoration of radiation damages in polyimide implanted with rare gas ions. Surf Coat Technol **159–159**, 391–394 (2002)
46. Hnatowicz V, Vacík J, Červená J, Švorčík V, Rybka V, Popok V, Fink D, Klett R, Doping of ion implanted poyletyhylene with metallocarborane. Nucl Instrum Methods **B105**, 241–244 (1995)
47. Fink D, Vacík J, Klett R, Chadderton LT, Hnatowicz V, Doping of 20 MeV fullerene ion tracks in polyimide. Nucl Instrum Methods **B119**, 591–595 (1996)
48. Fink D, Vacík J, Červená J, Hnatowicz V, Kobayashi Y, Hishita S, Ghosh S, Klett R, On the uptake of aqueous tracer solutions by pristine and ion irradiated PEEK. Nucl Instrum Methods **B134**, 61–72 (1998)
49. Fink D, Chadderton LT, Cruz SA, Fahrner WR, Hnatowicz V, Te Kaat EH, Melnikov AA, Varichenko VS, Zaitsev AM, Ion track doping. Rad Eff Defects Sol **132**, 81–90 (1994)
50. Fink D, Müller M, Petrov A, Klett R, Palmetshofer L, Vacík J, Červená J, Hnatowicz V, Aqueous Penetration into low and high energy ion irradiated polyimide. Nucl Instrum Methods **B191**, 662–668 (2002)
51. Guimarães RB, Livi RP, de Souza JP, Zawislak FC, Fink D, Biersack JP, Mass and energy dependence of implanted ion profiles in the AZ111 photoresist. J Appl Phys **60**, 1322–1324 (1986)
52. Hnatowicz V, Vacík J, Švorčík V, Rybka V, Popok V, Jankovskij O, Fink D, Klett R, Iodine diffusion and trapping in PE implanted with 150 keV F^+ and As^+ ions to different fluences. Nucl Instrum Methods **B114**, 81–87 (1996)
53. Hnatowicz V, Vacík J, Červená J, Švorčík V, Rybka V, Fink D, Klett R, Doping of polyethyleneterephthalate modified by keV Ar^+ ion bombardment with anorganic agents. Rad Meas **28**, 777–780 (1997)
54. Davenas J, Boiteux G, Jardin C, Electron and mass transport in ion beam nanostructured polymers: role of irradiation. Nucl Instrum Methods **B131**, 91–96 (1997)
55. Fink D, Klett R, Müller M, Omichi H, Hosoi F, Vacík J, Hnatowicz V, Chadderton LT, Influence of energetic ions on grafting to polyethylene. Appl Phys **A63**, 441–446 (1996)
56. Torrisi L, Radiation damage in polyvinyltoluene (PVT) induced by 50–400 keV Helium beams. Rad Eff Defects Solids **143**, 19–32 (1997)
57. Fink D, Klett R, Behar M, Sánchez G, Kaschny JR, Hertlein WG, Changes in the photoresist inhibitor distribution after ion irradiation and thermal treatment. Nucl Instrum Methods **B132**, 660–670 (1997)
58. Fink D, Omichi H, Hosoi F, Tamada M, Hnatowicz V, Vacík J, Chadderton LT, Klett R , Solid and liquid phase doping of energetic ion tracks in polymers. Trans Mater Res Soc Jpn **17**, 581–583 (1994)
59. Okayama Y, Hashimoto T, Yoguchi T, High dose ion implantation into photoresist. J Electro Chem Soc **125**, 1293–1298 (1978)
60. Sias US, Sánchez G, Kaschny JR, Amaral L, Behar M, Fink D, Polymer thermal stability enhancement induced by high energy ion beam bombardment. Nucl Instrum Methods **B141**, 187–192 (1998)

61. Sias US, Sánchez G, Kaschny JR, Amaral L, Behar M, Fink D, Polymer thermal protection induced by ion beam irradiation. Nucl Instrum Methods **B134**, 35–45 (1998)
62. Behar M, Guimarães RB, Grande PL, Amaral L, Biersack JP, Fink D, Kaschny JR, Zawislak FC, Range and thermal-behavior studies of Au and Bi implanted into photoresist films. Phys Rev **B41**, 6145–6153 (1990)
63. Maltez RL, Amaral L, Behar M, Zawislak FC, Thermal behaviour study of Sb implanted into photoresist film. Nucl Instrum Methods **B80/8**, 1316–1319 (1993)
64. Zhang G, Thomas JK, Photoionization in polymer films. In: RL Clough and SW Shalaby (eds.), Irradiation of polymers: Fundamentals and technological applications, Am Chem Soc Symp Ser 620, Washington, USA, 1996
65. Davenas J, Xu XL, Thomas JP, Comparative Study of the stability towards ionic irradiation of $(CH)_x$ and LiF. Presented at the 2nd Intl. Conf. on Radiation Effects in Insulators, Albuquerque, N.M., May 30–June 3, 1983 (unpublished)
66. Fink D, Biersack JP, Städele M, On the performance and lifetime of solar mirror foils in space. Space Solar Power Rev **5**, 91–100 (1985)
67. Hnatowicz V, Kvítek J, Peřina V, Švorčík V, Rybka V, Popok V, Anomalous diffusion of Pb atoms into polyethylene implanted with F^+ and As^+ ions to different doses. Nucl Instrum Methods **B93**, 282–287 (1994)
68. Švorčík V, Rybka V, Micek I, Popok V, Jankovski O, Hnatowicz V, Kvítek J, Structure and properties of polymers modified by ion implantation. Eur Polym J **30**, 1411–1415 (1994)
69. Hnatowicz V, Vacík J, Peřina V, Švorčík V, Rybka V, Jankovskij O, Popok V, Fink D, Klett R, Doping of latent tracks in polyethylene by iodine diffusion. Proc. 17th Conf. on Nucl. Tracks in Solids, Dubna; JINR E13-94-290 (1994), pp. 113–114
70. Fink D, Chung WH, Klett R, Döbeli M, Synal HA, Chadderton LT, Wang L, On the dyeing of ion tracks in polymers. Nucl Instrum Methods **B108**, 377–384 (1996)

3 Transport Processes in Tracks

D. Fink, V. Hnatowicz, and P.Yu. Apel

The linearly extended overall radiation-induced change in free volume – either resulting in a net increase or a decrease – together with the extended distribution of newly formed radicals, is one of the most characteristic properties of ion tracks in polymers, and therefore deserves special attention. According to findings on tracks in, e.g., polyimide [1], the free ion-track volume by no means just forms a small empty cylinder of nanometric dimension in the ion-track center, unless the tracks are extremely short [2]. Rather, the polymeric ion tracks appear to have a sponge-like consistency of free volume intermixed with carbon-enriched radiochemical reaction products with a fractal structure, which therefore often enables an enhanced transport of matter along them.

By etching the as-produced so-called "latent" tracks can be transformed to extended pores, the so-called "etched tracks".[1] In the following we consider the motion of matter along both latent and etched tracks. The moving matter can, in principle, be a solid, a liquid, a gas, a charge carrier such as an electron, or an energetic ion.

As *latent tracks* are known to be regions rich in radicals (which may act as traps for moving ions), newly formed free volume (stemming from degassing of volatile products) and compaction (stemming from shock waves during the ion impact, and from cross-linking), the motion of matter will be both retarded and enhanced at the same time (in comparison with pristine matter), so that the net effect cannot be predicted straightforwardly, but rather depends on the system's peculiarities.

As in etched tracks, the modified chemical products have been largely removed by etching so that essentially a cylindrical (or conical, hyperboloidal, etc.) pore remains, predictions of transport of matter are much easier than in latent tracks. However, we show below that even the interaction of matter with the etch-track walls was found to be still good for many surprising discoveries.

[1] Strictly speaking, "etched tracks" are not tracks but pores, as the characteristic feature of ion tracks – the radiation-damaged matter along the ion trajectories – has been removed by etching. As, however, it has become commonplace for half a century to denote these pores as "etched tracks", we adopt this custom.

3.1 Transport of Solids in Ion Tracks

3.1.1 Latent Tracks

Radiation-Enhanced Penetrant Diffusion

Experiments have been performed where metals (Cu, Au) evaporated onto polymer (teflon) layers showed enhanced intermixing after irradiation with 1.1-GeV Bi ions [3]. Whereas RBS showed just a broadening of the Au interface, ESCA in combination with sputtering revealed at least in one case [4] a long-range transport of the metal into the polymer down to several micrometers depth. These experiments were done in connection with metal–polymer adhesion studies to prove that the metal adhesion is strongly correlated with metal–polymer intermixing. The adhesion was found to scale well with the electronic stopping power, so that one concluded that the long-range metallic diffusion into the polymer should also be an effect introduced by the electronic-energy transfer. On the other hand, many careful examinations revealed that the system: {boron on teflon} showed neither intermixing nor any long-range diffusion into the polymer under otherwise else identical conditions, and the system remained unchanged even after subsequent annealing at high temperatures.[2]

Unfortunately, these experiments were not continued, and no other experiment was undertaken on the radiation-enhanced mobility of metals in polymers. However, recent findings of Avasthi et al. [5] on swift heavy-ion irradiation of Fe/Si and CuO/glass bilayers showed an enhanced mixing effect, and the results also pointed to the electronic-energy transfer as being responsible for this. As during the track formation considerable structural rearrangement processes take place that store considerable energy, this localized target activation may be responsible for the long-range mobility of Au in teflon reported above. The absence of any radiation-enhanced mobility of B in teflon is also understandable, as chemically reactive penetrants are known to bond readily to polymers and not to participate in diffusional processes.

Recently Bolse [6] found for ceramic/ceramic and ceramic/metal systems that the radiation-enhanced transport is usually (with the exception of the FeO/SiO_2 system, here isotropic diffusion was found) a one-dimensional diffusion directed along the track direction, with diffusion coefficients of the order of 10^{-3} to 10^{-4} $cm^2\,s^{-1}$, hence typical for liquid-state diffusion.

Formation of Metallic Nanowires along Latent Tracks?

The motion of solid matter along tracks has received only little attention up to now. As it has been found recently that metal atoms (such as Au [7], Ag [8], Pd [9], and others) dispersed in a polymeric matrix (photoresist and plexiglas, respectively) undergo regular thermal diffusion after low-energy

[2] Hnatowicz V, Vacík J, Fink D, Červená J (1993), unpublished work

low-fluence ion implantation or after in situ formation at low concentration by chemical reactions, one might think of inserting these atoms or metallic nanoclusters into polymer foils subject thereafter to swift heavy ion irradiation and thermal treatment, expecting that the moving metal atoms or clusters should then be trapped at the swift heavy-ion tracks, thus forming metallic nanowires. The latter could be detected by TEM or by an increase in conductivity along the polymer foil, where characteristic quantum electronic features should show up.

Such experiments are in progress, and first results have already been obtained, indicating that the decoration of ion tracks by at least metallic clusters is indeed possible. For this, ∼3- to 10-nm sized clusters with a volume filling of ∼1 to 20% were prepared by vapor-phase tandem evaporation of, e.g., Ni [10] or Au and teflon monomeric precursors onto a carbon-covered Cu TEM grid. Upon 120-MeV Au ion irradiation, several changes were observed. First, large clusters grew at the expense of small ones (Ostwald ripening) so that the overall cluster density decreased. Secondly, the irradiated clusters assumed nearly perfect spherical shapes. Third, hints for the spatial alignment of clusters along the ion tracks were found at a fluence of 5×10^{11} cm^{-2}. And fourthly, these strings of clusters appear to also align parallel to the beam direction so that curtain-like structures emerge [11].

3.1.2 Etched Tracks

Attempts to put nanoparticles (e.g., $LiNbO_3$ nanocrystals) as dry powder into etched tracks fail due to their great tendency to cluster into larger units and thus to block the track openings. However, when cladding the crystallites with organic ligands (e.g., amyl acetate), they can be dissolved in water and transported along with the liquid into the tracks. This will be treated below in Sect. 3.2.7.

3.2 Transport of Liquids in Ion Tracks

3.2.1 Penetration of Liquids into Hydrophobic and Hydrophilic Polymers

Many liquids are electrolytes that dissociate into ions. The solubility of ions in hydrophobic polymers is extremely small [12], which is a result of the polarization of the medium around electrical charges. The "self-energy" of a monovalent ion in a medium with the dielectric constant ε can be estimated using the Born formula [13]:

$$E = e^2/2\varepsilon r_\mathrm{i}, \tag{3.1}$$

where e is the elementary electrical charge, and r_i is radius of the ion. Calculations for hydrophobic polymers ($\varepsilon \approx 3$) and the ion radius of 0.2 nm show that the concentration of the ions in the polymer phase should be approximately 10^{20} times smaller than that in the aqueous phase [14]. The adsorption of ions in aqueous solutions by hydrophobic surfaces is normally negative. In the aqueous phase very close to the polymer surface, the concentration of ions is lower than in the bulk solution – due to the proximity of the medium with low ϵ. If the polymer surface is neutral, the negative adsorption is poor and can be neglected in considering the penetration kinetics, and the thermodynamic parameters of penetrant in the bulk and at the polymer/solution interface can be assumed to be identical. However, it is not the case when the surface has electrical charge due to dissociating chemical groups. An example is PET treated with an alkaline agent. Its surface acquires negatively charged carboxylate groups rigidly bound to the surface. An electric field results in increasing the local concentration of positive ions (counterions) in the adjacent layer of the aqueous phase. Accordingly, the local concentration of negative ions (co-ions) decreases. This phenomenon produces a considerable effect on the penetration kinetics, especially in the case of diluted etchants [15].

In contrast to hydrophobic matrices, diffusion of electrolytes in hydrophilic polymers is appreciable. Hydrophilic polymers (PVA, cellulose, etc.) are characterized by a high content of polar groups and readily take up water, i.e., swell. The presence of water in the polymer significantly reduces the ion self-energy [16]. Water molecules play the role of carriers promoting the transport of ions through the polymer. Other hydrophilic polymers are of an ion-exchanging nature (such as polyamides). In such systems functional groups are preferred resting positions for diffusing ions, with energetically unfavorable regions in between. The ions hop from one preferred site to another. Protonated amino groups can serve as active sites for the diffusion of anions, whereas carboxylic groups can provide the route for the transport of protons or other cations. These considerations have important implications for the penetration of degrading agents into ion tracks, which leads to their etching, see Chap. 4 of this volume.

3.2.2 Depth Profiles of Penetrating Liquids

General

Suppose a penetrant moves in a swift heavy-ion-irradiated polymer. Then two principle possibilities exist for the penetrant transport: (a) the penetrant moves preferentially along the ion tracks, or (b) it moves preferentially through the remaining unirradiated matter. In the latter case the tracks can be regarded as impermeable as the result of an excessive free-volume reduction and/or a high density of radicals acting as traps. In the first case tracks can be regarded as regions of excess free volume with little trapping.

If the penetration of a solution through the unirradiated zones of a polymer dominates (this is the case for, e.g., DMF in PET [17]) and the transport through the ion tracks is negligible, then the amount of penetrant uptake will not depend, of course, on the track direction. However, if the penetrant proceeds preferentially along the tracks (this is given, e.g., for the transport of chromophores dissolved in DMF or of aqueous salt solutions in irradiated PET [17]), one finds a dependence of the penetrant uptake on the track direction. This implies that here tomographic reconstructions for 3D profiling of penetrants are no longer possible as they require equal penetrant concentrations for all input profiles. The reason for this is that now the transport processes determine the penetrant distributions, and not the damage distributions.

Rapid and deep penetration of liquids is found in the case of hydrophilic polymers, and in the case of those with a nanoporous structure. The penetrant profiles often follow curves that are flatter than exponential ones (see Figs. 3.1 and 3.3 as examples), and that consequently cannot be described by simple Fickian diffusion without or with a homogeneous background of traps. Therefore it has been suggested that the initial penetration phase is laminar rather than diffusive, so that the penetrant occupies the available accessible free volume rapidly. That is, it is assumed that the penetrant first proceeds into the polymeric bulk through pre-existing interconnected nanopores that act as "irrigation channels", and thereafter continues penetrating via diffusion into the neighboring less accessible regions. That secondary diffusion step may well be of Case II type, which combines diffusion, penetrant sorption, solid-to-gel transition, and the opening of new pathways by crazing. Whatever polymeric matter has been transformed into the gel state will be dissolved in the case of etchant penetration so that in this case the nanocapillaric pathways will gradually increase with exposure time. As polymers that favor scissioning usually enable a more rapid solid-to-gel transition in their irradiated areas than in the pristine material, the opening of new pathways is even more efficient here. Both the nanopore density and distribution and the polymer's radiochemical reactivity determine how far the initial overall penetrant kinetics is dominantly nanocapillaric or just diffusional.

In Sect. 1.1 of this volume it had been shown that both the models of pure nanocapillaric penetration and of pure diffusional penetration of liquids yield the same $t^{1/2}$ dependence of the penetration depth with penetration time. However, they differ from each other in respect to the angular dependence of the etchant uptake C_{tot}. Whereas the nanocapillaric model roughly leads to proportionality of C_{tot} with $\cos(\Theta)(\Theta =$ angle of ion-track direction towards the surface normal), the diffusion model scales reciprocally with $\cos(\Theta)$. So there is a possibility given to derive from experimental evidence whether the etchant penetration proceeds dominantly via diffusion or via capillarity into the tracks.

For example, the initial etchant penetration into PSiO [18, 19] proceeds preferentially via nanocapillarity, whereas etchants have been found to penetrate into PI dominantly via diffusion [20]. In PC the etchant penetration is found to be nearly isotropic [21], which was interpreted as a combined action of both penetration mechanisms.

Some Individual Cases

Penetration of Liquids into Tracks in PET

In agreement with the picture of PET being an essentially hydrophobic polymer, Lück [22] reported on β-spectroscopic examinations that had shown that the alkaline degradation of PET foils by NaOH or KOH takes place only at the surface, or at least within very few molecular layers. By contrast, attack by acids or organic solvents is said to lead to deep diffusive penetration into the bulk PET by the undissociated compounds. These earlier hints could only partly be reconfirmed by recent depth-profiling studies of the penetrant distributions using NDP and a Li marker technique. Figure 3.1 shows the penetrant depth profiles in pristine and ion-irradiated PET after 5 min, for room-temperature penetration. Marker penetration was measured in tracks tilted at various angles towards the surface normal. Ambient temperature was chosen to slow down the penetration mechanism, so that "snapshots" during the initial phase of penetration could be taken. (Earlier studies [23] on, e.g., PI have shown that the transition from 45°C to room temperature retards the penetration process by about a factor of 3.)

In fact, it turns out that the Li marker does not penetrate strongly into either the pristine or the irradiated PET foils after a short time (typically only 500 nm after 1 min), as the etchant concentration profiles drop rapidly to negligible values at greater depths, see Table 3.1. Furthermore, one can conclude

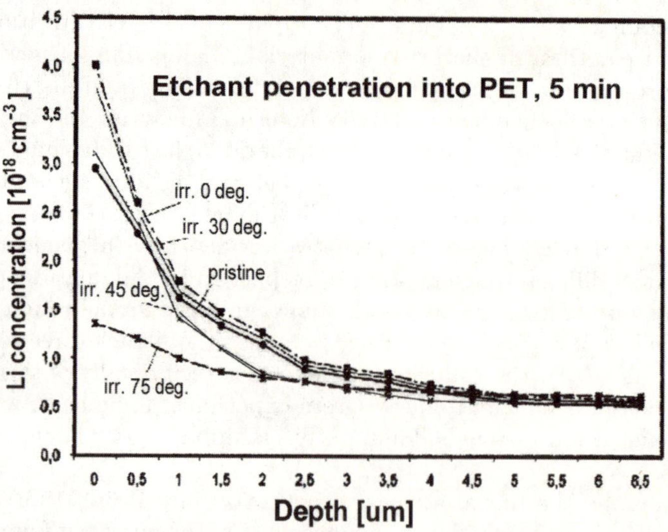

Fig. 3.1. Li marker profiles in irradiated and pristine PET for various track directions, after penetration of LiCl solution at ambient temperature for 5 min [20]. Raw depth profiles, not yet background subtracted (background here $\approx 4 \times 10^{17}\,\mathrm{cm}^{-3}$)

from this study that the penetrant uptake of PET is limited to low concentrations due to its relative hydrophobicity and compactness. On the other hand, the penetration depth definitely exceeds the order of a few molecular layers as was anticipated previously [23]. The initial penetrant uptake is somewhat higher in pristine than in irradiated PET, which indicates that ion irradiation leads to some material compaction in the vicinity of the tracks. This is consistent with the finding of Apel et al. [24] on the radial etchant speed that also indicates a zone of compaction around the ion-track core. After prolonged etching the radiochemical changes appear to make the ion tracks in PET more prone to solvent uptake than pristine PET, so that the etchant concentration within the tracks soon exceeds that outside them, Table 3.1.

Also, the initial penetrant distribution (after 1 min sample exposure) is nearly independent of the ion-track direction, i.e., the liquid penetration proceeds preferentially through the unirradiated bulk rather than through the tracks. In fact, other experimental findings [21] support this assumption (Table 3.1).

After longer exposure time (5 min at ambient temperature) the penetrant uptake is slightly lower in tracks at tilted angles than in tracks parallel to the sample surface normal (Fig. 3.1), which points to the emergence of dominant nanocapillaric penetration in PET in this case. The explanation is thought to be found in a reduction of the bulk diffusional penetration with time, once the polymer starts swelling by the action of the penetrant.

According to the model derived in [19,25], surface tension in nanopores is the driving force to incorporate the penetrant at high pressure – typically at some 10^9 dyn/cm^2, whereas sorption at the capillary walls, as well as friction and the penetrant's viscosity (the latter greatly enhanced by radiochemical debris within the penetration channels) are the effects retarding and limiting the penetration.

Penetration of Liquids into Tracks in PC

The penetrant depth distributions in PC look, in principle, similar to those shown for PET [22]. However, the general penetration depth in PC appears to be larger, see Table 3.1. The distributions in PC differ as a function of penetration time essentially only in their absolute magnitude but hardly in their shapes. The total penetrant uptake first increases rapidly (stage I), then decreases slightly (stage II), and finally increases again (stage III). The dominance of nanocapillarity in PC is supported by the slight angular dependence of penetration. As the near-surface penetrant concentrations C_{total} are the sum of etchant penetration into pristine C_{prist} and irradiated (C_{tracks}) zones:

$$\begin{aligned} C_{\text{total}}(x=0,\Theta) &= C_{\text{tracks}}(x=0,\Theta) + C_{\text{prist}}(x=0) \\ &= C_{\text{total}}(x=0,\Theta=0)[Q\,F_{\text{tracks}} + (1-F_{\text{tracks}})], \end{aligned} \tag{3.2}$$

with $F_{\text{tracks}}(\Theta) = \Phi t \pi r_{\text{eff}}^2/\cos(\Theta)$ being the sum of all track intersection areas with the surface, and Q being the ratio of the near-surface penetrant uptake in tracks to that in the pristine material, one can deduce an effective radius of the penetrant uptake $r_{\mathrm{eff}} = 0.5$ to $1\,\mu\mathrm{m}$. This means that a cylindrical region of this radius is accessible along each ion track to the penetrant via nanocapillaric mobility [21]. This radius being of the order of typical etched tracks, it is speculated that the irradiation produced a system of excess nanopores (cracks) eventually extending laterally this far.

Penetration of Liquids into Tracks in PI

It was shown above that there is, in general, a competition between the radiation-induced production of free volume along the tracks due to chain scission and a compaction due to cross-linking and shock waves. In fact, there seems to be a change in the ratio of free-volume production / free-volume annihilation when going from low to high transferred energy densities. Whereas for light low-energy projectiles ($< 1\,\mathrm{MeV}$) PI is preferentially compacted, and also for 135-MeV Ar irradiation of PI at fluxes up to $1 \times 10^{13}\,\mathrm{cm}^{-2}$ (corresponding to $3\,\mathrm{eV\,nm}^{-3}$ deposited energy density for $(\mathrm{d}E/\mathrm{d}x)_{\mathrm{el}} = 30\,\mathrm{eV\,nm}^{-1}$) compaction still dominates, irradiation with swift heavy ions such as 3 GeV U (at fluxes of $1 \times 10^{10}\,\mathrm{cm}^{-2}$, corresponding to $0.02\,\mathrm{eV\,nm}^{-3}$ with $(\mathrm{d}E/\mathrm{d}x)_{\mathrm{el}} = 180\,\mathrm{eV\,nm}^{-1}$) leads to a strongly enhanced marker uptake as compared with the value of bulk PI, pointing to dominant scission at this high transferred energy density [26–28].

The angular dependence of penetrant uptake (Fig. 3.2) indicates the increasing dominance of diffusional penetration over capillaric penetration for increasing ionic-energy transfer. This might be interpreted as a progressive destruction of the irrigation channels by ion impact at increasing energy. From the directional dependence of the penetrant profile shapes in 3-GeV U irradiated PI after 1 min exposure time one can conclude that the penetrant motion transversal to the swift heavy-ion tracks is about twice as large as the penetrant diffusion along the tracks, which amounts to about $1 \times 10^{-10}\,\mathrm{cm}^2\,\mathrm{s}^{-1}$.

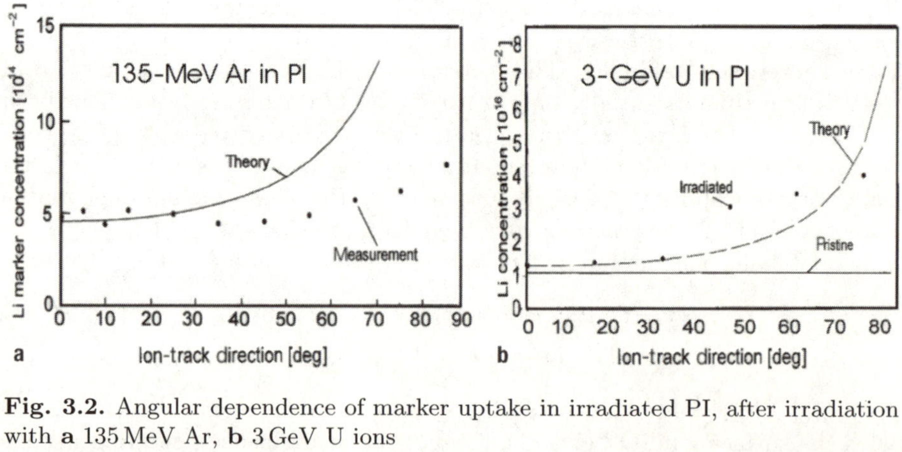

Fig. 3.2. Angular dependence of marker uptake in irradiated PI, after irradiation with **a** 135 MeV Ar, **b** 3 GeV U ions

Remeasurements after 5 min yield qualitatively the same behavior, however, the marker profiles extend deeper due to progressive penetration.

The absolute amount of the near-surface marker uptake increases with increasing exposure time of the marker solution. This points to some sol-to-gel transition. However, the marker penetration comes to an end before extended gel regions are formed as the penetrant mobility strongly decreases with penetration time – an effect that is not treated by the classical Case II diffusion theory [29].

Also the penetration of $\{C_{60}/\text{toluene}\}$ solutions into pristine and swift heavy-ion-irradiated PI has been studied by means of a specially developed technique (FTP, see Sect. 1.3 of Vol. I). Pristine PI allows little or no fullerene solution to enter. This contrasts with the behavior of a solution in THF of smaller entities such as hydrated Li^+ ions [28] or butyllithium $(Bz^{2-}\ Li^+)_2$ [30]) that both readily penetrate into pristine PI. The reason is thought to lie in different migration mechanisms. Whereas small penetrants can migrate through intrinsic voids in the polymer matrix, fullerene molecules cannot transverse them that easily due to their larger sizes (about ~ 0.75 nm diameter). However, already irradiation with, e.g., 500-MeV I^{22+} ions at a very low fluence enables C_{60} penetration. The additional empty space provided heterogeneously in each ion track brings about new efficient pathways for C_{60} penetration.

Irradiated PI can host typically 10^{19} C_{60} molecules per cm^3, independent of the fluence. In other words, whereas each track can accomodate 1.8×10^7 molecules per track at areal track densities as low as 1×10^9 tracks per cm^2, the uptake capacity of a track drops to $3.2 \times 10^4\ cm^{-2}$ in the case of 3×10^{11} tracks per cm^2. Such an effect has been observed earlier [28,30] and was ascribed to ion-induced compaction that counteracts the free-volume production by the energetic ions. An effective penetrant capture radius of 5 nm can be ascribed to each track in the case of high-fluence irradiation.

The C_{60} depth distribution decreases exponentially from the surface to the bulk, with as much as 5 to 10 at.% fullerene molecules embedded in the near-surface layer (similar concentrations are also found for butyllithium penetration into PI). Within the limited sensitivity of FTP, the C_{60} penetration can be followed up to at least 3 μm depth. The depth-profile shapes enable one to estimate the room-temperature diffusion coefficient of the fullerene solution in the tracks to be around 2×10^{-12}–$1 \times 10^{-13}\ cm^2\,s^{-1}$. This should be compared with the diffusion coefficient of aqueous LiCl solution at $\sim 100°C$ ($5 \times 10^{-13}\ cm^2\,s^{-1}$ after 1 h, and $1 \times 10^{-13}\ cm^2\,s^{-1}$ after 4 h penetration time) and of butyllithium in THF at ambient temperature (2×10^{-13}–$1 \times 10^{-14}\ cm^2\,s^{-1}$). The low migration speed of this organic salt solution is attributed to the strong interaction of this liquid with the irradiated PI, whereas the astonishingly high migration speed of C_{60} in solution is thought to be a consequence of the high symmetry, the minimization of the surface free energy, and the absence of dangling bonds of the buckyballs.

Penetration of Liquids into Tracks in PSiO

The penetration into both pristine and ion-irradiated PSiO is characterized by a very fast and very deep penetration already in the first seconds (Figs. 3.3–3.5), pointing to the capillaric penetration into a porous matrix rather than to diffusional penetration, especially in swift heavy ion-irradiated PSiO. The idea of capillaric penetration is also supported by the peculiar angular dependence of the penetrant uptake, Fig. 3.6. In all observed cases, the ion irradiation of PSiO creates excess free volume (Fig. 3.3). As a consequence for the case of etchant penetration, this signifies the simultaneous etching of the tracks over their whole lengths, and even of the adjacent zones beyond the ion range. Therefore – in contrast to the liquid penetration into, e.g., PET or PC – the initial penetration depth distribution is very deep and pronounced for perpendicular tracks and negligible for tracks at tilted angles. The penetration can be understood by assuming the existence of a sponge-like free-volume structure along the track. The resistance of the matter along the tracks increases the effective viscosity η_{eff} of aqueous penetrants from that of pure water $\eta_{\text{H}_2\text{0}}$ to about $\eta_{\text{eff}} \approx 3\times10^4\eta_{\text{H}_2\text{O}}$, thus making the penetration process comparable with heavy motor oil or honey penetrating into completely empty pores. Once the penetrant has been incorporated in the nanoporous network, the penetrations stagnate, and the penetrant concentration along the tracks (as given by the difference between the penetrant distributions in irradiated samples and in in pristine PSiO) becomes stationary.

After typically 10^3 s, when the depth profiles of Fig. 3.3 have approached the saturation concentration, a second stage of penetration is recorded that in the case of pristine PSiO leads to the motion of a penetrant concentration front into the bulk (see the plateaus and maxima in Fig. 3.4a), typically for

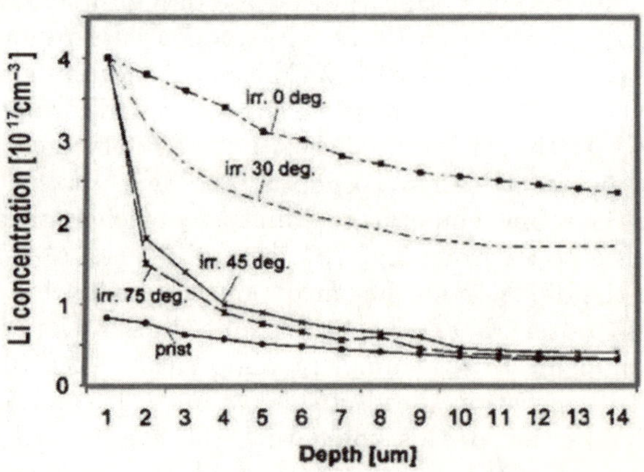

Fig. 3.3. Li marker profiles in irradiated and pristine PSiO for various track directions, after penetration for 1 min [19]. Raw depth profiles, not yet background subtracted (background here $\approx 5 \times 10^{16}$ cm^{-3})

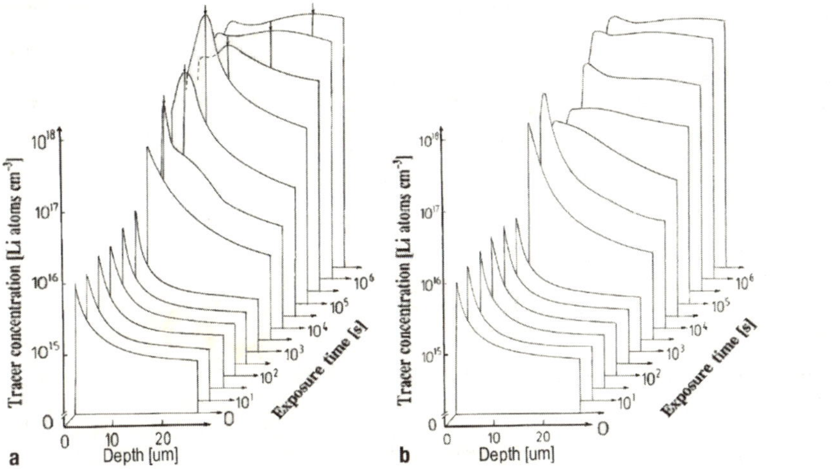

Fig. 3.4. Background-subtracted spectra of marker distribution in PSiO after different exposure times to an electrolyte; NDP measurement exploiting the ^{3}H spectra of the ^{6}Li(n,α)^{3}H reaction to obtain an overview for large depths with limited depth resolution; (**a**) for pristine, (**b**) for ion-irradiated PSiO.[3] Pristine samples even decreases somewhat in the near-surface zone due to the onset of swelling [19]

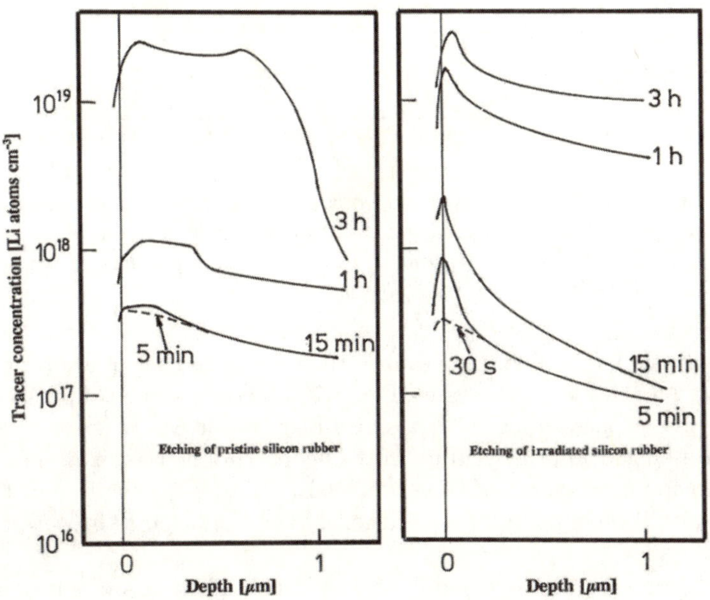

Fig. 3.5. Background-subtracted spectra of marker distribution in PSiO after different exposure times to an electrolyte; near-surface details of Fig. 3.4, exploiting the α spectra of the ^{6}Li(n,α)^{3}H reaction. (**a**) for pristine; (**b**) for ion-irradiated PSiO

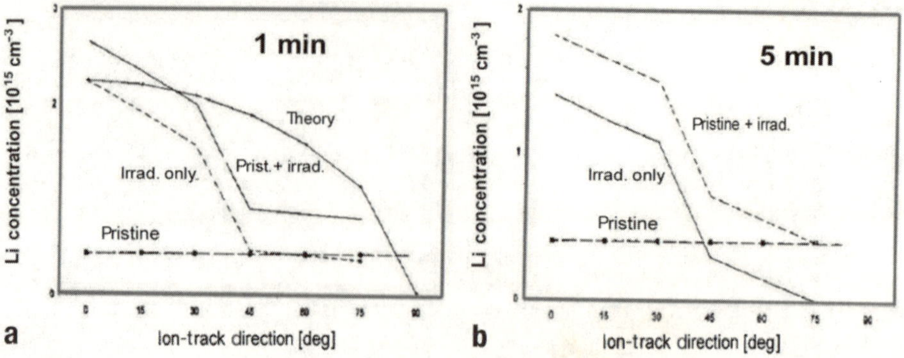

Fig. 3.6. Angular dependence of penetrant uptake in PSiO (**a**) after short time, (**b**) after long time [19]

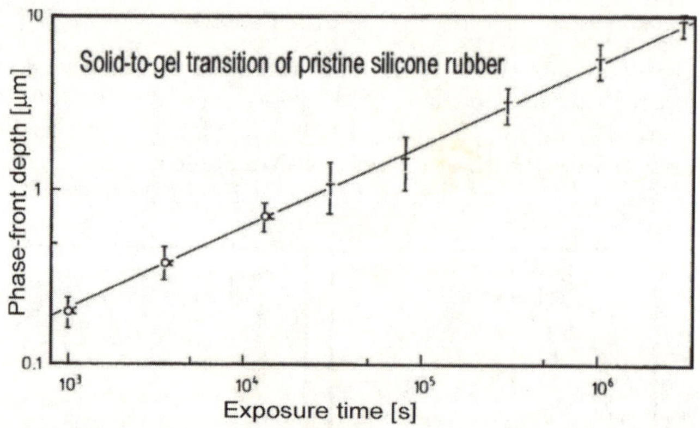

Fig. 3.7. Correlation between phase-front depth and exposure time for marker penetration into pristine PSiO. Result of evaluation of previous figures for pristine PSiO. T: Evaluation of ^{3}H, and α: evaluation of α-particle spectra after the ^{6}Li(n,α)^{3}H NDP reaction

Case II diffusion. The front moves into the bulk with the square root of time, as expected for diffusive processes, Fig. 3.7. The coexistence of both capillaric and diffusive processes in PSiO has also been reported on another occasion, when the penetration of pyridine into silicone rubber was examined. In this case, diffusion coefficients of the order of 10^{-9} cm^2 s^{-1} were found, the same order as would be expected for diffusion of small molecules in water through membranes [31].

Ion-irradiated PSiO is capable of uptaking slightly more penetrant in an intermediate stage, which indicates larger transient free volume. The Case II diffusion mechanism is, however, prevented here, which indicates that the swelling and crazing are largely inhibited in this case, for an as yet unknown reason.[3]

[3] Fink D, Müller M (1999), unpublished work

Summary

Tables 3.1 and 3.2 summarize the results of the penetration of aqueous liquids into different pristine and ion-irradiated polymers. In the above observations, the pH value of the penetrant does not yet play any role in the initial stage of liquid penetration, i.e., whether it is an etchant or not. The peculiar etching effects show up only in later stages of liquid penetration, see Chap. 4.

Table 3.2 shows that swift heavy-ion impact modifies the nanostructure of different polymers in different ways. Also, swelling – as revealed by the time dependence of liquid uptake – often plays an important role for the penetrant behavior. Furthermore, the type of penetrant appears to be of importance: whereas the experiments summarized in the table indicate that aqueous salt solutions move partly via a nanocapillaric mechanism into PET, experiments performed with dyes dissolved in dimethyl formamide (DMF) indicate diffusional motion along the tracks [17].

Table 3.1. Summary of the liquid-penetration experiments. The data refer to ion impact parallel to the sample normal. Given here are the marker concentration at the surface C_{surf}, the penetration depth $x_{1/2}$, and the total areal marker concentration C_{tot}. Numbers without brackets indicate the values for the pristine material, numbers with brackets are the data for penetration into irradiated polymer foils. Irradiation conditions: 1- to 3-GeV heavy ions, 10^7 to $10^8\,\mathrm{cm}^{-2}$. Penetration by 3 mole/l LiCl or LiOH solution at ambient temperature

	Penetration time	PET	PC	PI	PSiO
C_{surf} [cm^{-3}]	1 min	8×10^{20} (5×10^{20})	3×10^{20} (6×10^{20})	2.7×10^{20} (5.1×10^{20})	2.7×10^{19} (1.2×10^{20})
$x_{1/2}$ [µm]		0.55 (0.45)	10 (3)	6.4 (1.75)	9.6 (~26)
C_{tot} [cm^{-2}]		1.8×10^{18} (1.7×10^{18})	4.5×10^{17} (5.4×10^{17})	3.3×10^{18} (3.9×10^{18})	1.2×10^{17} (8.1×10^{17})
C_{surf} [cm^{-3}]	5 min	8.7×10^{21} (1.2×10^{21})	6×10^{20} (8×10^{20})		4.3×10^{19} (6.6×10^{19})
$x_{1/2}$ [µm]		1.3 (0.8)	10 (5)		1.1 (~28)
C_{tot} [cm^{-2}]		3×10^{18} (3.5×10^{18})	9×10^{17} (7.2×10^{17})		1.2×10^{17} (5.6×10^{17})

Table 3.2. The dominant mechanisms of liquid penetration in tracks in different polymers

Polymer type	PET	PC	PI	PSiO
Pristine	Isotropic	Isotropic	Isotropic	Isotropic after longer penetration times: Case II diffusion
Irradiated, general description	After 1 min: Near-isotropic; dominant penetration through pristine matter + some additional capill. mobility. After 5 min: pronounced capill. mobility	Dominant penetration through pristine matter, and some additional capillaric mobility	Diffusional penetration through both prist. matter and tracks; track permeation Enhanced with increasing deposited energy	Dominant capillaric penetration through tracks. No Case II diffusion after longer exposure
Effective track radii r_{eff} [nm] for capillaric penetrat.; effective zone of diffusion r_{d} [nm] along tracks	$r_{\text{eff}} = 4$	$r_{\text{d}} =$ 500–1000	$r_{\text{d}} = 50$	$r_{\text{eff}} = 10$
Effective diffusion coefficients [$\text{cm}^2\,\text{s}^{-1}$]	–	1.7×10^{-10}	90° irradiat.: 1×10^{-10} 75° irradiat.: 4×10^{-17}	

Time Dependence

In general, the uptake of aqueous liquids by swift heavy-ion-irradiated (SHI) polymers proceeds in three consecutive distinct stages (described here for penetration at ambient temperature):

I Rapid and deep penetration within typically a few seconds, until saturation in liquid uptake is achieved after typically half a minute at ambient temperature;

II For PI, PC, PET: decrease in liquid concentration for typically half an hour. These two stages had been denoted as the "incubation time" of penetrant uptake in earlier studies. The incubation time depends strongly on the polymer, the type of ion tracks and the temperature. For PSiO and polysilanes, there is no decrease of the penetrant concentration observed in stage II, but the penetrant concentration remains constant for a while.

III Uptake of new liquid. Whereas in the case of nonreactive penetrant solution, the liquid uptake comes to a permanent end after typically a few hours, in the case of chemically reacting penetrants (i.e., in the case of etching) the uptake continues until complete disintegration of the foil has been achieved. The onset of this stage can be observed after drying the sample in the microscope by visible etch-track formation.

The times given here are approximate values for the case of 5 mole/l aqueous Li marker penetration into SHI irradiated polymers at ambient temperature, according to [20, 21, 23, 32, 33]. For an example, see Fig. 3.8.

For the interpretation of stage I two possibilities show up. Either one might deal with a swift penetration of the liquid into the polymer's nanocapillaric system, or one deals with a rapid diffusion process, see Table 3.2.

In stage II and for PI, PET, and PC, both the intrinsic and radiation-induced free polymeric volume are reduced by the onset of swelling that leads to expulsion of a major part of the previously incorporated penetrant essentially from smaller microvoids. Apparently, for PSiO and polysilanes this free-volume reduction plays a minor role due to the more open structure of these materials.

In stage III the formation of new free volume sets in by means of crazing, i.e., by formation of nanocracks due to osmotic pressure differences in the material, and finally, continuous dissolution of the polymer's gel phase takes

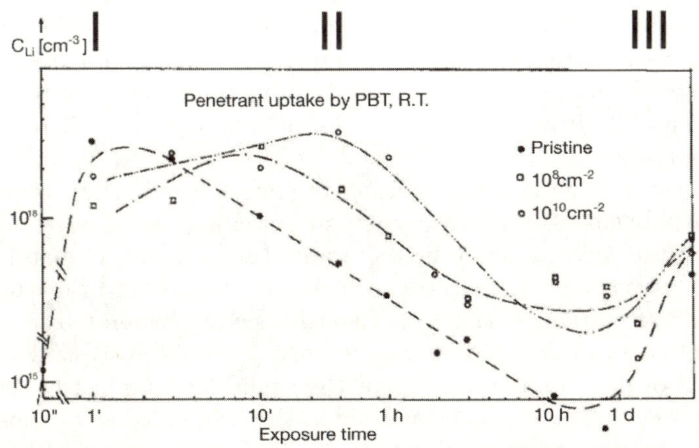

Fig. 3.8. Time dependence of the marker uptake at ambient temperature in pristine PBT, and PBT irradiated with 500-MeV Xe^{17+} ions at different fluences. Lines are to guide the eye [32]

place in the case of alkaline attack [23,32,33]. In stage III the overall penetrant uptake increases with increasing exposure time [1].

To interpret the time dependence of penetrant uptake by PP [34], it was earlier discussed that the initial rapid uptake may be due to the liquid's capillaric penetration into the polymer's amorphous phase, and that the subsequent slow penetrant uptake could be assigned to the penetrant's diffusion into the crystalline phase. This simplified model must be discarded, however, as it would not explain the intermediate decrease in penetrant concentration, and as the stepwise penetrant uptake is found in both crystalline and amorphous polymers.

As in the case of pristine and ion-irradiated organic polymers and silicone rubber, penetrants also move into polysilanes rapidly and deeply, indicating an initially nanocapillaric penetration process into an interconnected nanoporous network. The latter appears to be different for different polysilanes, as irradiated PMPySi samples uptake about three times more penetrant than the corresponding pristine ones for all exposure times, whereas in the case of PDHSi this difference is found only after exposure times of a few weeks. In contrast to most organic polymers, polysilanes do not swell (i.e., do not form a gel) so that the penetrant concentration remains rather constant for the first hour of penetration. Thereafter, when the penetrant is an etchant, its concentration increases proportional to the exposure time, due to the formation of new free volume by etching. For PDHSi, this new free volume is twice as large at elevated temperature (80°C) than at room temperature, which indicates that the ratio of cross-linking to chain scissioning slightly shifts in favor of chain scissioning with temperature in this special case, in agreement with Seki's findings for low-energy ion irradiation of PDHSi [35].

The majority of the results reported in this section have been obtained by studying the motion of Li^+ markers in aqueous 5 mole/l LiCl solutions at ambient temperature by NDP. The overall marker uptake along SHI tracks follows in this case that of the electronic-energy transfer for both low- and high-fluence SHI implantation. This has been found not only for aqueous LiCl solutions but also for methylene blue solution, and even for fullerene solution in toluene [1,36].

In the case of penetration of suitable organic solvents (such as DMF) into tracks, the solvent induces strong polymer swelling, and matter eventually dissolved in the solvent then diffuses rapidly through the gel phase [17], i.e., the solvent "sensibilizes" the track to facilitate the uptake of matter.

Leaching of markers that had previously been incorporated in irradiated organic polymers or silicone rubber is only partially successful. Typically around 50% or more of aqueous Li^+ or BO_3^- markers, and nearly 100% of incorporated dyes are bound permanently at the tracks, for any exposure time of the corresponding original solution [1,34,36]. It appears that at least within the first 1000 s the amount of marker leached from ion tracks is proportional to the leaching time, Fig. 3.9. Both the leachable and the immobile penetrant fractions may be correlated with bonding to O centers and radiation-induced

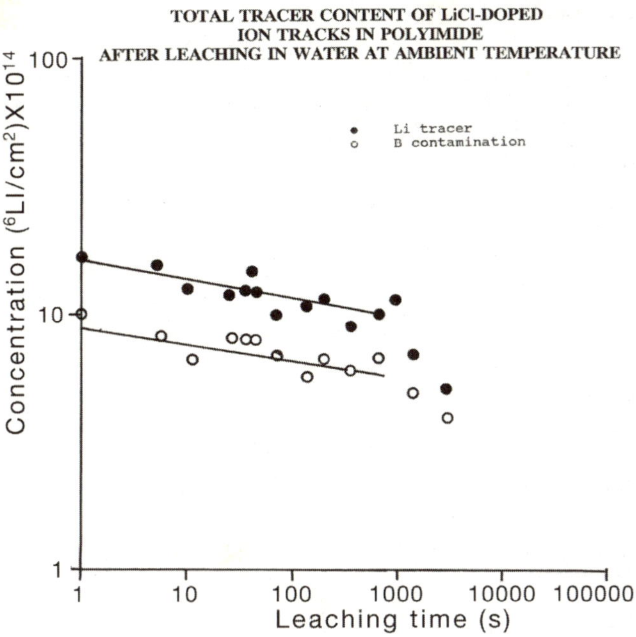

Fig. 3.9. Retained marker (Li from 5 mole/l LiCl and natural boron contamination) along swift heavy-ion tracks in PI, during leaching, as a function of leaching time

defects, respectively – see below (Sect. 3.2.3). The simple correlation for the leaching of penetrants from PI is surprising as leaching of incorporated penetrants is usually quite a complex phenomenon. For example, for porous matter (e.g., porous Si or C_{60}) leaching of incorporated aqueous ionic Li^+ markers reveals two different components,[4] a "volume component" of markers that stem from the interior of the pores and therefore are not bound so that they can be released rapidly, and a "surface component" of markers that were absorbed to the pore walls and hence show retarded release. In the case of PPy – a conducting polymer with aligned fibers and abundant free volume in between – even a third stage shows up[3] that could be assigned to the gradual dissolution in the water of the polymer itself (including the bound marker ions). No hint has yet been found for comparable processes in ion-irradiated polymers.

In nonoverlapping tracks, one finds typically up to 10^5 radicals per μm track length in each ion track. This high number unambiguously points to some spatial expansion of the accessible track region. Assuming "close packing" of these radicals along the tracks, one derives track diameters of 10 to 50 nm, which is in accordance with other information about penetration processes. With increasing fluence the radical density decreases from 10^5 down

[4] Fink, D, Vacík, et al. (1998), unpublished work

to 10^2 radicals per ion track and per μm track length, which points to overlapping of the outer track zones whereby the radicals can recombine.

In PP, the fraction bound to irradiated material exceeds the fraction bound in pristine matter by about one order of magnitude, i.e., ion-track bonding is a quite efficient process. By contrast, bonding to tracks in PI is less efficient, which could point to shorter lifetimes of the radicals in this polymer. As both the depth profiles of total penetrant uptake and of uptake of bond penetrants after leaching usually follow the electronic-energy transfer, one can conclude that both free volume and radical formation are consequences of the projectile's electronic-energy loss.

Comparing the bonding of Li^+ in aqueous solution to tracks with that of C_{60} in toluene, one finds that in average one C_{60} molecule is bound in every 30 nm along an individual track, which signifies that the capability of fullerene to bond (graft) to latent tracks is quite poor [34].

In the case of polysilanes leaching readily decreases the etchant concentration by factors of between 3 and 30. This poor penetrant bonding shows that polysilanes and their destruction products are chemically more inert than other polymers.

Fluence Dependence

For low-fluence SHI impact (corresponding to individual ion tracks) and when the marker uptake has not yet come to saturation, the marker uptake depth distributions drop from surface concentrations of typically some 10^{17} to 10^{18} marker atoms/cm^3 (depending on the polymer and the marker concentration) to concentration values of $\sim$10 μm depth that are lower by at least up to one order of magnitude. From these concentrations effective ion-track radii for marker penetration of the order of 0.1 to 0.3 nm can be derived.

After exposure of, e.g., irradiated PI to a tracer solution for a prolonged time at elevated temperature, the tracer penetrates very deeply along the tracks – at least up to some 5 to 15 μm. Due to the predominantly diffusive motion in PI, the decrease of bulk concentration with depth can be described approximately by straight lines in {log(concentration) vs. (depth)^2} plots, Fig. 3.10a. At sufficiently low fluences, occasionally a slight decrease in marker uptake capability is observed with increasing fluence, Fig. 3.10b. By averaging over the whole exposure time, an effective marker diffusion coefficient can be derived that turns out to be rather fluence independent [28, 37], Fig. 3.10c. However, in the case of PC, the diffusion coefficient steadily decreases with fluence, even for low-fluence ion irradiation, see Fig. 3.11.

When, however, ion-track overlapping sets in, both the polymer's capability for marker uptake in the polymeric bulk and the effective marker diffusion coefficient increase strongly with increasing fluence, see Figs. 3.10c, 3.12 [28] and 3.13 [36].

Dividing the penetrant uptake by the fluence, i.e., considering the marker uptake per ion track, one finds, however, that the latter is not constant but

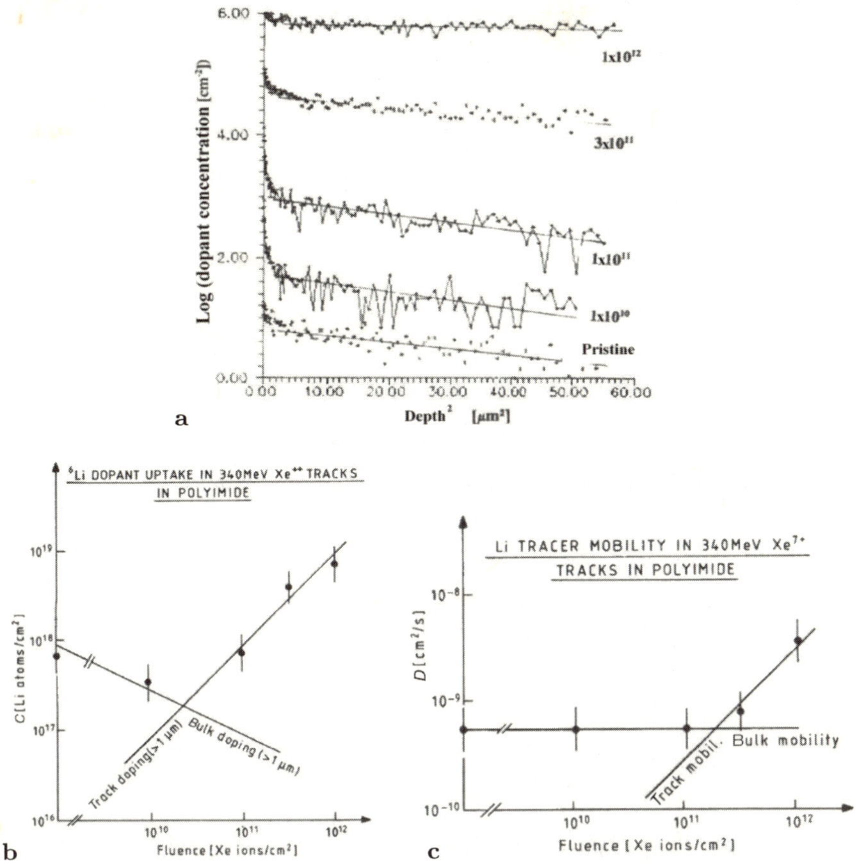

Fig. 3.10. Penetration of LiCl marker solution into 340-MeV Xe^{7+}-irradiated PI [38]. (**a**) The depth distributions are depicted as log(concentration) vs. $(\text{depth})^2$. The numbers refer to the fluence [cm^2]. Most distributions are seen to consist of two parts, a surface contribution stemming from the near-surface distortions of the track structure, and the bulk region described by regular (Fickian) diffusion. (**b**) Fluence dependence of uptaken Li concentration C at 100°C, and (**c**) of the corresponding Li diffusion coefficient D

roughly reciprocal to the track density. This is attributed to compaction due to pressure waves and cross-linking. In PI, the proportionality factor is two orders of magnitude smaller for methylene blue markers than for LiCl markers, indicating a strong dependence on the marker size, Fig. 3.14. For overlapping tracks, the penetrant mobility also increases with increasing fluence.

The fluence dependence of both the marker uptake in overlapping tracks of different ions in PI, and the effective marker diffusion coefficient (defined here by $D_{\text{eff}} = x^2/(4t)$, with x = average marker penetration depth and t = total exposure time) can be approximately generalized by scaling the marker

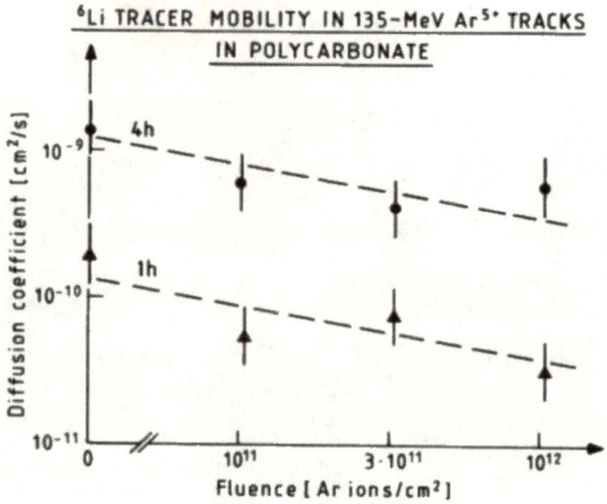

Fig. 3.11. Fluence dependence of total marker uptake; example: LiCl penetration into PC

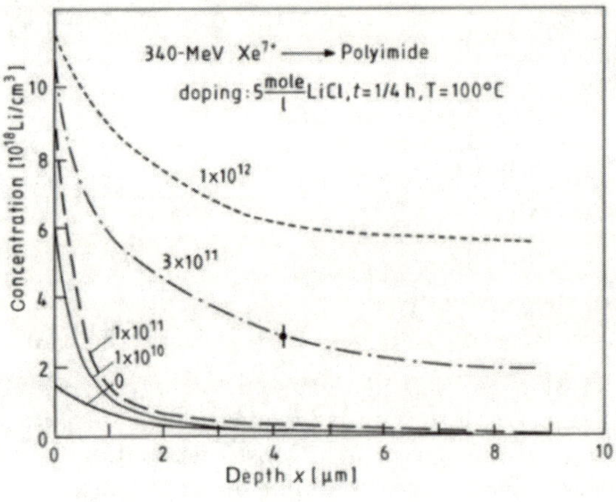

Fig. 3.12. Fluence dependence of hot penetration of LiCl marker solution into PI. Fluences given in [cm^{-2}], [1, 30]

uptake with the totally deposited electronic energy density ΦS_e (Fig. 3.13), or even better by scaling with ΦS_e^2 – where Φ is the SHI fluence. Interestingly, even the marker uptake in low-energy ion-irradiated PI and in energetic cluster-ion (C_3^+–C_{60}^+)-irradiated PI scales approximately according to that rule, for whatever marker [1, 36, 38, 39] – see Vol. I, Chap. 5.

The increase of marker uptake and mobility with fluence at the onset of track overlapping in most polymers is understood by the continuous free-volume production as a consequence of the polymer's progressive destruction. In this connection, it may be worthwhile noting that the formation of new voids and/or cracks after high-fluence ion irradiation of PI can also

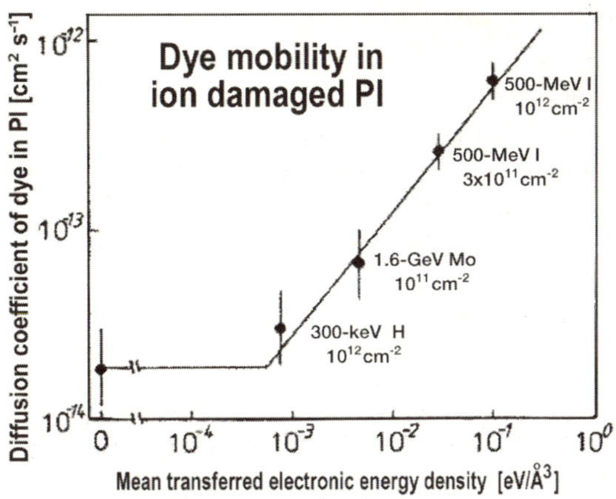

Fig. 3.13. Dependence of diffusion coefficient in PI on the electronic-energy density, for methylene blue dye-marker solutions

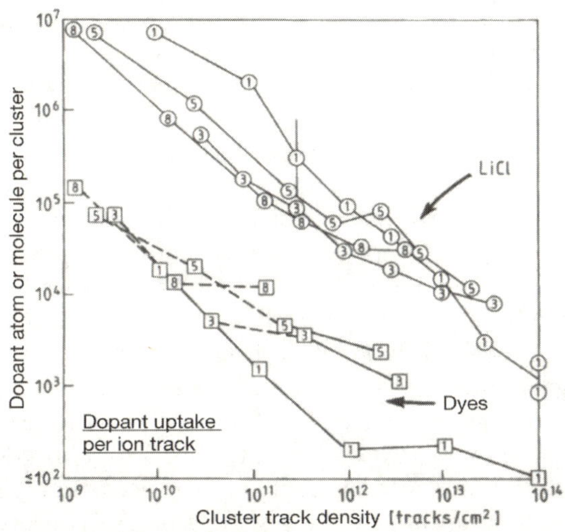

Fig. 3.14. Penetrant uptake per track as a function of fluence, for dye and LiCl solutions penetrating into tracks of C_n^+ clusters in PI. The numbers in the graph refer to the cluster size n

be detected by ion transmission spectrometry (ITS) [40]: In the case of $2 \times 10^{15}\,\mathrm{cm}^{-2}$ 5-MeV Li^+ irradiation the energy-loss spectra are seen to reflect the superposition of both the nuclear and electronic damage, with the electronic damage being several hundred times less efficient in void production than the nuclear one. For 20- to 30-MeV fullerene ion impact, apparently long-ranging cracks are formed in the polymer that enable the marker's long-range diffusion far beyond the ion range [39].

In the case of pristine PI the accessible free volume V_{free} – as determined from measured marker concentrations – is very low, of the order of some 10^{-4}–10^{-5} vol.% only (depending on the foil production parameters). This contrasts with many other polymers where free volumes of up to several vol.% are found, and it illustrates the necessity to perform these examinations with a highly sensitive measuring technique such as NDP.

In the case of penetration of chemically reactive liquids into latent tracks, polymeric modifications may take place. The most important ones are swelling, sensibilization, grafting and etching. They will be treated in the subsequent Sects. 3.2.4, 3.2.5, and 3.4, respectively.

Diffusion of Chromophores in Organic Solvent along Latent Tracks

The mobility of 53 chromophores along tracks has been examined by means of UV-Vis or fluorescence spectroscopy [17]. Table 3.3 gives an extract of some representative data. All hitherto examined cases point to diffusional penetration.

Nearly all diffusion coefficients lie in between those ones of pure DMF in pristine and in ion-irriadiated PET. The dye mobility is always lower than that of the pure solvent by factors between 4 and 32, and the dye mobility along tracks amounts to typically {0.5 to 1} times the mobility of the solvent itself. There is some tendency for decreasing dye diffusion coefficients with increasing dye molecule length, but this correlation is not very pronounced. Incorporation of cationic chromophores with amino groups is favored, which is probably due to the anionic nature of the host polymer used (in this case: PET) with its acidic groups. (Dyes that are used as pH indicators, such as methyl green, bromophenol blue, and others, could not be analyzed by the optical techniques applied in the cited work with sufficient accuracy.)

For low track densities, the dye uptake scales with the square of time within 20%, which points to regular diffusion. From the deviation of this correlation at elevated fluences the radius of the accessible track zone can be determined to be 18 ± 0.1 nm for 11.4-MeV/amu Au tracks, independent of the type of dye used.

By means of polarized fluorescence spectroscopy, the orientation of the chromophores along the tracks can be determined to be of the order of $\langle P_2 \rangle = 0.12$, which has to be compared with $\langle P_2 \rangle = -0.022$ for the unirradiated host polymer ($\langle P_2 \rangle$ is the 2nd Legendre coefficient of the dye's orientation distribution). The orientation has been shown to exist only in single ion tracks and to vanish when tracks overlap. The degree of chromophore orientation does *not* scale with the chromophore length or the degree of polymer destruction along the track.

Once incorporated, the oriented molecules do not relax with time and temperature, which indicates a considerable rigidity in the central track zones [17]. The latter observation is supported by examinations of ultrathin

Table 3.3. Some representative diffusion coefficients of dyes in DMF solution through pristine and swift heavy-ion-irradiated PET (irradiation data: 7-MeV/amu Se to 14-MeV/amu U ions, fluence 1×10^8 to 5×10^{11} cm^{-2}, flux 1.5×10^7 cm^{-2} s^{-1}

DMF + dye	D_{prist} [cm^2 s^{-1}]	D_{irrad} [cm^2 s^{-1}]
No dye, only DMF	1.5×10^{-10}	2.43×10^{-9}
Flouresceine	below detection limit	below detection limit
Rhodamine B	below detection limit	2.45×10^{-10}
Methylene blue	8.2×10^{-11}	3.67×10^{-10}
Methyl violet	1.05×10^{-10}	1.62×10^{-10}
4-Di-2-ASP	8.7×10^{-11}	2.45×10^{-10}
RH 180	8.8×10^{-11}	3.10×10^{-10}

polymer films where it was found that the polymer-chain conformation differs from that in the bulk by greater ordering, as expressed by parallel chain orientation, chain swelling and slightly larger chain dimensions.

Some experiments on thin high molecular mass polymer films point to the influence of long-range order (up to 10 µm) from the substrate. Corresponding systematic observations on polymers embedded within etched tracks are missing.

Noncircular Track Cross-Sections

It is generally assumed that track cross-sections are circular. There are, however, also cases where this rule is violated. Vetter et al. [41] found by TEM observations of ion-irradiated and stained HDPE that track cross-sections in crystalline regions of HDPE exhibit pronouncedly elongated elliptical or wedge-like shapes. The reason for this is found in the structure of that material. The ~20-nm thick lamellae in the crystalline zone of HDPE consist of hydrocarbon chains that are folded back and closely packed with distances of 0.254 nm. Once the stretched macromolecules inside the lamellae are broken due to the impact of a swift heavy ion, hereby affecting typically 25 macromolecule segments in the damaged zone with an assumed diameter of roughly 5 nm, the remaining part of the macromolecular segments relaxes and is transformed into an amorphous state. Consequently, they are more susceptible to staining by compounds such as OsO_4 and, therefore, become visible in the TEM. As a result, stained spots appear with an elliptical cross-section, with axes of about 6–7 nm and 20 nm, respectively. If the ion irradiation affected the crystalline/amorphous interface region, these elliptical shapes change to wedge shaped.

3.2.3 Transport of Liquids into Latent Tracks: Corrosion Effects and Aging

The marker uptake along tracks in PI increases with increasing sample age, possibly due to oxidation by penetrating ambient oxygen. Another, perhaps better, explanation may be to correlate the increase of dopant uptake with the increase of reacted O centers, due to the progressive oxidation depth, as is well known for the oxidation of γ-irradiated polymers [42]. These extrinsic centers might act as traps for the dopants. There exist two different extrinsic trap populations in aged irradiated polymers with different bonding strengths to dopants, namely the ion-induced defects and the O centers, the latter increasing in number with age. Indeed, the first leaching experiments of dopants from aged tracks point to the existence of two different dopant components – a leachable, easily mobile fraction that increases with track age, and another immobile one that is rather independent of the track age. The mobile component could be related to the O centers. On the one hand, the O centers provide the necessary free volume for additional dopant uptake, and on the other hand they act as shallow traps from which the dopants can readily be released during leaching [34]. The immobile component is attributed to deep radiation-induced traps with lifetimes $\gg$ 1 month. Assuming that each such deep trap bonds just one penetrant atom or molecule, then the density of these traps can be estimated to be of the order of some 10 traps per ion track and per 1 µm length. EPR measurements on aged tracks yield similar densities of unpaired spins per track.

In air, the increase in free volume by about an order of magnitude comes to an end after about half a year [1, 2, 45], Fig. 3.15. The corrosion depends slightly on the intensity of long-term illumination with visible light.

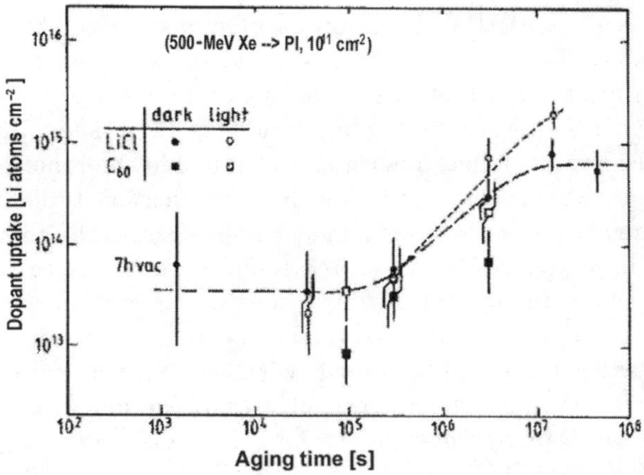

Fig. 3.15. Illustration of aging: increase of penetrant uptake capability of ion-irradiated PI

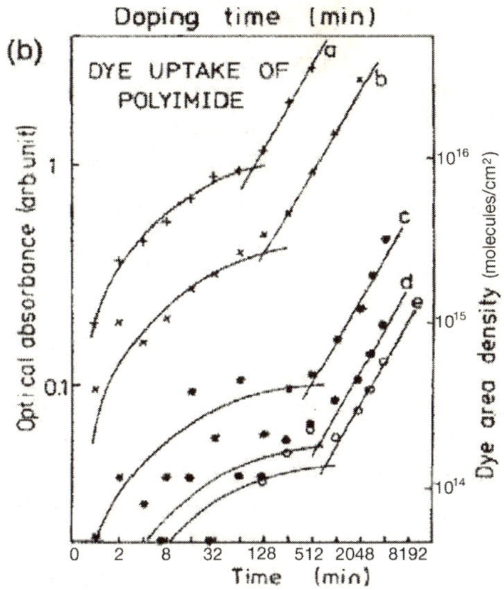

Fig. 3.16. Time dependence of methylene blue incorporation in ion-irradiated PI. After a first stage during which regular diffusive penetration is observed (as verified by the parabolic correlation in the log/square plot), catastrophic dye incorporation (verified by an exponential increase) takes place due to the onset of corrosion effects. The curves **a** to **e** refer to the different systems examined in Fig. 3.13

In the case of track doping with methylene blue markers, a dramatic increase of marker uptake is observed after typically 2 to 10 h (depending on the track type), due to the onset of corrosive reactions of the dye with the ion tracks that then act as unsaturable traps for the dye uptake [36], see Fig. 3.16.

In contrast to the above observations, Steckenreiter [44] reported on the basis of FTIR examinations that no aging effect occurs in swift heavy-ion-irradiated polymers (PC, PET, PVDF, PMMA, PEEK, PSU, PI), once the CO_2 desorption has been completed. Possibly this disagreement results from different commercial additives (e.g., antioxidants) to the polymers used by the different authors.

3.2.4 Swelling after Penetration of Liquids into Latent Tracks

Polymers can uptake liquids via hydration with alkyl groups, hydrogen bondings, and through bonding with carboxylate groups. This solvent uptake is accompanied by swelling, i.e., volume expansion. Swelling results, in general, from loose dipole–dipole bonding of water molecules at the polymeric chains. Additionally, in alkaline liquids, hydrolysis processes may bind hydroxide ions at the polymeric chains, which at a later stage support the chemical decomposition of the latter.

The incorporation of a solvent by a polymer often induces a phase transition from the solid into a highly viscous gel, and the latter can frequently be dissolved in the solvent. Depending on the system under consideration, solvent molecules are either attached weakly to the molecular chains so that they can be removed again by annealing, or they bond with the polymer permanently and cannot be removed any longer. It is speculated that in the case of PET soaking with solvents such as DMF (n,n-dimethyl formamide), the swelling is accompanied by polymer-chain transport processes from the outer track region into the inner ones and by recrystallization processes [44].

Irradiated matter is often more prone to swelling, hence the volume expansion is higher there. This is due to both the track amorphization and the formation of excess free volume along the track that facilitates swelling. The swelling is essentially restricted to the track region, as the rigid unirradiated track walls largely prevent swelling outside the track. The only possibility for expansion is towards the polymer's intrinsic and radiation-induced (by chain scissioning and degassing) free volume. Therefore, the time dependence of the total penetrant incorporation in a (pristine or irradiated) polymer, in general, shows the tendency that, upon the onset of swelling, the already-incorporated etchant is largely driven out again from the polymer, see Fig. 3.8 [32, 45].

Swelling continues until the forces arising from cohesion of the polymer chains balance the osmotic pressure. These forces act the earlier, the more the polymer is cross-linked. Crystalline polymers swell less than amorphous ones, due to less intrinsic free volume being available. In the case of nanoporous tracks, equilibrium of track swelling will occur when the pressure buildup by swelling equals the capillaric pressure inside the tracks, i.e., when the pressure in the swollen zones is of the order of some 10^9 dyn/cm^2.

Only in the very near-surface region does the polymer have the possibility to expand towards the outside. Therefore the study of hillocks forming at the place of each ion track during solvent penetration (see the corresponding Fig. 4.2 in Chap. 4) gives one the opportunity to quantify the swelling – and hence also to estimate the pressure within the ion tracks experimentally. If one supposes that the depth up to which the near-surface track region can expand towards the outside is roughly equal to the visibly expanding track radius r according to the AFM images, then one can estimate the volume ν of the near-surface track zone $\nu \approx \pi r^3$ to which the observed volume expansion $\mathrm{d}\nu$ (i.e., the hillocks in the AFM pictures) corresponds. The volume expansion is in the order of 1% initially, and becomes as large as some 5% in the last stages of ion-track etching (after 10 min exposure time at elevated temperature). According to Hooke's law, this expansion is directly correlated with the pressure P that initiates the swelling:

$$P = \kappa * \mathrm{d}\nu/\nu, \qquad (3.3)$$

with κ the compression modulus. Inserting the measured $\mathrm{d}\nu/\nu$ values, one arrives at pressures of the order of some 10^9 dyn/cm^2. This is consistent with the capillaric pressure that balances this pressure in the polymeric bulk.

3.2.5 Latent-Track Sensibilization (Sensitization) and Grafting

Suitable organic solvents such as DMF, DMSO (dimethyl sulfoxide), methanol, or trichlorethane, may sensibilize the tracks, i.e., facilitate subsequent uptake of matter or chemical reactions. The interaction of organic solvents with tracks is highly complex as processes occur such as:

- Track swelling [46],
- aminolysis of the polymer [47],
- dissolution of oligomers [48, 49],
- solvent-induced recrystallization of the host matrix [50], and
- decrease of the host polymer glass temperature [51].

The decay of DMF to methyl amine and dimethyl amine in the tracks induces an aminolytic degradation of the host polymer [47], which leads to an enhancement of the ion-track radius [17]. The sensibilization with DMF depends sensitively on impurities, especially on amines [47]. Sensitization is most important for ion-track etching (Chap. 4 of this volume). According to [44], sensibilization with DMF has been performed sucessfully only in irradiated PET samples. The sensibilization increases with temperature according to an Arrhenius law [44]. After storage in air for a few months after the irradiation the sensibilization efficiency of the irradiated polymers decreased strongly.[5]

If ion tracks in suitable polymers exhibit sufficient free volume so that matter can penetrate into them, the latter can be grafted onto the tracks due to their chemical reactivity. Thus, for low ion fluences, ion-irradiated polymers can be transformed into inhomogeneous composite materials. In this way, composites such as polystyrene-polyvinylidene fluoride (PS-PVDF) and polyethylene terephthalate – N* isopropylacrylamide (PET-NIPAAm) have been produced, transforming the original homogeneous polymer films into new ones with highly anisotropic properties [53]. Depending on external parameters such as temperature, pH, pressure, electric field, etc., the foil's permeability, electrical conductivity and optical properties can be strongly altered. An example is the anisotropic PS-PVDF structure that generates a functionalized polymer membrane with submicroscopic channels down to 10 nm diameter [53], which is applicable especially in biomedicine due to high haemocompatibility. First applications of this material as prostheses to regulate drug release inside the human body have been reported [54].

SAXS and SANS spectra of pristine PVDF are anisotropic due to the orientation of the crystalline lamellae after the thin-film preparation. After ion irradiation and grafting at low levels ($< 10\%$), an intense scattering at very low angles appears and the SANS peak positions shift to smaller q values. This indicates an increase of the long period between crystallites from 11.5 to 18 nm due to swelling accompanying the grafting process. No difference

[5] Chen J (2001), personal communication

is observed when grafting 10-MeV/amu oxygen- and xenon-irradiated PVDF with styrene. Cross-sectional SEM images clearly indicate a heterogeneous structure with the existence of styrene-enriched latent ion tracks. By contrast, γ irradiation of PVDF enables homogeneous PS grafting into the amorphous zones without disturbing the crystalline texture [55].

In these cases, the degree of chemical penetrant absorption, the so-called grafting yield Y [wt%], can be simply determined by the weight difference of the polymer before and after the grafting process, according to the relation: $Y[\%] = (W_g - W_0)/W_0$ where W_0 and W_g are the weights of the samples before and after grafting, respectively.

To graft PS onto PVDF, ion irradiation is performed in an oxygen atmosphere [53]. This interaction produces radicals of the type R–O–O–R, i.e., peroxide bonds. At 60°C the O–O bonds are broken and radicals R–O• initiate graft polymerization of styrene in which the PVDF is bathed after the irradiation. On this occasion, the PS enters the reactive microchannels. In this way, up to 10^9 submicroscopic filaments have been produced per cm^2, which gives the resulting material highly anisotropic properties. The grafting process increases quickly at low doses to about 8 kGy, and approaches saturation at about 15 kGy; [53, 56].

In contrast, no peroxy radicals are formed in PI [57]. Moreover, it was found that, e.g., for gamma irradiation of PI in air, oxygen inhibits grafting of styrene. However, irradiation at a pressure below 0.01 Pa allows the formation of some radicals capable of meeting with styrene (free of oxygen). The same holds for ion irradiation. Also, in the case of heavy ion induced grafting PI could be subsequently etched. However, the etching rate decreases with growing graft yield, leading to reduced etch track diameters [57]. Finally one should keep in mind that uncritical application of grafting is not always beneficial; frequently the excellent properties of commercial polymers are only deteriorated by grafting [57].

3.2.6 Transport of Liquids through Etched Tracks

With the removal of matter from latent tracks by etching, the accessible free volume increases. It has been shown in Sect. 1.1 that influence this has on the transport properties of liquids. Liquid transport in etched tracks is always governed by capillarity and viscous flow according to the Hagen–Poiseuille law. This implies that the main precondition for transport of liquids in ion tracks is that the liquids are wetting and that their viscosity is not too high. The wetting properties can be modified by adequate chemical surface treatment of the etch-track walls [58].

In the above considerations on etch tracks we treated "large" micropores only, by neglecting the surface flow along the pore walls, i.e. by assuming that $R_{eff}\zeta$, with ζ being the distance of interaction between penetrant molecules and the pore walls. In this case, the motion of a specie in an electrolyte will be essentially determined by the electrolyte motion, as its behavior is

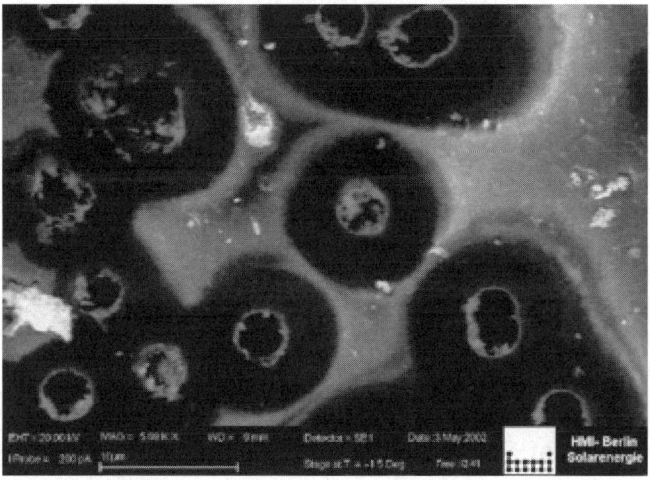

Fig. 3.17. SEM image of etched tracks in PET that were subject first to penetration of NaOH etchant, and later to penetration of a silver complex salt solution. A concentric zone of enhanced conductivity around each track shows up (black in this image), due to the contrast with the unaffected bulk regime that charges up highly by the electron beam impact in the SEM (bright in this image), and thus alters its electron-emission properties. Some white spots indicate precipitation of metallic silver on intrinsic surface defects of the polymer, and on the inner walls of the etched tracks. Please note the astonishingly sharp transition between conducting and insulating zones. The circular etched tracks are partly slightly deformed in this image, due to charging effects [59]

determined by the specie–electrolyte interaction. This point is treated below under the topic of charge transport through etched tracks (Sect. 3.4.2).

It has been shown above (Chap. 1) that liquids are capable of penetrating somewhat through polymeric surfaces into their bulk. This is not only restricted to the planar surface of polymeric foils, but it also applies to inner surfaces of, e.g., pores or etched tracks. In other words, etched tracks serve as "irrigation pipes" for the neighboring matter upon passage of liquids through them. This has been verified by Petrov [59] and Fink et al.[6] via changes in the conductivity of the polymer regions surrounding the etched tracks. A major fraction of, e.g., salts dissolved in carrier liquids cannot be leached out again from the polymeric matrix into which the solution had penetrated, but is firmly bound to the polymeric medium. These salts, being ionically bound, exhibit some ionic conductivity that increases the overall conductivity of a previously highly insulating polymer in the region of their presence. As a result, scanning microscopy observes concentrical regions of enhanced conductivity around etched tracks, and well-insulating zones in regions further

[6] Fink D, Wilhelm M, Petrov A, unpublished results (2002)

away. The latter show up in SEM pronouncedly by being charged up by the analyzing electron beam, Fig. 3.17.

The transport of liquids and gases through etched tracks depends on the smoothness of their inner walls. AFM examinations have revealed that PI has smoother surfaces after etching than PET, due to the rigidity of the rings in the first, and due to dangling bonds in the latter case [60].

As the transport of liquids through etched tracks is governed by the Hagen–Poiseuille law, one can use that flow through the pores to determine the pore radius if the viscosity of the liquid is known. As the flow rate of a given solution depends on the concentration of the matter dissolved therein, one also can use the penetration of solutions through micropores to determine that concentration [61].

3.2.7 Transport of Suspended Particles through Etched Tracks

Transport of Colloids in Etched Tracks

For the transport of colloids in a liquid medium, Stoke's law applies. Particles suspended in water (e.g., negatively charged cells) are surrounded by are surrounded by the so-called electrical double layer comprising the Stern layer (consisting of counter- and co-ions in immediate contact with the surface) and a diffusive layer (with charge equilibrium). This enhances the effective size of the particles during the flow through narrow capillaries, and hence lowers their mobility. The penetration of suspended particles through capillaries is hindered by the need to replace the pre-existing fluid [62]. Colloids transported into tracks can be deposited there permanently by evaporating the carrier solvent.

Figure 3.18 shows some typical depth distributions of colloids deposited in tracks.[7] It is seen that the highly viscous colloidal solution used in this experiment slowly penetrates into the pores until they are filled homogeneously.

Transport of Suspended Particles through Narrow Channels

In the case of medical blood purification, one encounters the problem of passage of micrometer-sized particles (the red blood cells) embedded in the serum, through the thin pores of the filter ("haemofiltration") [63]. It has been found that in this case, the application of pulsative blood flow at 2–4 Hz through the porous etched-track filters (with typical pore radii of 0.5 to 1 μm) is quite advantageous to prevent the blocking of the narrow channels, as this causes a tumbling of the red blood cells due to the repetitive acceleration and deceleration. In this case, one can write for the diffusion coefficient of the solute:

$$D = D_0[1 + 0.06(a^2\gamma\omega/D_0)^{0.886}], \tag{3.4}$$

[7] Fink D, Chemseddine A, Müller M, Heitjans P, Borck D, to be published (2005)

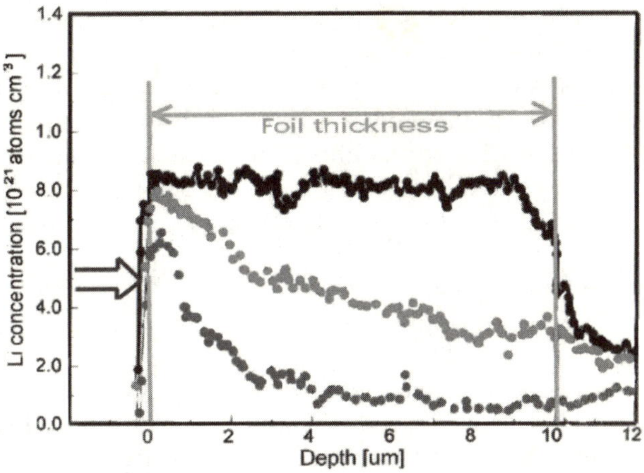

Fig. 3.18. Distribution of colloidal $LiNbO_3$ nanocrystals along etch tracks in PET as determined by NDP by using ^6Li as the marker isotope, for different penetration times (see the arrow for the penetration direction) until a homogeneous filling (*top distribution*) is obtained after sufficiently long time. The smooth decline to nonzero values at the *right side* is due to poor depth resolution at these great depths and due to NDP background

with D_0 being the diffusion coefficient in the stagnant solution (i.e., without pulsating), ω being the pulsation frequency, a being the sphere equivalent radius of the red blood cell, and γ being the wall shear rate [63].

3.2.8 Transport Processes in Irradiated Biological Matter

Cell membranes are vital constituents of living matter, as they control the cell's metabolism and functionality. Their malfunction often having fatal consequences, a great number of studies have been devoted to the consequences of their irradiation. The studies published up to recently have been restricted to irradiation effects of X- or γ-rays on the membrane fluidity, permeability and – in the case of erythrocyte membranes, the haemolysis. In the meantime, the first papers have also been published on the effects of particle irradiation of cell membranes, ranging from high-energy electrons [64] to swift heavy ions such as 17.8-MeV/u^{12}C, 3.9–13.3-MeV/u ^{20}Ne, 8.2-MeV/u ^{40}Ar, 8.4-MeV/u ^{84}Kr, 7.4–9.8-MeV/u^{131}Xe, and 5.6- to 7.9-MeV/u ^{197}Au, [65]. (Considerably more work has already been invested to study swift heavy-ion-induced changes of DNA, chromosomes, cell cycle delays and cell inactivations, which will not be reported here.)

Cell membranes form a two-dimensional fluid matrix of various lipids and proteins. Upon irradiation, their fluidity and microviscosity may change considerably leading to alterations in membrane permeability and protein function [66]. Earlier results on this topic have been controversial; also an increased and a reduced fluidity have been reported.

Irradiation effects of swift heavy ions on erythrocyte membranes can be well characterized by the haemolysis of red blood cells, that is the loss of integrity of the cells leading to the liberation of intracellular haemoglobin. In this way it could be demonstrated clearly that the swift heavy-ion-induced haemolysis of erythrocytes is of osmotic nature, i.e., that the cell membranes develop increased permeability for intra- or extracellular solutions – probably upon lipid–lipid cross-linking.

The haemolysis starts immediately upon the onset of particle irradiation with good proportionality to the LET, in contrast to X-ray irradiation, which shows a subproportional effect upon low doses. Ten ion hits per cell let typically 50% of all cells die 200 min after the exposure to ion irradiation. The lysis cross-sections for different ions (ranging from $3.0 \times 10^{-2}\,\mu m^2$ for 17.8-MeV/u ^{12}C to $20.1\,\mu m^2$ for 6.2-MeV/u ^{197}Au) follow a power-law relationship in the {[log(lysis cross-section)] vs. [log(LET)]} plot, with the exponent being slightly higher than unity [66].

3.3 Transport of Gases in Ion Tracks

3.3.1 Latent Tracks

Transport of Nonreactive Gases

The diffusion process of gas in a polymer occurs by small displacements of side groups or segments of chains of the polymer. If a polymer cross-links, the coordinated segment motion available for permeation is reduced, with a significant reduction in the diffusion coefficient (see, e.g., the diffusion of nitrogen in vulcanized natural rubber [67]). However, the radiation may also induce brittleness of the polymer, which leads to microcracks, the latter enhancing the gas transport in the polymer dramatically. In general, one may say that light-ion impact below a critical fluence preferentially induces polymer densification that leads to the sealing of the surface against penetrants, hence has a protective action.

If a polymer is, however, particularly sensitive to chain scissioning, the emerging excess free volume will enhance the gas mobility. This is given for many swift heavy-ion-irradiated polymers. There exist numerous gas-diffusion experiments through latent tracks in polymers. To give some representative results, at least a few of them are reported here. For example, Remmert [68] has studied the permeation through 11.6-MeV/amu ^{238}U tracks in PET and PI, respectively. Table 3.4 summarizes some of the derived permeabilities at 25°C. For the diffusion coefficient through pristine PET, a value of $1 \times 10^{-9}\,cm^2\,s^{-1}$ was found, and the solubility coefficient is given as

Table 3.4. Permeabilities of noble gases through tracks, after [68]

Polymer	Track density [cm^{-2}]	Penetrant gas	Permeation coefficient [$cm^2\,s^{-1}\,hPa^{-1}$]
PET	0	Ar	$\sim 5 \times 10^{-14}$
	1×10^{11}	Ar	$\sim 4 \times 10^{-12}$
	1×10^{11}	Kr	$\sim 2.5 \times 10^{-12}$
	1×10^{11}	Xe	$\sim 4 \times 10^{-13}$
	5×10^{11}	Ar	$\sim 1 \times 10^{-11}$
	5×10^{11}	Kr	$\sim 7.5 \times 10^{-12}$
	5×10^{11}	Xe	$\sim 9 \times 10^{-13}$
	5×10^{11}	Kr	$\sim 7.5 \times 10^{-12}$
PI	5×10^{10}	Ne	$\sim 1 \times 10^{-10}$
	5×10^{10}	Ne	$\sim 1.6 \times 10^{-10}$

$\sim 5 \times 10^{-5}\,hPa^{-1}$. Remmert found that the diffusion coefficients of permeating gases through tracks could not be derived unambiguously from permeation measurements, as the diffusion paths are not well defined. It appears that the penetrants also follow the path straight through the latent tracks, as do other paths that partly follow the tracks and partly run through the neighboring pristine bulk. As a consequence, the permeation measurements yield two different diffusion coefficients for the two different penetrant fractions. Gas flow measurements through latent tracks were also performed by, e.g., Komaki [69] and Packard et al. [70].

Backmeister et al. [71] found that the results measured for gas permeation through latent tracks remained constant for more than one year. They did not find, however, any reasonable correlation between the diffusion coefficients of permeating gases through latent tracks and the energy transfer dE/dx of the track-forming projectile ions. For low-fluence ($4 \times 10^{10}\,cm^{-2}$) heavy-ion impact (11.4- to 13.4-MeV/amu Se, Pb, Au, and U ions) with an energy transfer exceeding some $20\,MeV/(mg\,cm^{-2})$, the gas diffusion (about $2 \times 10^{-8}\,cm^2/s$) was about twice that of the pristine material ($\sim 1.25 \times 10^{-8}\,cm^2/s$), whereas for higher-fluence ($> 6 \times 10^{10}\,cm^{-2}$) Au and Se ion irradiation, the diffusion coefficients were observed to fall below the pristine values. The same was observed for electron-irradiated materials. This decrease could be explained by a three-phase model [72].

In a one-phase model that describes the permeation of a gas through a polymer foil irradiated at low fluence (i.e., below overlapping of the tracks with effective radius r_t), the total diffusion coefficient was assumed to be given by the sum of the diffusion coefficients for motion through the bulk D_{bulk} and for motion through the tracks $D_{single\,tracks}$, so that:

$$D = D_{bulk}(1 - \mu) + D_{single\,tracks}\,\mu, \tag{3.5}$$

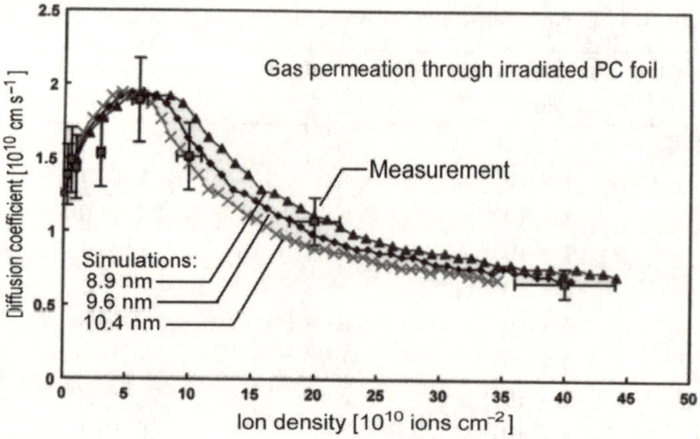

Fig. 3.19. Fluence dependence of the diffusion coefficient of Ar through 11.6-MeV/amu ion tracks in PC, measurement and best-fit simulation according to Turowski [72]

with $\mu = \Phi \pi r_t^2$. The change of the mobility $D_{\text{single tracks}}$ along tracks is attributed to chain scissioning. The strict proportionality predicted by (3.5) holds up to fluences of typically $2.5 \times 10^{10}\,\text{cm}^{-2}$. In a two-phase model it was assumed that a second hit of an already once-irradiated polymeric area does not alter its diffusive properties any longer, so that the overall diffusion coefficient comes to saturation:

$$D = D_{\text{bulk}} \exp(-\mu) + D_{\text{single tracks}}[1 - \exp(-\mu)]. \tag{3.6}$$

This model holds up to typically $7 \times 10^{10}\,\text{cm}^{-2}$. Assuming now that, in contrast to the two-phase model, multiple hits lead to a deterioration of the diffusivity through a track, one obtains:

$$\begin{aligned} D = {} & D_{\text{bulk}} \exp(-\mu) + D_{\text{single tracks}}\, \mu \exp(-\mu) \\ & + D_{\text{multiple tracks}}[1 - (1 + \mu) \exp(-\mu)], \end{aligned} \tag{3.7}$$

where $D_{\text{multiple tracks}}$ is the reduced diffusion coefficient in the overlapping tracks [72]. The reduction of the diffusion coefficient due to the material's densification, resulting from both shock waves and cross-linking, may even reach values near 100%. Assuming that the local diffusion coefficient $D_{\text{comp}}(r)$ at the distance r from an ion track is proportional to the material's free volume, changes with the square of the distance r from a track: $D_{\text{comp}}(r) = D(1 - \chi/r^2)$ with χ being the compaction coefficient [76], one can describe the fluence dependence of the diffusion coefficient by means of a Monte-Carlo program [73]. From a best fit between simulation and measurement (Fig. 3.19), Turowski [72] found $\chi = 0.8$, $r_t = 9.1$ to $9.7\,\text{nm}$, $D_{\text{single tracks}} = (4$ to $10)D_{\text{bulk}}$, and $D_{\text{multiple tracks}} = (0.2$ to $0.3)D_{\text{bulk}}$, depending on the position along a 11.6-,MeV/amu U track in PC. The effective track radii scale with dE/dx_{el} according to two different linear relations for high and low energies, respectively. This may account for the influence of the projectile velocity on the track damage.

Transport of Reactive Gases

CO_2 permeation through etched tracks in polymers leads to an apparent shrinking of the pores, for a yet unknown reason[8]. Most studies on the transport of reactive gases in polymers have been performed with iodine, due to the importance of this element and the ease of iodine depth profiling with RBS. To our knowledge these examinations have been restricted, however, until now to the behavior of reactive gases in low-energy ion-irradiated polymers. It is expected that the permeation of such gases through latent tracks is greatly hindered by strong sorption at chemically active sites along the tracks. One should also mention at this point the penetration of oxygen and of humidity, which alone or in combined action lead to ion-track corrosion, see Sect. 3.2.3.

3.3.2 Etched Tracks

Due to the narrow distribution of radii of etched tracks in polymer foils (for both $r_{track} = 0.105\,\mu m$ and $r_{track} = 0.20\,\mu m$; the FWHM of the track-radius distribution was measured to be only $0.02\,\mu m$ [74]), they are suitable for separating gases. For example, Sudowe et al. [74] showed that CO and CO_2 can be separated to some extent. Whereas for large pores ($r \geq 0.1\,\mu m$) and small pressure differences, the gas permeation is proportional to the latter, for small pores ($r \leq 0.2\,\mu m$) a larger pressure difference is necessary to feed the gases through the foils. This indicates the transition from a laminar gas flow according to the Hagen–Poiseuille law to a molecular diffusion process as described by Knudsen with increasing ratio of mean free path of the gas molecules to the pore diameter. As, according to Knudsen's law, the flow rate of the penetrant is inversely proportional to the square root of the gas molecular weight M_g, the separation factor f of a permeating gas mixture (M_1, M_2 with $M_1 < M_2$) is proportional to $(M_2\ /\ M_1)^{1/2}$ in the case of CO and CO_2: $f = 1.25$. For tracks with $r \approx 0.1\,\mu m$, Sudowe et al. [74] found $f = (1.00 \text{ to } 1.12) \pm 0.09$. Ovchinnikov et al. [75] studied the problem of gas separation on small track-etched pores in more detail. They applied PET track membranes with pores as small as several nm and observed that the penetration comprised both the Knudsen flow and surface diffusion of adsorbed gases. The adsorption and the surface diffusion had a significant influence on the separation factors.

[8] Backmeister GU, personal communication (1998)

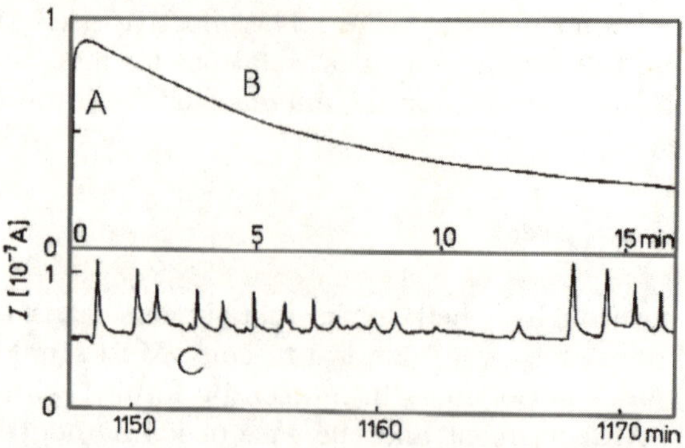

Fig. 3.20. Current flow through latent swift heavy-ion tracks (of 135-MeV Ar) at high areal density ($1 \times 10^{11}\,\mathrm{cm}^{-2}$) in 2-µm thick PET foils; applied electric field: 75 kV/cm [76]. In the first stage (A), the current increases rapidly due to the electrolyte (here: aqueous 5 mole/l LiCl) penetrating along the tracks. In the second stage (B), swelling along the tracks reduces the free volume and hence ejects a major part of the electrolyte, which results in a current decrease. In a final stage (C), rather regular current fluctuations are observed that are attributed to pulsating surface currents through very narrow ion channels. This operation state was found to be stable for at least several days

3.4 Current Transport through Ion Tracks Embedded in Liquids

3.4.1 Electrolytes in Latent Tracks

Ionic conduction is likely to be the normal surface-conduction mechanism of polymers. It is known to be enhanced by absorbed water. Ionic conduction may also be intrinsic due to dissociation of the main chain or side groups followed by proton or electron transfer through hydrogen-bonded networks.

Immediately upon contact of pristine or swift heavy ion irradiated polymer foils with an aqueous solution of 5 M LiCl the conductivites of the irradiated polymer foils increase by many orders of magnitude from their intrinsic (dry) polymeric values of around 10^{-12} to $10^{-15}\,\Omega^{-1}\,\mathrm{cm}^{-1}$ [1] to some $10^{-6}\,\Omega^{-1}\,\mathrm{cm}^{-1}$, as could be verified by the application of a DC potential across the two sides of the foil. After a few minutes the conductivity comes to saturation, its value being most pronounced for high-fluence irradiated samples. Thereafter one observes a decrease in current as a function of time until a steady-state value is reached after a few hours, the ratio of maximum conductivity to steady-state conductivity being 2.2 ± 1.1 for 8-µm thick pristine PI, and up to 10 ± 2 for high-fluence irradiated 8 µm thick PI foils.

After prolonged time, regular current spikes are observed under the influence of the applied electric field as shown in curve C, due to the following two probable reasons: (i) as the electrolyte enters the latent tracks, it swells the polymer, thereby hindering the further passage of charge carriers, and (ii) the DC potential polarizes the charge carriers and the final residual current is only due to electrons, their counterions remaining outside the tracks. The hindered migration of charge carriers results in spikes of decreasing amplitude in the current (curve C) due to the hopping of the charge carriers from one site to the other before the polarization process is complete. Note that the continuous decrease in the amplitude of the current spikes is also due to the fact that some of the Li^+ ions form a permanent bond with the dangling bonds of the ion tracks thereby decreasing the number of mobile charge carriers.

What is also remarkable is that this effect is not limited to individual isolated tracks only, but that a multitude of tracks can switch coherently with each other. This collectively pulsed charge carrier movement through the tracks is possibly mediated by the cloud of counterions accumulating on the polymer surface.

It should be noted at this point that, similarly as was found for the marker uptake of latent tracks [76], the conductivites also vary strongly (within about ±50%) for samples subject to the same preparation conditions. Therefore a multitude of measurements must be performed with samples prepared under identical conditions, to obtain reasonable average values.

In contrast to penetrant uptake results [76] in no case has a continuous second increase in electrical current been observed, once the low steady-state value has been reached [77], Fig. 3.20. The reason for this behavior is found in the explanation of the penetration mechanism in Stage II. New free volume that is formed by crazing branches off more or less perpendicularly from the open main pores (i.e., the main penetration channels) and hence is not necessarily connected with other penetration channels, so that it need not contribute to the overall permeability through a polymer foil – hence also need not contribute to the electrical current through the irradiated polymer. In other words, whereas the penetrant concentration yields the *totally accessible free polymer volume*, the conductivity measurements yield the *permeable free polymer volume*.

Please note that the conductivity after penetrant uptake is slightly higher for pristine PI foils than for those ones irradiated by low-fluence and low energy ions, reflecting (as for the free volume) the ion-induced reduction of the overall free polymeric volume. Only when the ion energy becomes large and ion-track overlapping becomes marked, does the conductivity exceed that of the pristine case.

In order to check for the influence of the driving electrostatic force onto the protons and hydroxide ions, the voltage was applied either continuously or switched on only briefly for the measurement, allowing the system to diffuse otherwise thermally. As all these experiments yielded identical results within the experimental variability, it is concluded that driving forces by applied voltages contribute only negligibly to the overall water mobility within the latent tracks.

Replacement of the distilled water in the above-described experiments by 5 mole/l LiCl solution leads to an enhancement in conductivity by a factor of 30 or so, within a few seconds. About an order of magnitude difference from the experiments with distilled water still remains when the saturation conductivity was reached after a few hours. In situ replacement of the LiCl solution by pure water, and vice versa, showed that the conductivity changes are only partly reversible. A major fraction of the penetrating Li^+ ions remains bound permanently to the ion tracks, thus contributing steadily to an enhanced polymer conductivity – a result that coincides with the penetrant-leaching behavior as determined by NDP, see Sect. 3.2.2.

3.4.2 Etched Tracks

The electric current through aqueous liquids embedded in etched tracks is governed by ionic transport. It is correlated with the effective track diameter, see (4.24). Several factors influence the current transport through liquids in tracks:

- The bulk ionic conductivity,
- a surface conductivity component,
- phenomena in microcapillaries such as steric hindrance [78].

Whereas the surface conductivity is negligible for large etched tracks and highly concentrated etchants, it becomes important for neutral electrolytes. As a result of track etching, polymers such as PET contain carboxylic groups (COOH) in their inner track surface. For $pH > pH_{crit}$ ($pH_{crit} \approx 4$ to 7; average 5.5), the COOH dissociates, leaving the immobile COO^- in the etch-track surface and the mobile H^+ entering the liquid. Consequently, a mobile counterion layer close to the surface is formed in the solution – the so-called "electrical double layer", compensating the surface charge. Thus the ion concentration is locally increased near the track walls, associated with increasing conductivity: $\sigma_{tot} = \sigma_{volume} + \sigma_{surf}$. As a first approximation, the latter is independent of the concentration of the solution and depends only on the charge density of the surface [79]. More accurate measurements of the surface charge density of etched tracks in PET and PC show, however, its dependence on the electrolyte concentration, probably due to adsorption of ions onto the surface. The thickness of this double layer is typically around 0.2 to 10 nm; it can be decreased by increasing the electrolyte concentration passing through the capillary. The electrical potential falls off exponentially with distance from the pore wall. The surface current is proportional to the pore diameter. Therefore, the current–voltage characteristics is composed of two components, a constant one due to surface currents (i.e., $I_{surf} = \text{const.}$), and a current I_{volume} that is proportional to the volume conductivity σ_{volume}. The transition from the constant current to the proportionality current regime is around 2 mS/cm for the example reported in [79] (30 to 150 nm pore diameter, PET foil, 19 μm thickness).

The transport of counterions adjacent to the charged capillary wall upon an applied electrical field along the track axis induces a mass transport, so-called electro-osmosis. As electro-osmosis is an important effect that is exploited in advanced chromatography, many details are known in this field [80]. As one deals in this field essentially with capillaries of fused silica, we summarize here only those factors that may be of interest in connection with etched tracks.

As capillaries (of fused silica or etched tracks in PET) are charged up negatively, the counterions are positive and hence drawn to the cathode. The negative charges bound in the capillary walls can be as large as 10^{-6} C/cm^2 in fused silica [80], and even 10^{-4} C/cm^2 in PET [84]. This surface charge density can be controlled by:

1. The adsorption of small cationic molecules for neutralization of the latter.
2. Covalently blocking the charged groups (COO^- in the case of PET, and Si-OH^- (silanol) in the case of fused silica).
3. Simple titration of the charge on the surface walls. In the case of fused silica walls, a maximum surface charge is obtained for pH = 10, whereas pH = 2 is the point of zero charge.

These well-known recipies may also be applied to etched tracks to optimize their transport behavior.

The migration of single-stranded DNA strands through voltage-biased nanometer-sized pores produces a unique and reproducible blockage current that depends on the composition of the DNA. This phenomenon was observed first in a-hemolycin, a biological nanopore where up to ten individual DNA bases could be distinguished in this way. Possibly the use of etched ion tracks will allow improved resolution, to enable the formation of a rapid DNA sequencer [81].

It has been found that ion-track membranes with nanoscopic conical pores embedded in an electrolyte show both asymmetric current–voltage characteristics and pronounced current rectifying [82, 83]. The degree of this rectification depends on the electrolyte concentration and the presence of divalent cations in the solutions. For etched cones with small openings this diode-like effect is more pronounced than for large openings. The asymmetric I–V characteristics are more pronounced in dilute solutions and at higher pH. They are also more pronounced for PI than for PET. The shape of the diode characteristics is practically the same for one individual and for a multitude of parallel diodes. When changing from LiCl to NaCl or KCl electrolyte, the rectification effect and its direction hold but the shape of the I–V characteristics changes in accordance with different transference numbers of the cations.

These observations show unambiguously that the electrokinetic asymmetry is associated with the presence of electric charges on the pore walls that are formed due to, e.g., dissociation of carboxylic groups on the etched surface of PET. Though the detailed mechanism of this diode-like behavior is not yet clear it is speculated that the etch-track wall is transformed into a thin gel layer that covers the cone tip on the polymer foil surface. Such gel layers are thought to arise from the interaction of hydroxide ions with the PET chains that leads to the production of a partly hydrolyzed polymer volume with a certain density of carboxylic end groups. As the latter are highly hydrophilic, this layer absorbs water creating a gel. This gel may shrink or swell depending on pH and electric-field conditions, and thus lead to the observed electrical diode-like switching behavior.

In fact, in an earlier study [84] hints were obtained that etched surfaces of PET might host a surface density of electrical charge as high as 10^{-4} C/cm^2, corresponding to about seven elementary charges per nm^2. (In [79], it is reported that the cleavage of polymer chains during the etching releases about one carboxylate group per nm^2 on the inner track surface.) Assuming that every monomer unit of PET can only carry one elementary charge, one is led to the conclusion that this charge is stored in-depth, i.e., is distributed within a gel layer on the pore walls.

Associated with the rectifying property of asymmetrically etched tracks is the finding that negatively charged micrometer-sized colloid beads can flow through asymmetric polymer channels [85], and also the capability of conical tracks to act as an ion pump if subject to a low-frequency alternating voltage [86]. In a narrow conical pore that is covered with negative charges on its inner surface, a driving force acts on positive ions within this pore towards the larger side of that pore. Consequently, a mass transport takes place upon application of an AC field, even against the concentration gradient of these ions. Numerically evaluated equations have been established by Siwy et al. [86] to describe the motion of ions in such pores. This pumping phenomenon is determined by the size of the narrow opening, the surface charge of the cone, and the frequency of the applied field.

This pumping effect is a small-size phenomenon only. For cones with small radii above ~15 nm, the current–voltage characteristics become linear, indicating that the pore loses both its rectifying and pumping properties. Also, when compensating the pore surface charges (e.g. by buffering at pH ~ 3) no pumping occurs. The pumping is restricted to frequencies below ~1 Hz, possibly due to the large aspect ratio of the tracks examined up to now. This ion-pumping phenomenon is also known in biology where K^+, Rb^+, and Na^+ are pumped in the direction of their increasing concentration gradient in the absence of other energy sources through biological membranes, mediated by electric-field-sensitive conformations of ATPases [87].

When for $R_{\text{tot}} \to b$ (or even $R_{\text{tot}} \leq b$; R_{tot} = total etch-track radius, b = thickness of surface layer) the surface flow starts dominating, the specific electrostatic action of the pore walls onto the penetrant molecules becomes decisive for the penetrant motion. In general, the pore wall will be covered with fixed charges so that the vicinity of the walls will be a region of high electric field strength, which determines the behavior of neighboring ionic species. As the water dipoles align along the electric field perpendicularly to the pore walls, the water molecules are essentially frozen in place, and this leads to a decrease of the dielectric constant from the bulk value of 78 to a much lower value. The decrease depends on the degree of solvation. The higher their degree of solvation in aqueous solution, the less they are allowed to approach the charged wall, so that the charges are separated by a minimum distance. Of all the alkalines Li is most solvated, hence in Li salt solutions the charges are separated at a greater distance as, e.g., in K salt solutions; the difference amounts to about 5%. The consequence of this is a higher current density of Li salt solutions through membranes by a factor 1.72 in comparison with Na salt solutions, and by 2.22 for K salt solutions. Another consequence of the pore charging up is that the viscosity of fluids penetrating through these pores increases as compared with their bulk value [88]. Please take into consideration that the chemical properties of nanoscopically confined water in nanotubules are also different from those of bulk water, as the spatial limitations prevent the existence of the extended hydrogen-bonding network of bulk water.

The fixed negative charges (carboxylate groups) that are generated, e.g., in sufficiently narrow ($R_{\text{tot}} \leq b$) ion tracks in PET by the hydrolysis of some of the bonds of PET during the ion-track etching, makes these tracks about twenty times more selective for the cationic transport than for the anionic one. If the carboxylic groups are methylated the cation permselectivity is reduced; if they are substituted by positive charge carriers (e.g., by grafting polycations such as polymethylvinylpyridine (PMVP) onto the track walls) the selectivity can be even reversed [89]. A specific feature of many induced pores – whether pores in biological membranes, narrow cylindrical ion tracks in synthetic polymers, or conically shaped etched tracks immediately after the etching breakthrough – is their enhanced sensitivity to divalent cations and protons. The presence of Ca^{2+} or Zn^{2+} ions, or of low pH values decreases the passage of low molecular weight metabolites through the tracks. The ion current is also decreased, and the fluctuations from low- to high-conductance states are diminished [89,90]. The relative efficiency of the cations in all pores is found to decrease in the sequence: $H^+ > Zn^{2+} > Ca^{2+} > Mg^{2+}$.

The rapid change in the flow of ions such as Li^+, Na^+, K^+, or Ca^{2+} through specific protein-lined ion channels from "open" to "closed" states is the backbone of signal transport in nerves. In fact, such fluctuations have also been observed in micropipettes of 0.1 to 1 μm radius, and in membranes with about 1 pore/cm^2 areal density and 1 to 10 nm pore radius [92]. Applications of this effect have been suggested, such as devices for desalination and sensors. Current fluctuations have also been observed in swollen latent ion tracks in PET, see Fig. 3.20.

If Au is electrodeposited onto etch-track walls (see Sect. 4.2) and subsequently Cl^- ions are adsorbed on the newly formed Au nanotubule walls, then these tubules have an ideal cation permselectivity, provided the radius of the tubule is small relative to the thickness of the electrical double layer within the tubule (being, e.g., −30 nm for 0.1 mmole/l KCl and −0.3 nm for 1 mole/l KCl) [91]. In this case the double layer accumulates cations and rejects anions due to the fixed negative wall charges. By contrast, larger Au tubules with radius ≈ double-layer thickness reject anions only at low electrolyte concentrations, but become inefficient at elevated concentrations, due to the reduction of the double-layer thickness.

If the adsorption of ions from the electrolyte on the Au tubules is blocked, then even narrow Au tubules do not exhibit any transport selectivity. However, by applying an external potential to them in this case, they can be switched from anion to cation permselective, and vice versa. Such a blocking of the Au surface against anion adsorption is made possible by chemisorption of a thiol to the Au nanotubule walls. 1-propanethiol has been found to be adequate to enable, on the one hand, this Au surface blocking, and, on the other hand, to leave the Au surface still wetting for water.

Furthermore, it is possible to expand this recipe by applying hydrophobic thiols (e.g., $HS{-}C_{16}H_{33}$) to yield membranes that preferentially transport hydrophobic molecules (thus treated Au tubules are not wetted by water), or by applying hydrophilic thiols (e.g., $HS{-}C_2H_4OH$) to yield membranes that preferentially transport hydrophilic molecules (e.g., pyridine) [91]. Derivatizing the Au nanotubules with $HS{-}C_{16}H_{33}$ is also good for separating hydrophobic species from hydrophilic ones [92]. If chemisorbing cystein, one can obtain membranes that exhibit pH-switchable ion-transport selectivity, by either allowing cation transport (with anion rejection), non-ion-permselectivity, or anion transport (with cation rejection). Depending on the solution pH value, the membrane can have positive charge (low pH, protonated amino group), no net charge (isoelectric point, protonated amino and deprotonated carboxyl), or excess negative charge (high pH, deprotonated caroboxyl). With such membranes, anionic and cationic species can be separated from solution [91]. In this way, Au nanotubules treated by adequate chemisorption can be considered as model systems for naturally occurring membranes.

Another interesting point is that the flux of, e.g., toluene across Au nanotubules *increases* with carbon chain length of the thiol chemisorbed onto the metal, as the partition coefficient (see Sect. 1.1) increases with the chain length of the thiol employed (as verified for propyl, decyl, and hexadecyl thiol). Apparently the toluene is transported here through the 0.5-nm thick thiol layer. (The fact that the thiol layer does not have the theoretically expected thickness of 2.6 nm indicates that the thiol layer on Au is neither well ordered or well packed.) The higher the penetrant molecular weight, the *higher* the penetrant flux, as the partition coefficient between the feed solution and the thiol is expected to increase with the penetrant molecular weight [92].

3.5 Transmission of Energetic Ions through Tracks in Vacuum

3.5.1 Latent Tracks

This case is closely connected with ion transmission spectrometry (ITS) through latent tracks for probing their radial density distributions. Until now, only one such experiment has taken place on light-ion transmission through latent tracks in collodium [93, 94]. The radiochemical background has been treated in Vol. I, Chap. 8.3.3. The basic idea is that ions transmitted along different radial zones of a latent track will suffer different energy losses, due to different radial track densities. The energy dispersion of monoenergetically transmitted ions will be the larger, the higher the density difference between track and pristine polymeric ambient. Therefore collodium is a favorite target material, as $\sim 96\%$ of the ion-irradiation products are gaseous and escape, thus leaving the latent track as a zone with near-zero density. From the energy spectrum of the transmitted particles, the radial distribution can be reconstructed.

3.5.2 Etched Tracks

For the transmission of ions through etched tracks in vacuum, two different cases have to be distinguished. On the one hand, light energtic ions are used routinely to probe the shape of etched tracks. This special case of ion transmission spectrometry (ITS) has been treated exhaustively in Vol. I, Chap. 3. The other case is restricted to low-energy highly charged particles that show during their passage through etched tracks a peculiar effect denoted as "ion guiding".

Ion Transmission Spectrometry of Highly Charged Ions to Probe the Ion-guiding Effect

The possibility of an easy transmission of energetic ions through crystalline matter along the main crystal axes and planes has been commonplace since the discovery of ion channeling about 30 years ago. The basic mechanism in this case is the glancing angle reflection of the projectile ions along atomic chains or planes. As long as a certain angle of impact is not exceeded, the ions will move in the narrow channels defined by the potential walls of the adjacent crystal atom rows or planes. This "channeling" movement comes to an end either when the projectile ion's energy is gradually used up due to the multiple small energy transfers to the channel wall atoms, or if the projectile "dechannels", i.e., encounters at a large angle a displaced atom stemming, e.g., from intrinsic or radiation-induced defects. (Please note that it is the discrete distribution of atoms along rows or chains that leads to the channeling effect. If the atomic charges were smeared-out homogeneously all

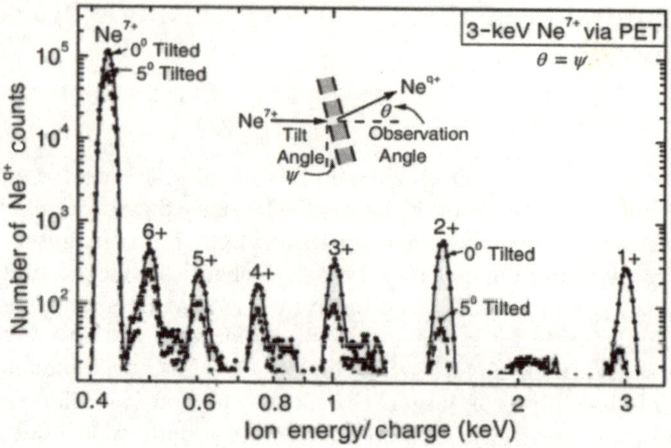

Fig. 3.21. Charge-state spectra of Ne^{q+} ions produced after the passage of Ne^{7+} through capillaries in PET. The charge state q is given at each peak. The tilt angles are $\Psi = 0°$ and $5°$ as indicated at the spectra. The-ion observation angle is $\Theta = \Psi$. The angles are defined by means of the diagram shown as an insert in the figure [95]. Copyright (2003) by the American Physical Society

over the channel, the latter would then be field free as in a Faraday cup, and no projectile deflection would take place at all.)

It was a great surprise to find just at the end of the past millienium an effect along etched tracks in polymers that, regarded superficially, shows some similarity with channeling [95]. If, e.g., 3-keV Ne^{7+} ions are shot into a 10-μm thick PET foil (coated with 5-nm Au on both foil sides to suppress the charging-up of the foil) containing 2×10^8 etched tracks per cm^2 with 80 to 100 nm radius each (i.e., which have an aspect ratio of ~ 50), one finds ion transmission even if the foil is tilted by an angle towards the beam direction that exceeds the geometrical opening angle ($\sim 1°$) by far. Furthermore, for the great majority of the transmitted ions, their high charge state is retained (Fig. 3.21). The transmitted particles are found to distribute within a narrow cone that is always directed nearly perfectly along the track direction, whatever the direction of the impinging ion beam. The latter can deviate from the track direction by as much as at least 25°, however, larger deviations in impact angle give rise to considerably lower transmission yields (Fig. 3.22). The transmission cone width depends on the transmitted particles' charge state, it ranges from $\sim$ 5.8° FWHM for Ne^{7+} ions to $\sim$ 2.2° FWHM for Ne^{1+} ions, Fig. 3.23. The angular width is proportional to the square root of the final charge state.

The angular transmission also depends strongly on the nanocapillary radius. Whereas for pores with 200 nm radius the maximum angle of incidence for recordable transmission is $\sim$ 15°, it amounts to $\sim$ 25° for pores with 100 nm radius. Also the angular distribution of the transmitted beam de-

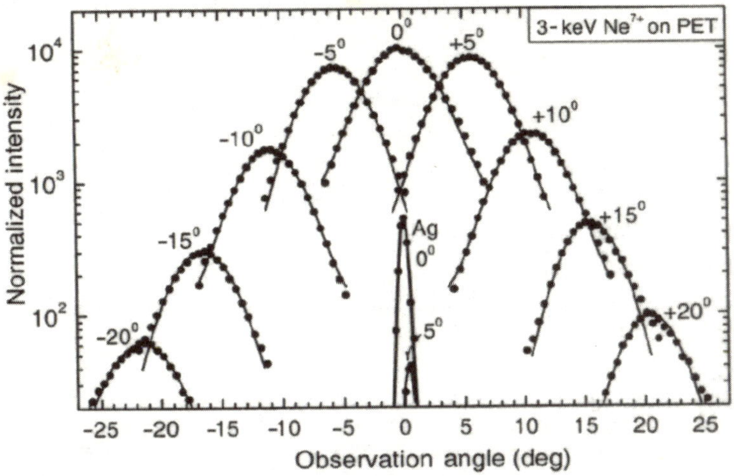

Fig. 3.22. Angular distributions of Ne^{7+} ions transmitted through capillaries in PET. The foils were tilted by the angles as indicated in the figures. The *solid lines* represent Gaussian functions fitted to the data. The distinct peaks near 0° were obtained using conducting capillaries, i.e., etched tracks covered with Ag [95]. Copyright (2003) by the American Physical Society

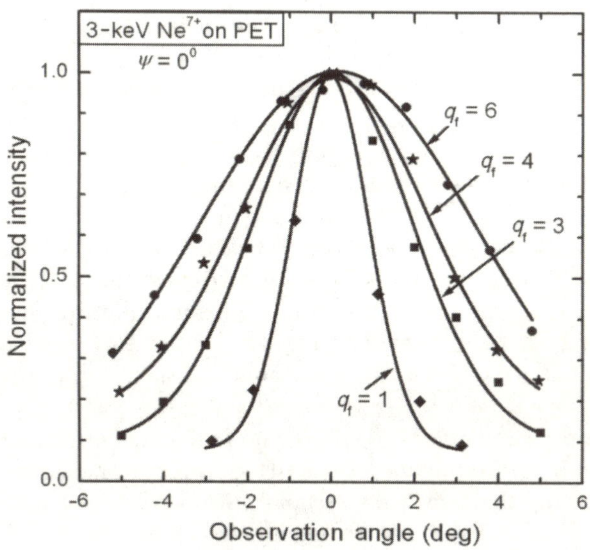

Fig. 3.23. Normalized angular distribution of transmitted Ne^{q+} ions with final charge $q+ = 6$ that emerge from the tracks for a tilt angle $\Psi = 0°$, when Ne^{7+} ions are injected [95]. Copyright (2003) by the American Physical Society

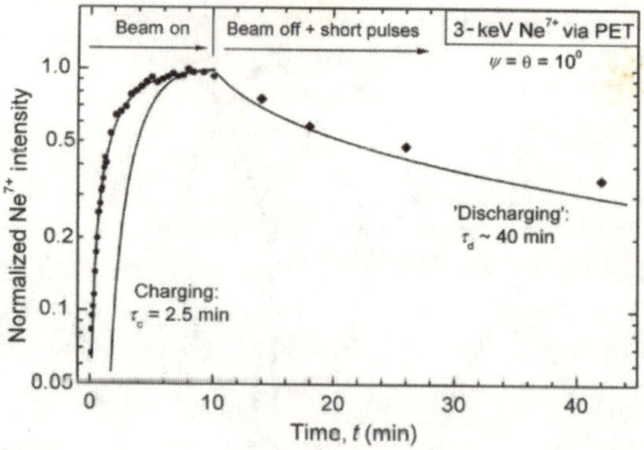

Fig. 3.24. Time dependence of the transmitted Ne^{+} intensity showing the charging and discharging phenomena of capillaries in PET. A beam of 1.3-nA Ne^{7+} is directed onto the PET foil tilted at 10°. The transmitted Ne^{7+} intensity, measured at 10°, increases exponentially with a time constant τ of 2.5 min. Short beam pulses probe the initial decrease of the transmission with $\tau_d \sim 40$ min [95] under the given experimental conditions. Copyright (2003) by the American Physical Society

pends strongly on the pore radius: whereas it amounts to 0.5° and 0.7° for pores with radii of 100 and 200 nm each at 0° incidence angle, it reaches values of 0.024° and 0.0037° at 15° incidence angle, respectively. This means that the narrower the pores, the more the incoming ions can be deflected, but the higher is their angular divergence.

This ion transmission does not set in immediately after the ion's impact into the tracks. Initially being completely impermeable (except for the negligible geometric transmission), the foils allow the first ions to be transmitted only after typically half a minute. Thereafter the foil's permeability rises steeply with a time constant of 0.75 to 2.5 min (depending on the chosen experimental conditions), and it arrives at a stationary transparency value only after typically 3 to 10 min. If the beam is switched off, this transparency decays with a half-lifetime τ_d around 40 min. This half-lifetime increases steadily with decreasing residual track charging up, finally reaching values of $\tau_d \sim 55$ days. Some enhanced capillary transmission can still be recorded after several months.

When a short time after the end of the first transmission experiment the ion beam is switched on again, transmitted ions are registered immediately. The transmission efficiency rises steeply within a few tens of seconds only, and reach values that exceed the original ones by about 10%. Thereafter this overshoot in transmission vanishes within ~3 min, and the original transmission is restored.

For an explanation of this "ion-guiding" effect one might assume that the first impinging highly charged ions continue their original straight flight path inside the etched tracks until they hit the track walls. At a distance closer than 1 nm charge exchange will take place, i.e., the ions will capture polymeric elec-

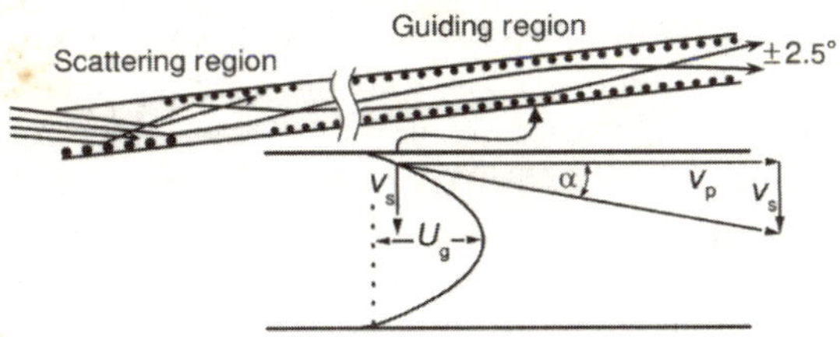

Fig. 3.25. Guiding of highly charged ions in a capillary. The graph shows results of the calculated potential inside the capillary. The influence of neighboring capillaries increases the absolute value of the potential by nearly a factor of 10, whereby the potential depth is unchanged [95]. Copyright (2003) by the American Physical Society

trons thus charging up the polymer wall positively. The following ions will feel these charges on this wall region, which scatters them towards the opposite track wall side, to be annihilated there. The subsequent ions will then be able to already undergo two reflections inside the ion track, before they contribute to the further track charging-up at greater depths by sacrificing themselves. With continuous ion impact, gradually the whole track becomes charged up until it is transformed to an ion-guide tube after a characteristic incubation time τ that depends on the ion flux, the track inner surface area, and the surface conductivity. Due to multiple ion deflection from the charged areas of the inner track walls, these walls become charged up everywhere. (This incubation effect is missing in ion channeling through crystals, as here the whole crystal channel is available for transmission from the first moment on.)

As a result, one has to distinguish between two zones, a near-surface scattering region (left-side of Fig. 3.25) where the ions are fed into the tracks by means of an inhomogeneous charge distribution, and a guiding region where the surface charges are distributed homogeneously on the track walls. Once the guiding is optimized, a stationary state is reached between charge accumulation by the impinging ions and charge removal by surface-charge diffusion. Transmission experiments performed in this steady-state mode are fully reproducible. Consequently, switching off the beam signifies an exponential decay in transmissivity. The latter can serve to estimate the charge mobility on the track walls. The sinks for the moving charges being beyond the track entrances, one expects a preferential surface-charge depletion where the tracks end. Upon resuming the irradiation, the newly impinging ions will preferentially recharge the entrance region. It appears that this can happen with transient charge excess until charge equilibrium is re-established along the tracks.

The fact that most of the highly charged ions are transmitted practically without any change of their charge state indicates that the minimum distance of most ions to the track walls never falls below ~ 1 nm, i.e., that the repulsive force along the track is considerable.

The negative outcome of the corresponding experiment with metal-clad inner track walls clearly demonstrates that ion guiding is based on Coulombic repulsion. If discrete charges were not fixed (at least transiently) on the track walls (as is the case for the insulating track) but were removed immediately (as is the case for metallized tracks), then the guiding effect does not take place.

There is an important difference between channeling and ion-track guiding. Channeling takes place due to repulsion from atomic potentials (i.e., by the Coulombic repulsive force between the two electron clouds of projectile and target atoms, respectively). Guiding, however, needs as a prerequisite dispersed isolated fixed charges, and therefore continuous metallized etched track walls do not allow for guiding, no matter whether the walls are charged up or not. What differs between guiding and ion transmission spectrometry (ITS) is the fact that ITS is interested only in the geometrical transmission process and not in the guided oscillatory motion. Therefore ITS works with energetic ions (e.g., 5.5-MeV α particles), whereas guiding favors low-energy highly charged projectiles.

The time dependence of the deposition of the charge Q along the tracks is governed by the differential equation:

$$\mathrm{d}Q(t)/\mathrm{d}t = J_\mathrm{r} - (1/\tau_\mathrm{p} + 1/\tau_\mathrm{d})Q(t), \tag{3.8}$$

with J_r – residual beam current surviving the scattering of the primary impinging beam at the channel entrance, $\tau_\mathrm{p} = Q_\mathrm{max}/J_\mathrm{r}$ being the time constant associated with the building up of the charge distribution and τ_p being the time constant associated with the discharge of the capillary wall. From the solution, $Q(t) = Q_\infty(1 - \exp(t/\tau_\mathrm{c}))$ with $Q_\infty = Q_\mathrm{max}/(1 + \tau_\mathrm{p}/\tau_\mathrm{d})$ and Q_max being the maximum charge uptake of a capillary, one can estimate by comparison with the experiment that $Q_\infty \approx Q_\mathrm{max} = 3200\ e$ (e = elementary charge). That is, the mean distance between two neighboring charges on the etched-track walls is about 30 to 40 nm. From this one can calculate the potential distribution within a track. It turns out that, for 3-keV particles incident into 200-nm wide etch tracks in PET, the calculated potential has a depth of 2 eV, i.e., exceeds the 0.8 eV by far that are required to guide the particles in these mesoscopic channels. A future more elaborate theory should also include the image charges formed when the ion passes along the charged track walls, and also the angular deflection at the track entrances and exits should be taken into account.

The approach according to (3.8) assumes that both the discharge current due to the bulk and surface conductivity J_d and the propagated current passing through the capillary J_p are proportional to the deposited charge Q: $J_\mathrm{d} = Q(t)/\tau_\mathrm{d}$ and $J_\mathrm{p} = Q(t)/\tau_\mathrm{p}$, respectively. This simple model does not, however, explain the strong decrease in intensity of the transmitted ions at increasing incident angles Ψ and the exponential increase of the capillaric conductivity with increasing deposited charge. Therefore an advanced nonlinear model was developed recently [96].

This model is based on the fact that field-enhanced thermal excitation of trapped electrons into the conduction band is governed by tunneling processes in which J_d depends exponentially on $Q(t)/Q_d$, with Q_d being the charge constant governing the discharge. Assuming that this is the process governing the capillary conductivity, and inserting the corresponding correlation into (3.8), one in fact obtains the desired exponential increase of the capillaric conductivity with increasing deposited charge. Also, the fraction J_p/J_r of transmitted ions is then seen to scale with the $\exp(\sin^2 \Psi)$, consistent with the observations.

Though it is still premature to speculate about future applications of this track-guiding effect, one should keep in mind that one has here a technique to deflect ion beams of nanometric diameter by at least 25°. Consequently, one might think of using this effect for constructing future deflection elements and self-aligning ion focusing devices, with a minimum of space requirement (as compared with conventional large-sized magnets and ion lenses) and no extra energy consumption (apart from some first self-sacrificing ions). In this way it may be possible to obtain "etch-track magnets" and "etch-track lenses" with μm- to mm-sized foci by irradiation and etching of suitably curved foils, which could be useful for future lightweight and smallest-scale ion-analytic devices, especially for space-technology applications.

Cladding such tracks with magnetic layers might eventually lead to the easy production of polarized ion beams, and, furthermore, the guiding effect may serve as a new probe to determine the electric nature of the inner etched track-wall surfaces.

References

1. Klett R, Charakterisierung von hochenergetischen Schwerionenspuren in Polyimid. PhD. Thesis, Humboldt-University, Berlin 1996 (in German)
2. Eyal Y, Gassan K, Observation of latent heavy-ion tracks in polyimide by means of transmission electron microscopy. Nucl Instrum Methods **B156**, 183–190 (1999)
3. Wang L, Trautmann Ch, Vetter J, Quan Z, Cohen D, Fladry H, Adhesion enhancement by GeV heavy ion irradiation. Radiat Eff Defects Solids **126**, 403–407 (1993)
4. Wang L (1996) personal communication, and Wang L, Angert N, Trautmann C, Vetter J, Effect of ion irradiation and heat treatment on adhesion in the Cu/Teflon system. J Adhes Sci Technol **9**,1523–1529 (1995)
5. Avasthi DK, Assmann W, Nolte H, Mieskes HD, Huber H, Subramaniyam ET, Tripathy A, Ghosh S, On-line study of ion-beam induced mixing at interface by swift heavy ions. Nucl Instrum Methods **B156**, 143–147 (1999)
6. Bolse W, Atomic transport in hot ion tracks. Presented at the 5th Intl. Symposium on "Swift Heavy Ions in Matter", May 22–25, 2002, Giardini Naxos, Italy

7. Soares MRF, Kaschny JRA, dos Santos JHR, Amaral L, Behar M, Fink D, Diffusion and solubility of Au implanted into the AZ1350 photoresist, Nucl Instrum Methods **B166–167**, 615–620 (2000)
8. Soares MRF, Kaschny JRA, dos Santos JHR, Amaral L, Behar M, Fink D, Diffusion and solubility of Bi implanted into the AZ1350 photoresist. Nucl Instrum Methods **B191**, 690–694 (2002)
9. Fink D, Müller M, Nakao Y, Hirata K, Kobayashi Y, Behar M, Kaschny JR, Vacík J, Hnatowicz V, Ion-induced redistribution of palladium in polymethyl methacrylate, Nucl Instrum Methods **B166–167**, 610-614 (2000)
10. Biswas A, Marton Z, Kanzow J, Kruse J, Zaporojtchenko V, Faupel F, Controlled generation of Ni nanoparticles in the capping layers of teflon AF by vapor phase tandem evaporation. In print (2003)
11. Biswas A, Avasthi DK, Kanzow J, Ding SJ, Fink D, Gupta R, Zaporojtchenko V, Faupel F, Nanostructural modifications in Au cluster arrays distributed in teflon AF layers upon MeV heavy ion impact. In print (2003)
12. Moiseev YuV, Zaikov GE Chemical Stability of Polymers in Aggressive Media. Khimia, Moscow (1979) (in Russian)
13. Born M, Volumen und Hydratationswärme der Ionen. Z Phys **1**, 45–48 (1920)
14. Markin VS, Chismadzhev YuA, Induced Ionic Transport. Nauka, Moscow (1974) (in Russian)
15. Samoilova LI, Apel PYu, Etching of small pores in PETP by different alkalis. Radiat Meas **25**, 717–720 (1995)
16. Parsegian A, Energy of an ion crossing a low dielectric membrane: solutions to four relevant electrostatic problems. Nature **221**, 844–846 (1969)
17. Eßer M, Lösungsmittel-induzierte Delegation molekularer Sonden in latente Kernspuren und ihre photophysikalische Analyse, PhD. Thesis, Technical University Clausthal, 1996 (in German)
18. Fink D, Müller M, Petrov A (2002) Etching kinetics of swift heavy ion irradiated polymers with insoluble additives or reaction products. Proc 5th Intl. Conf. on Swift Heavy Ions in Matter, May 22–25, 2002, Giardini Naxos, Italy
19. Fink D, Müller M, Capillaric penetration of etchant solution into swift heavy ion irradiated silicone rubber. Nucl Instrum Methods **B170**, 134–144 (2000)
20. Fink D, Petrov A, Müller M, Hnatowicz V, Vacík J, Červená J, Marker penetration into high energy ion irradiated polymers. Surf Coat Technol **158-159**, 228–233 (2002)
21. Fink D, Dwivedi KK, Müller M, Ghosh S, Hnatowicz V, Vacík J, Červená J, On the penetration of etchant into tracks in polycarbonate. Radiat Meas **32**, 307–313 (2000)
22. Lück HB, Kinetik und Mechanismus der Bildung und Ätzung von Teilchenspuren in Polyethylenterephthalat. PhD. Thesis, TU Dresden, published as report ZfK-473 of the "Zentralinstitut für Kernforschung Rossendorf bei Dresden"(1982), and references: Be51, Be60, LL56, RM72, RM75, GJ78, DB75, SB78, and RA60 therein (in German)
23. Ghosh S, Klett R, Fink D, Dwivedi KK, Vacík J, Hnatowicz V, Červená J, On the penetration of aqueous solutions into some pristine and heavy-ion irradiated polymers, Radiat Phys Chem **55**, 271–284 (1999)
24. Apel PYu, Schulz A, Spohr R, Trautmann C, Vutsadakis V, Tracks of very heavy ions in polymers. Nucl Instrum Methods **B131**, 55–63 (1997)
25. Ferry JD, Viscoelastic Properties of Polymers, 3rd. edn, Wiley, New York

26. Davenas J, Xu XL, Diffusion of iodine into polyimide films modified by ion bombardment. Nucl Instrum Methods **B81**, 33–38 (1992)
27. Fink D, Chaderton LT, Cruz SA, Fahrner WR, Hnatowicz V, TeKaat EH, Melnikov AA, Varichenko VS, Zaitsev AM, Ion track doping. Radiat Eff Defects Solids **132**, 81–90 (1994)
28. Fink D, Hnatowicz V, Vacík J, Chadderton LT, On the lithium uptake of MeV ion irradiated polymer foils from a LiCl solution. Radiat Eff Defects Solids **132**, 1–10 (1994)
29. Thomas NL, Windle A, A theory of Case II diffusion. Polymer **23**, 529–542 (1982)
30. Fink D, Klett R, Mathis C, Vacík J, Hnatowicz V, Chadderton LT, Depth profiles of fullerene in ion irradiated polyimide. Nucl Instrum Methods **B100**, 69–79 (1995)
31. Wood E, Sutton C, Beezer AE, Creighton JA, Davis AF, Mitchell JC, Surface enhanced Raman scattering (SERS) of membrane transport processes. Intl J Pharmaceut **154**, 115–118 (1997)
32. Fink D, Ghosh S, Klett R, Dwivedi KK, Kobayashi Y, Hirata K, Vacík J, Hnatowicz V, Červená J, Chadderton LT, Transport processes during the incubation time of ion track etching in polymers. Nucl Instrum Methods **B146**, 486–490 (1998)
33. Fink D, Ghosh S, Hirata K, Klett R, Dwivedi K, Vacík J, Hnatowicz V, On the Interaction of penetrant solutions with pristine and ion-irradiated polyimide. In: Yu.Ts. Oganessian, R. Kalpakchieva (eds.): Heavy Ion Physics. World Scientific, Singapore 1998, pp. 784–791
34. Fink D, Klett R, Hnatowicz V, Vacík J, Mathis C, Omichi H, Hosoi F, Chadderton LT, Wang L, Bonding of dopants to irradiated polymers, Nucl Instrum Methods **B116**, 434–439 (1996)
35. Seki S, Kanzaki K, Yoshida Y, Tagawa S, Shibata H, Asai K, Ishigure K, Positive-negative inversion of silicon based resist materials: poly (di-n-hexylsilane) for ion beam irradiation. Jpn J Appl Phys **36**, 5361–5364 (1997)
36. Fink D, Chung WH, Klett R, Döbeli M, Synal HA, Chadderton LT, Wang L, On the dyeing of ion tracks in polymers. Nucl Instrum Methods **B108**, 377–384 (1996)
37. Fink D, Omichi H, Hosoi F, Tamada M, Hnatowicz V, Vacík J, Chadderton LT, Klett R, Solid and liquid phase doping of energetic ion tracks in polymers. Advanced Materials '93 / Laser and Ion Beam Modification of Materials, I. Yamada et al. (eds.). Trans Mater Res Soc Jpn, Vol. **17**, Elsevier B.V., 1994, pp. 581–583
38. Fink D, Klett R, Chung WH, Grünwald R, Döbeli M, Ames F, Chadderton LT, Vacík J, Hnatowicz V, Doping of C_n^+ ($n = 1, 3, 5, 8$) cluster ion tracks in polyimide. Radiat Eff Defects Solids **140**, 3–20 (1996)
39. Fink D, Vacík J, Klett R, Chadderton LT, Hnatowicz V, Doping of 20 MeV fullerene ion tracks in polyimide. Nucl Instrum Methods **B119**, 591–595 (1996)
40. Vacík J, Červená J, Hnatowicz V, Posta S, Fink D, Klett R, Strauss P, Simple technique for characterization of ion-modified polymeric foils. Surf Sci Technol **123**, 97–100 (2000)
41. Vetter J, Mickler GH, Naumann I, TEM observation of latent tracks of heavy ions in semicrystalline polymers. Radiat Eff Defects Solids **143**, 273–286 (1998)

42. See, e.g., Munro HS, ESCA studies on the photooxidation and the gamma-radiation-induced oxidation of low density polyethylene, Polym Degrad Stab **12**, 249–259 (1985)
43. Fink D, Klett R, Hu X, Müller M, Schiwietz G, Xiao G, Chadderton LT, Wang L, Mathis C, Hnatowicz V, Vacík J, Characterisation of aged latent ion tracks in polyimide. Nucl Instrum Methods **B116**, 66–71 (1996)
44. Steckenreiter TH, Charakterisierung von Spuren energiereicher Ionen in Polymeren. PhD. Thesis, TH Darmstadt (1997) (in German)
45. Ghosh S, Klett R, Fink D, Dwivedi KK, Vacík J, Hnatowicz V, Červená J, On the penetration of aqueous solutions into pristine and radiation damaged polyimide. Radiat Phys Chem **55**, 271–284 (1999)
46. Eßer M, Apel PYu, Brüchle W, Fuhrmann J, Heinrich B, Remmert G, Spohr R, Trautmann C, Vetter J, Solvent induced sensitization, GSI-Nachrichten, 03-93 11–16 (1993)
47. Eßer M, Fuess H, Spohr R, Steckenreiter T, Trautmann C, Solvent induced track sensitization, role of amines. Nucl Instrum Methods **B107**, 393–396 (1996)
48. Apel PYu, Angert N, Brüchle W, Hermann H, Kampschulte U, Klein P, Kravets LI, Oganessian YuTs, Remmert G, Spohr R, Steckenreiter T, Trautmann Ch, Vetter J, Solvent induced sensitization, extraction of oligomers. Nucl Instrum Methods **B86**, 325–332 (1994)
49. Apel PYu, Track regression effects in polyethylene terephthalate after sensitization. Instrum Exp Techn, translated from Pribory I Tekhnika Experimenta No 5, 71–75 (1992)
50. Steckenreiter Th, Dimethylformamid-Sensibilisierung latenter Teilchenspuren in Poly(ethylenterephthalat)-Folien. Diplomarbeit, TH Darmstadt (1994) (in German)
51. Zachmann HG, Kinetik der Kristallisation von gequollenem Polyethylenterephthalat. Makromol Chem **118**, 189 pp. (1968) (in German)
52. Tamada M, Yoshida M, Asano H, Omichi H, Kakakai R, Spohr R, Vetter J, Thermo-response of ion track pores in copolymer films of methacryloyl$_{-L}$-alanine methyl ester and diethyleneglycol-bis-allylcarbonate. Polymer **33**, 3169–3172 (1992)
53. Torrisi L, Percolla R, Ion beam processing of polyvinylidene fluoride. Nucl Instrum Methods **B117**, 387–391 (1996)
54. Renardy M, Planck H, Trauter J, Zschocke P, Siebers U, Zecorn T, Federlin K, In: Heime G, Soltész U, Lee AJC (eds.): Clinical implant material. Adv Biomater **9**, 633 (1990)
55. Gebel G, Ottomani E, Allegrand JJ, Betz N, LeMoel A, Structural study of polystyrene grafted in irradiated polyvinylidene fluoride thin films. Nucl Instrum Methods **B105**, 145–149 (1995)
56. Duraud JP, Le Moel A, Le Gressus C, Aging of fluoropolymers irradiated by X-rays, low energy electrons and enegetic heavy ions. Radiat Eff **98**, 151–157 (1986)
57. Friese K, Plack V, Mehnert R, Angert N, Spohr R, Trautmann Ch, Radiation-induced grafting of styrene onto polyimide ion track membranes. Nucl Instrum Methods **B105**, 139–144 (1995)
58. Martin CR, Nanomaterials: A membrane-based synthetic approach. Science **266**, 1961–1966 (1994)

59. Petrov A, Production of micro- and nanoelectrotechnic devices by help of ion tracks in insulators. PhD Thesis, Fernuniversität Hagen (2004)
60. Neumann R, Ion induced modifications in solids: basic aspects and applications in nanoscience. 21st Int. Conf. on Nuclear Tracks in Solids, New Delhi, 21–25 Oct. 2002
61. Singh S, Sinha D, Srivastava A, Ghosh S, Dwivedi KK, Some chemical applications of nuclear track microfilters, to be published, 2002
62. Chin VI, Ozkan M, Bhatia S, Rapid three-dimensional arraying of single cells, Proc MRS Boston, 27.11.–1.12.2001, Contribution Y5.8
63. Jaffrin MY, Innovative processes for membrane plasma separation. J Membr Sci **44**, 115–129 (1989)
64. Barber DJW, Thomas JK, Radiat Res **74**, 51–65 (1978)
65. Schön W, Gärtner H, Kraft G, Radiat Envir Biophysics **33**, 233–242 (1994)
66. Leyko W, Bartosz G, J Radiat Biol **49**, 743–770 (1986)
67. J Guillet, Polymer Photophysics and Photochemistry: An Introduction to the Study of Photoprocesses in Macromolecules. Cambridge University Press, Cambridge, 1985
68. Remmert G, Transporteigenschaften und Geometrie von Schwerionenspuren in Polymerfolien. PhD. Thesis, Johann Wolfgang Goethe Universität, Frankfurt am Main (1994) (in German)
69. Komaki Y, Growth of fine holes by the chemical etching of fission tracks in polyvinylidene fluoride. Nucl Tracks **3**, 33–44 (1979)
70. Packard RE, Pekola SP, Price PB, Spohr RNR, Westmacott KH, Zhu YQ, GSI Scientific Report 1985, Darmstadt 1986
71. Backmeister GU, Enge W, Observation of the latent track structure in polymers by diffusion measurements. Nucl Instrum Methods **B131**, 643–70 (1997)
72. Turowski T, Schockwellenmodell zur Beschreibung des ionendichteabhängigen Diffusionsverhaltens bestrahlter Polymerfolien. PhD. Thesis, Christian-Albrechts-Universität Kiel (2001) (in German)
73. Fink D, Biersack JP, Compactation of polymers by energetic ions. HMI Berlin, internal report (1994 and 1999), forwarded in 1999 to Turowski T. as the basis of his PhD. Thesis, see Ref. [72]
74. Sudowe R, Penzhorn RD, Vater P, Abu-Jaber S, Brandt R, Filters with small holes ($d < 1\,\mu m$) as a tool to separate gases. Radiat Meas **28**, 811–816 (1997)
75. Ovchinnikov V.V., Seleznev V.D., Surguchev V.V., Tokmantsev V.I. Investigation of separation efficiency for gases on nuclear membrane with hyperfine pores. J Membr Sci **55**, 311–323 (1991)
76. Ghosh S, Klett R, Fink D, Dwivedi KK, Vacík J, Hnatowicz V, Červená J, On the penetration of aqueous solutions into some pristine and heavy-ion irradiated polymers. Radiat Phys Chem **55**, 271–284 (1999)
77. Fink D, Klett R, Latent tracks in polymers for future use in nanoelectronics, an overview about the present state-of-the-art. Braz J Phys **25**, 54–75 (1995)
78. Deen WM, Hindered transport of large molecules in liquid-filled pores. AIChE Journal **33**, 1409–1425 (1987)
79. Wolf A, Reber N, Apel PYu, Fischer BE, Spohr R, Electrolyte transport in charged single ion track capillarities. Nucl Instrum Methods **B105**, 291–293 (1995)
80. Grossmann PD, Colburn JC, Capillary Electrophoresis: Theory & Practice. Academic Press Incl., San Diego (1992)

81. Deamer D, Westphal A, Progress towards the development of a rapid DNA sequencer using etched relativistic ion tracks. Proc of the Workshop on European Network on Ion Track Technology, Caen, France, 24–26.Feb. 2002
82. Baur D, Apel PYu, Korchev YE, Müller C, Siwy Z, Spohr H, Spohr R, Surface gel in ion track etching – observations and consequences. Proc of the Workshop on European Network on Ion Track Technology, Caen, France, 24–26.Feb. 2002
83. Apel PYu, Korchev YuE, Siwy Z, Spohr R, Yoshida M, Diode-like single-ion track membrane prepared by electro-stopping. Nucl Instrum Methods **B184**, 337–346 (2001), and references therein, and
Apel PYu, Tuchkin SV, Gotlib VA, Lebedeva NE, Lev AA, Rectification property of conical track etched pores and an explanation of the change of rectifying direction at some conditions. Proc of the Workshop on European Network on Ion Track Technology, Caen, France, 24–26.Feb. 2002
84. Berezkin VV, Kiseleva OA, Nechaev AN, Sobolev VD, Churaev NV. Kolloidn Zh **56**, 319–325 (1994) (in Russian)
85. Marquet C, Buguin A, Talini L, Silberzan R, Rectified motion of colloids in asymmetrically structured channels. Phys Rev Lett **88**, 168301/1–4 (2002)
86. Siwy Z. Fulinski, Fabrication of a synthetic nanopore ion pump. Phys Rev Lett **89**, 198103-1–198103-4 (2002)
87. see, e.g., Liu DS, Astumian RD, Tsong TY, J Biolog Chem **265**, 7260 (1990)
88. Pintauro PN, Verbrugge MW, The electric-potential profile in ion-exchange membrane pores. J Membr Sci **44**, 197–212 (1989)
89. Pasternak CA, Alder M, Apel PYu, Bashford CL, Korchev YE, Lev AA, Rostovtseva TK, Zhitariuk NI, Model pores for biological membranes: The properties of track-etched membranes. Nucl Instrum Methods **B195**, 332–334 (1995), and Pasternak CA, Bashford CL, Korchev YE, Rostovtseva TK, Lev AA. In: Colloids Surf, A: Physicochem Eng Aspects **77**, 119 (1993)
90. Lev AA, Korchev YE, Rostovtseva TK, Bashford CL, Edmonds DT, Pasternak CA, Proc Royal Soc **B252**, 187–192 (1993)
91. Martin CR, Nishizawa M, Jirage K, Kang M, Lee SB, Controlling transport selectivity in gold nanotubule membranes, Adv Mater **13**, 1351–1362 (2001)
92. Jirage KB, Hulteen JC, Martin CR, Effect of thiol chemisorption on the transport properties of gold nanotubule menbranes. Anal Chem **71**, 4913–4918 (1999)
93. Fink D, Müller M, Szimkowiak P, Klett R, Vacík J, Hnatowicz V, Chadderton LT, Rutherford backscattering of laterally heterogeneous structures: the determination of radial density distributions in ion tracks in collodium. Nucl Instrum Methods **B134**, 87–97 (1998)
94. Fink D, Müller M, Vacík J, Červená J, Hnatowicz V, Nanotomographic examinations of etched and latent ion tracks by ion energy loss spectrometry. Appl Phys **A68**, 87–91 (1999)
95. Stolterfoht N, Bremer JH, Hoffmann V, Fink D, Petrov A, Sulik B, Transmission of 3 keV Ne^{7+} through nanocapillaries etched in polymer foils: evidence for capillary guiding. Phys Rev Lett **88**, 133201/1–4 (2002)
96. Stolterfoht N, Hellhammer R, Pesic ZD, Hoffmann V, Bundesmann J, Petrov A, Fink D, Sulik B. Guiding of Ne^{7+} ions through nanocapillaries in a PET polymer: dependence on the capillary diameter. Presented at the Conf. IPMM-03, Sendai, Japan, May 2003

4 Ion-Track Etching

P.Yu. Apel and D. Fink

Chemical etching of the ion-irradiated polymers is a process that transforms every latent track into a hole that, depending on conditions, may have a wide variety of shapes – long cylinders, short cones, hemispheres – and many others. All uses of ion-track etching are aimed either at getting information on the particles that created the tracks or at modifying the structure of the pristine monolithic polymer. In the former case the polymer serves as a detector and, sometimes, as a spectrometer. From the number, shapes and sizes of the tracks, one can determine the particle fluence, the composition of the particle flux, the angle of incidence and the energy of the particles. In other words, etching develops "fingerprints" of the particles passed through the polymer. Numerous applications of polymeric track detectors in nuclear physics, radiography, cosmic-ray studies, applied radiochemistry, dosimetry, etc., are based on this principle. In the latter case a particle beam with known parameters in combination with chemical etching serves as a unique and powerful microtool for the creation of a micro- or nanoporous structure. These two branches of track-etch methodology have been presented in books by Fleischer et al. [1] and by Spohr [2] that were comprehensive at their time of publication.

4.1 Basics of Etching Physics and Chemistry

4.1.1 Etching Kinetics

In the general case, degradation of the polymer in aggressive media is a complex process including a few main stages [3]:

- Transport of etchant molecules to the polymer surface,
- adsorption of the etchant molecules on the surface,
- motion of the etchant molecules into the polymer volume,
- chemical reaction of the etchant with weak chemical bonds of the polymer,
- diffusion of reaction products from the polymer bulk to the surface,
- desorption of reaction products from the polymer surface and their transport into the etchant volume.

The mathematical description of such a multistage process is very complicated. What is worse, the values of some necessary parameters, such as diffusion coefficients in a swollen matrix, are hard to measure. Nevertheless, one or two stages are often much slower than the others. Thus, the overall rate of the process is determined by the slowest – limiting – stage. Degradation of the polymer throughout its volume is always undesirable when etching particle tracks. This is why the etching conditions should be arranged in such a way that the transport of the etchant into the polymer does not take place or proceeds at a rate that is much lower than the rate of the chemical reaction. In this case the chemical reaction occurs in a thin layer at the phase boundary. The reaction rate in terms of the number of monomer units undergoing transformation per time unit, $\mathrm{d}n/\mathrm{d}t$, is

$$\mathrm{d}n/\mathrm{d}t = k_{\mathrm{s}} C_{\mathrm{m,s}} C_{\mathrm{et,s}} C_{\mathrm{sol,s}}, \tag{4.1}$$

where k_{s} is the reaction rate constant, $C_{\mathrm{m,s}}$ is the surface density of the monomer (active sites), $C_{\mathrm{et,s}}$ and $C_{\mathrm{sol,s}}$ are the concentrations of etchant and solvent molecules, respectively.[1] The subscript "s" denotes that all the concentrations belong to the solution/polymer interface. The process of surface modification can be monitored by measuring the water contact angle which decreases rapidly upon the onset of etching until a new stationary value is obtained (e.g., for PET from 72° to 66° within 10 min of contact with 9 mole/l NaOH at room temperature)[2] Equation (4.1) describes a process that is widely met in practice – both solvent molecules and molecules of the etchant (alkali, acid) take part in the chemical attack of the polymer surface. An example is the alkaline hydrolysis of polyethylene terephthalate [3]. In such a reaction, a steady-state concentration of active sites on the polymer surface is reached very quickly after the process starts. The period needed to reach the steady state is the reaction induction time. The value of $C_{\mathrm{m,s}}$ depends on the type of macromolecules, on their conformation, on the cohesion energy, on the solubility of degradation products (monomers and oligomers) in the ambient medium [3].

Another example of etching kinetics is the reaction of alkaline hydrolysis of polycarbonate. Solvent (water) molecules do not take part in the cleavage of carbonate groups [4], and the corresponding equation becomes

$$\mathrm{d}n/\mathrm{d}t = k_{\mathrm{s}} C_{\mathrm{m,s}} C_{\mathrm{et,s}}. \tag{4.2}$$

The concentration of the etchant molecules at the solution/polymer interface, $C_{\mathrm{et,s}}$ may differ from the concentration in the bulk solution, C_{et}. The concentration gradient at the interface depends on the consumption of the etchant components due to the chemical reaction, on the migration speed of

[1] We use concentrations for simplicity. The rigorous approach is based on the concept of thermodynamic activities

[2] Akimenko SN, Apel PYu, unpublished

the etchant into the reaction zone, and on the removal of products from the reaction zone. Moreover, the attacking molecules and the reaction products can be adsorbed on the polymer surface thus changing the spatial distribution of the etch-rate-determining particles in the reaction zone.

Transport of etchant molecules to the surface occurs through a stagnant (also called laminar or diffusion) layer with the thickness Δx. Assuming that the concentration gradient at the surface is $(C_{\mathrm{et}} - C_{\mathrm{et,s}})/\Delta x$ and setting the diffusion flux equal to the etchant consumption rate in a steadystate, one can easily derive the equation for the resulting reaction rate.

$$D_{\mathrm{et}}(C_{\mathrm{et}} - C_{\mathrm{et,s}})/\Delta x = k_{\mathrm{s}} C_{\mathrm{m,s}} C_{\mathrm{et,s}}, \tag{4.3}$$

where D_{et} is the effective diffusion coefficient in the laminar layer. Therefore the etchant concentration at the interface is

$$C_{\mathrm{et,s}} = C_{\mathrm{et}}(D_{\mathrm{et}}/\Delta x)/(k_{\mathrm{s}} C_{\mathrm{m,s}} + D_{\mathrm{et}}/\Delta x). \tag{4.4}$$

Substitution of the right-hand term of (4.4) for $C_{\mathrm{et,s}}$ in (4.2) leads to the final result:

$$\mathrm{d}n/\mathrm{d}t = C_{\mathrm{et}}/[\Delta x/D_{\mathrm{et}} + 1/k_{\mathrm{s}} C_{\mathrm{m,s}}]. \tag{4.5}$$

If $\Delta x/D_{\mathrm{et}} << 1/k_{\mathrm{s}} C_{\mathrm{m,s}}$ the overall rate is determined by the surface reaction:

$$\mathrm{d}n/\mathrm{d}t = k_{\mathrm{s}} C_{\mathrm{m,s}} C_{\mathrm{et}}. \tag{4.6}$$

A concentration profile illustrating this type of etching kinetics is presented in Fig. 4.1a. In the opposite situation when $\Delta x/D_{\mathrm{et}} >> 1/k_{\mathrm{s}} C_{\mathrm{m,s}}$, the rate is limited by the diffusion in the solution near the polymer surface. The etchant concentration at the phase boundary is markedly lower than in the bulk solution, $C_{\mathrm{et,s}} < C_{\mathrm{et}}$ (see Fig. 4.1b). A particular case of the diffusion-controlled kinetics is the etching accompanied by the buildup of a layer of insoluble degradation products. Stirring may significantly affect the etching kinetics because of a decrease in the δ value and an increase in $C_{\mathrm{et,s}}$. Etching limited by external diffusion is often nonuniform since hydrodynamic parameters are not constant along the polymer surface. Turbulent convection mixes the solution in the depleted layer irregularly with the bulk solution. For this reason, the use of highly concentrated (large k_{s}) and highly viscous liquids (large Δx, small D_{et}) cannot be recommended for track etching.

4.1.2 Etching of Hydrophobic and Hydrophilic Polymers

Most etchants are electrolytes dissociating into ions, and therefore the considerations of Chap. 2 on the penetration into hydrophobic and hydrophilic polymers are applicable here. The solubility of ions in hydrophobic polymers is extremely small [3], which is a result of the polarization of the medium around the electrical charges. The "self-energy" of a monovalent ion in a medium with dielectric constant ε can be estimated using the Born formula [5]:

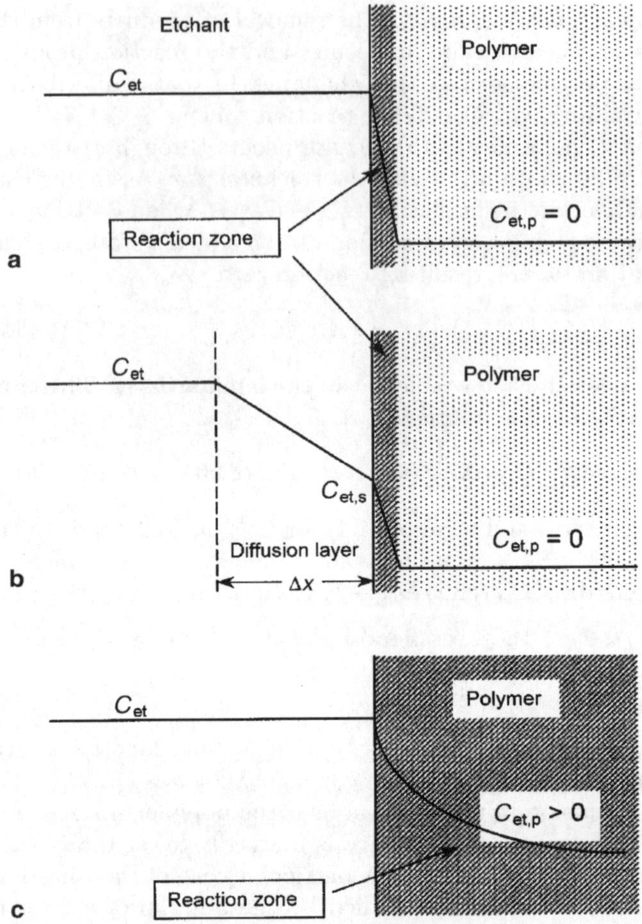

Fig. 4.1. Schematic etchant concentration profiles near the boundary between polymer and etch solution. (**a**) Surface reaction determines the rate of etching. (**b**) Diffusion of the etchant molecules through the laminar layer with the thickness D_x is the limiting stage of the process. (**c**) The etchant penetrates into a hydrophilic polymer. The rate of degradation depends on the internal mobility and the reaction rate in the polymer bulk

$$E = e^2/2\varepsilon r_\mathrm{i}, \tag{4.7}$$

where e is the elementary electrical charge, and r_i is the radius of the ion. Calculations for hydrophobic polymers ($\varepsilon \approx 3$) and the ion radius of 0.2 nm show that the concentration of the ions in the polymer phase should be approximately 10^{20} times smaller than that in the aqueous phase [6]. The adsorption of ions in aqueous solutions by hydrophobic surfaces is normally negative. In the aqueous phase very close to the polymer surface, the concentration of ions is lower than in the bulk solution – due to the proximity of the medium with low ε. If the polymer surface is neutral, the negative adsorption is poor and can be neglected in considering the etching kinetics. The thermodynamic parameters of the etchant in the bulk and at the polymer/solution interface can be assumed to be identical. However, this is not the case when the surface has electrical charge due to dissociating chemical groups. An example is polyethylene terephthalate treated with an alkaline agent. Its surface acquires negatively charged carboxylate groups rigidly bound to the surface. The electric field results in increasing the local concentration of positive ions (counterions) in the adjacent layer of the aqueous phase. Accordingly, the local concentration of negative ions (co-ions) decreases. This phenomenon produces a considerable effect on the etching kinetics, especially in diluted etchants [7].

In contrast to hydrophobic matrices, diffusion of electrolytes in hydrophilic polymers is appreciable. Hydrophilic polymers (polyvinyl alcohol, cellulose, etc.) are characterized by a high content of polar groups and an ability to take up water, i.e., swell. The presence of water in the polymer significantly reduces the ion "self-energy" [8]. Water molecules play the role of carriers promoting the transport of ions through the polymer. Other hydrophilic polymers are of ion-exchanging nature (such as polyamides). In such systems functional groups are preferred resting positions for diffusing ions, with energetically unfavorable regions in between. The ions hop from one preferred site to another. Protonated amino groups can serve as active sites for the diffusion of anions, whereas carboxylic groups can provide the route for the transport of protons or other cations.

For these reasons, degrading agents not only etch hydrophilic polymers at the surface but also diffuse into the depth and initiate overall degradation leading to a decrease in molecular mass, an increase in the number of end groups and to further swelling. By contrast, a hydrophobic polymer being in contact with a degrading agent retains its average molecular mass. Ideally, the etching is solely removing material in this case, i.e., changing the size and shape of a polymer sample without affecting its interior physical and chemical structure.

Figure 4.1c shows the schematic etchant concentration profile at the boundary between a hydrophilic polymer and a liquid phase. It is assumed that the decomposition of the polymer is slow, whereas the external diffusion is fast. The whole sample constitutes the reaction zone because of a nonzero etchant concentration, $C_{\mathrm{et,p}}$, in the polymer. Note that the three diagrams presented in Fig. 4.1 do not cover all possible situations. Various combinations can be met in practice. For instance, both the external and internal diffusion may influence the overall reaction rate. Finally, one should take into account that the classification of polymers by hydrophobic/hydrophilic properties is a matter of convention. Many of them are intermediates; accordingly, they behave neither like purely hydrophobic nor like purely hydrophilic solids. For realistic as-measured etchant depth distributions, please see Chap. 2 and 3.

A few years ago the idea emerged that a three-phase model (i.e., etchant/etchant-soaked polymer region/untouched polymer) eventually might be more adequate to describe the experimental observations during the etching process [9]. Recent observation on a number of polymers {polysilicone (PSiO) [10, 11], polycarbonate (PC) [12], polybutylene terephthalate (PBT) [13], polyethylene terephthalate (PET) [14], and PI [14, 15], and earlier observations on CN [93]} showed that this model indeed holds – even more, in a number of systems the intermediate etchant-soaked (swollen) layer turned out to be the most important one of the whole etching process.

4.1.3 Ion-Track Swelling

The pristine PET is known to be a rather hydrophobic material with a water contact angle of about 72°, and hence hardly uptakes water so that both solid-to-gel transition and swelling are hindered. However, after etching, at least a layer at the surface of unknown thickness appears to be hydrophilized due to its much higher concentration of carboxylic (−COOH) and hydroxylic (−OH) end groups in comparison with the pristine polymer. These types of end groups are hydrophilic and can be responsible for the water uptake.

Figure 4.2 reveals that, whereas during the onset of NaOH etchant penetration into SHI irradiated PET there is hardly any change in surface topography to be detected, considerable alterations in surface structure show

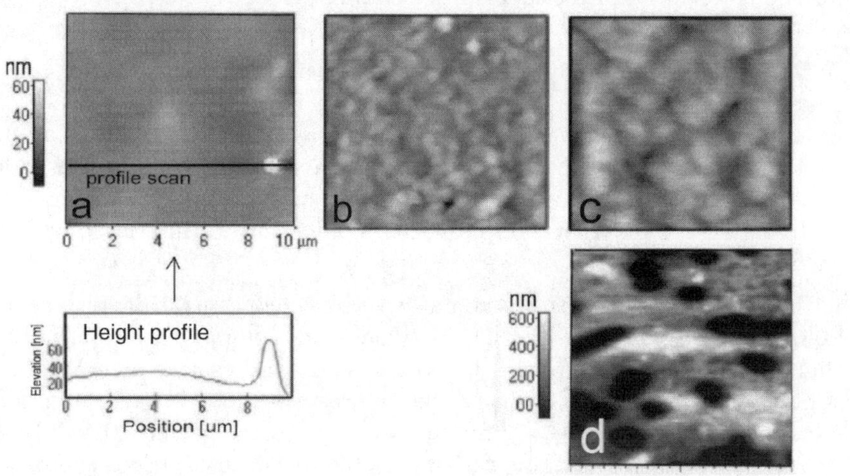

Fig. 4.2. SFM images of ion-irradiated PET (**a**) without etching, (**b**) after 10 s etching, (**c**) after 10 min etching at ambient temperature, as measured on wet samples. (**d**) Only samples dried after etching reveal pores. Irradiation by 3-GeV U ions at $2 \times 10^8\ \mathrm{cm}^{-2}$, [16]. The dust particle on the *right side* of (**a**) (see also the height profile) was incorporated for the sake of better illustration of the smoothness of the pristine foil. Sizes: (**a, b, c**) 10 µm × 10 µm, (**d**) 3 µm × 3 µm

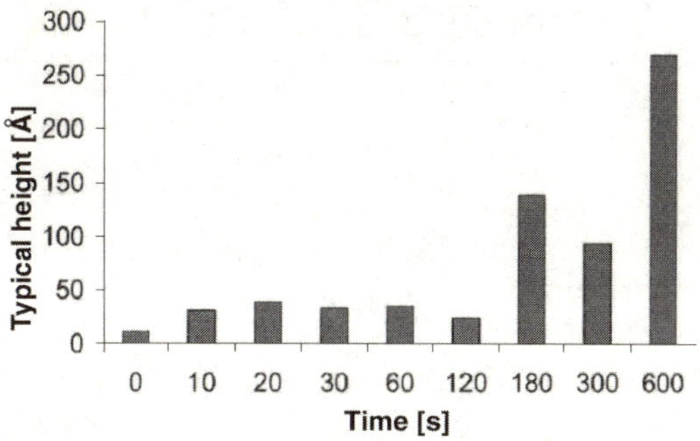

Fig. 4.3. Time dependence of surface roughening by swelling, as derived for the samples of Fig. 4.2 [16]

up after about 10 min exposure time at ambient, or 3 min at $\sim 45°C$. The time dependence of the surface roughness is quantified in Fig. 4.3. Even when continuing these SFM examinations up to the point when the samples disintegrate due to over-etching, SFM observations do not yet reveal any visible hole, Fig. 4.2c. Only when releasing the uptaken water after drying, does the shrinking polymer reveal its mass loss in the irradiated zones due to etching, Fig. 4.2d [16]. By contrast, recent STM observations on tracks in PC [17] have revealed the existence of open pores during etching, however, with diameters much smaller than in the dried state. The studies confirmed the existence of a highly conductive surface gel.

The transition from the latent tracks to etched tracks by the etching process can also be followed by ion transmission spectrometry, Fig. 4.4. Irradiated unetched PET foils show the same transmission spectra as pristine ones (not shown here), indicating that degassing after ion impact is a process of overall minor importance (i.e., less than a per cent loss in density) than is commonly assumed. Incorporation of etchant leads to a shift of the transmission spectra by as much as $\approx 50\%$ towards lower energies, if the foils are measured in the wet state (with only their surface dried by a cloth to remove drops of the etchant), Fig. 4.4. Subsequent drying after the etching reverses this process, leading finally to an ion-transparent foil, Fig 4.5 [16].

These observations indicate that a prerequisite for dissolution of radiochemical destruction products by etchants appears to be their transformation from the solid to the gel state by solvent uptake, associated with the material's swelling. This transformation, followed by new free-volume formation due to gel dissolution, sets in already at the beginning (stage I of Chap. 3, but becomes more important in stage III when sufficient etching has penetrated into the polymer due to sufficient newly created free polymeric volume.

Quite a number of experiments have been made to determine the diffusion coefficient of etchants in latent ion tracks, see Chap. 3. The kinetics of etchant penetration into ion tracks appears to be a highly complex process, where the penetration into charged gels plays a major role.

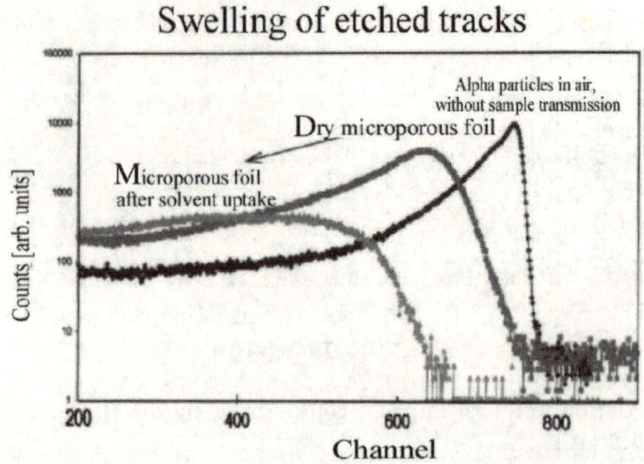

Fig. 4.4. ITS spectra of different stages of ion track etching. Solvent uptake during the etching leads to greater energy loss of the transmitted particles. Foil thickness: 10 µ*m*, final etch track diameter 200 nm [16]

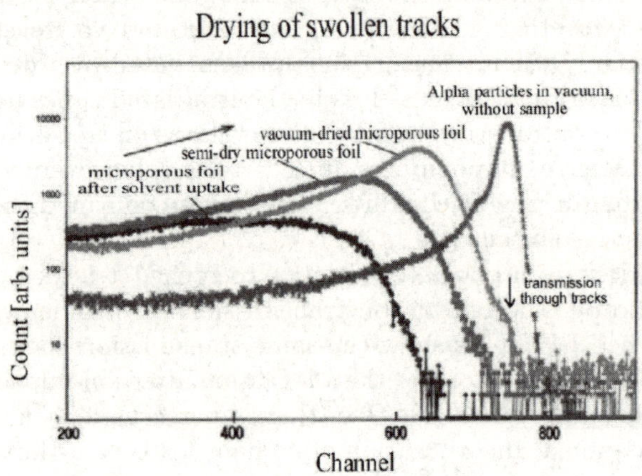

Fig. 4.5. ITS spectra of different stages of ion track etching. Solvent loss by drying leads to energy loss reduction and penetration through etched tracks with full projectile energy. The final spectrum reveals the porosity of the etched foil (right arrow). The measurement of the transmission spectra in air results in worse energy straggling and a slight peak shift [16]

Leaching of irradiated etched PC samples for ~ 1 h in distilled water does not remove much of the previously incorporated etchant. Experiments with a NaOH/LiOH mixture have shown that ~ 70% of the originally incorporated Li remain in the sample with unchanged depth distribution.[3] The halo of these adsorbed ions (see Fig. 3.17) enhances the conductivity along the tracks by up to a few orders of magnitude, thus worsening the polymer's insulating properties.[3]

[3] Fink D (2000), unpublished results

4.1.4 Etching of Polymers with Structural Inhomogeneities. Why Are Ion Tracks Preferentially Etched?

High polymers are always more or less heterogeneous. This is especially true for commercially available plastics of various kinds. Even in the case when a polymer constitutes a single phase, one can distinguish inhomogeneities of properties in its volume, so-called structure gradients [18]. For heterophase polymeric materials, the variety of the structure gradients is wider. Here, some of them are listed:

- Domains in the glassy state and in the state of high elasticity,
- differences in the degree of cross linking,
- oriented and not oriented domains,
- regions with different molecular mass distributions formed due to a spontaneous fractionation (e.g., during the shear flow).
- amorphous and crystalline domains,
- domains with different degrees of microheterogeneities in stereoregular polymers,
- domains with different spatial (*gauch–trans* isomerism) or chemical structures.

Moreover, industrial polymeric products often contain structural gradients due to various organic and inorganic admixtures and impurities. These are stabilizers, antioxidants, plasticizers, fillers, dyes, traces of catalysts, etc. All the structure gradients cause local anisotropy of penetrant mobility and chemical properties.

The availability and distribution of both the intrinsic and the radiation-induced free volume in a polymer determines the mechanism of penetrant migration that can either proceed via nanocapillarity or via diffusion (see Chaps. 1 and 3). The accessibility and, hence, reactivity of chemically active bonds in polymer chains (weak bonds, active sites) depend on the structural heterogeneity. For example, weak bonds are more accessible in the amorphous phase than in the crystalline phase. The free volume in amorphous domains gives more freedom for rearrangements of chain segments and facilitates every single chain scission event. From this point of view, selective etching of heavy-ion tracks in dielectrics, and particularly in polymers, is apparent from the properties of latent tracks, which were presented in the previous parts of this book. However, historically, the phenomenon of selective etching of ion tracks in polymers was discovered first [19]. Detailed studies of latent tracks in polymers were performed later and gave us a description of a track as an extended zone with changed (as compared with pristine polymers) physical and chemical properties. One can specify the following typical features that make the latent tracks preferentially etchable:

1. The material density in latent tracks is lower and, hence, the free volume is larger. A particular case are amorphous tracks in a crystalline or semicrystalline polymer.

2. The average molecular mass in the track core is lower. Smaller molecules are more mobile and reactive. Abstraction of a monomer unit at the chain end requires only one bond cleavage whereas two cleavage events are required to release the monomer from the middle of the chain.
3. Molecules in the track core often have polar groups formed in radiation-induced reactions. The polar groups attract water and promote the penetration of ionic species into the track.
4. Apparently, some new chemical bonds and groups may specifically react with a given etchant. In this case the selectivity of the etchant should be especially high.
5. Residual mechanical strain around the ion path makes weak chemical bonds even more susceptible to cleavage.

The reversal of some of the above-mentioned conditions may lead to a reversal of the etching process, so that the unirradiated material is etched more than the irradiated one, thus transforming ion tracks into hillocks or filaments (after prolonged etching), instead of pores. This is indeed found, e.g., for swift heavy-ion-irradiated PMPySi, see below. Here, the underlying radiochemical reason is that the polysilane is transformed into practically inert SiC_xH_y, whereas the pristine polymer is attacked [20].

Until now, it is not quite clear which one of the listed factors plays a crucial role for ion tracks in polymers. Very likely, the answer will be different for different combinations of polymers, ions, and etch recipes. It is difficult to "separate variables" in a real experiment and estimate individual contributions of the above mechanisms. As an example, we could mention studies performed with the aim of finding a correlation between etchability and the presence of radicals in the tracks [21, 22]. Parameters of etched tracks and concentration of radicals were measured as a function of annealing temperature. The presence of free radicals could not be correlated with the etchability of the latent tracks in Lexan polycarbonate [21]. For CR-39 and Kapton [22], a correlation seems to exist; however, the annealing could lead not only to radical recombination but also to other changes in the track structure that makes the conclusions ambiguous.

The concept of a free reaction volume, which is formed during etching from microvoids in latent tracks [23], seems to have a stronger justification. Heavily ionizing particles always produce gaseous products in polymers (CO_2 and CO in polycarbonates and polyesters [24], CO_2, CO and N_2O in cellulose nitrate [25], H_2 and hydrocarbons in polyolefines [26], etc.). The total yield of gaseous products significantly increases with increasing $\mathrm{d}E/\mathrm{d}x$ [26]. Escaping gas molecules form a trail of microholes along the particle trajectory. It is absolutely clear that such defects should have a strong effect on the etchability.

Similarly to low-LET radiation, heavy ions produce unsaturated chemical bonds in polymers. This was observed in the early stage of studies on polymeric track detectors [27]. At first, the accumulation of unidentified unsaturated bonds was confirmed by ultraviolet and visible absorption spectra [28, 29]. Later, the formation of double and triple bonds in various polymers under heavy-ion bombardment was evidenced from analysis of infrared spectra [30,31]. The unsaturated chemical bonds are often more reactive than ordinary ones, therefore they can act as active sites for selective etching.

The role of polar groups is seen from the fact that oxidation and photo-oxidation increase the track etch rate. A method of track sensitization is based on the strong correlation between the degree of oxidation of radiolysis products in latent tracks and the rate of alkaline hydrolysis. Below we will consider this subject in more detail.

Etching of dislocations in inorganic crystals can be considered as a demonstration of the potential role of the strained chemical bonds. A dislocation is a zone of a distorted crystal lattice. There are no small molecules, no different chemical groups and bonds, and no microvoids in it. Atoms (ions) in the distorted zone are characterized by a reduced binding energy because the distances between the ions and the angles between neighboring bonds are different in comparison with the undisturbed lattice. An appropriate solvent removes such atoms selectively. One could expect a similar behavior of chemical bonds in high polymers, see e.g. Vol. I, Fig. 1.1. Deformed organic molecules may have a higher reactivity, in particular, the molecules being under stress are oxidized at a higher rate [32].

Any track-sensitive etchant reveals other defects that existed in a given polymer before irradiation. Etched defects have the appearance of bubbles, craters (see Vol. I, Fig. 1.1), caverns, cracks, and recesses on the surface. Such "background" may conceal the etched tracks. Therefore a good polymer for track recording should be isotropic and as homogeneous as possible. Cartwright et al. [33] argued that only totally amorphous polymers could satisfy the requirement of homogeneity and isotropy.

4.1.5 Selection of Appropriate Etchants for Developing Ion Tracks

Some general principles of the selection of good etching conditions were recommended in [1, 33, 34]. Summarizing the accumulated experience, one can present guidelines as follows:

1. The etchant should be capable of "distinguishing" between regions of different chemical properties. Chemical polishes that tend to round corners are not suitable since they will obscure etched tracks.
2. The etchant should degrade the molecules at the liquid/solid interface slowly and carefully. The degradation should occur due to bond breaking and without swelling and dissolving of the matrix.
3. The reaction between the polymer and etchant should not lead to the formation of insoluble products. Such products are accumulated on the surface and impede further transport of reagents.

4. Avoid highly concentrated etch recipes since they, on the one hand, are often highly viscous and, on the other hand, etch very rapidly. This leads to the external diffusion-limited kinetics that strongly depends of the efficiency of convection in the liquid phase. The only exceptions are extremely resistant materials, such as fluoropolymers. They are etched very slowly even in concentrated etchants.
5. The etchant should be stable with time. Its spontaneous degradation at elevated temperature or under light exposure is undesirable.
6. The etchant should have a low surface tension and, thus, wet the polymer surface readily. Poor wettability may cause inhomogeneous etching because the process does not start simultaneously over all the sample.

In practice, the development of an "ideal" etchant is sometimes a hard problem. In many cases it turned out to be impossible to find a recipe that satisfies all the above requirements. So far the method of trial and error plays a major role in the selection of etching conditions. The general rules do help, however, only experimental tests may give the ultimate answer if the etchant is suitable or not. Examples are the etch recipes for polypropylene (PP). A solution of $KMnO_4$ (0.1–10%) in a concentrated sulfuric acid was known to be a good etchant for revealing the surface morphology of PP [35]. Surprisingly, this oxidizing etchant is not selective to heavy-ion tracks in isotactic polypropylene. Instead, the oxidizing systems based on chromic acid or its derivatives reveal the tracks in PP. Our knowledge of reaction mechanisms of polymer degradation in oxidizing multicomponent systems is still limited. The process of alkaline hydrolysis is not so intricate; relevant experimental data and a detailed analysis of its mechanism as applied to polymeric track detectors were presented in [36] for cellulose nitrate, in [37, 38] for polycarbonates and in [39, 40] for polyethylene terephthalate. Nevertheless, some empirical data on the alkaline etching – such as the maximum of selectivity at 20% concentration of NaOH in CR-39, cellulose acetate, and polycarbonate [41] – still remain unexplained.

The knowledge of the fundamental chemistry involved in the etching process enables one to understand and unify results obtained under a variety of conditions and assists in the search for more satisfactory and application-specific etches [38]. Equations of the relevant chemical reactions are of benefit when one uses stoichiometric relations to estimate consumption of etchant or determine the etch rate from the amount of etch products. In fact, all recipes used for track etching in polymers are either degradation catalyzed by base or oxidation promoted by base or acid. In the former case ether or ester bonds in the polymer backbone are destroyed. In the latter case the oxidative degradation provides cleavage of more stable carbon–carbon bonds in the main chain. Carboxylic end groups readily react with alkali and form carboxylates. Hydroxylic ones are less acidic but may also react at high alkali concentrations. Apart from COOH and OH groups, unsaturated bonds are important for oxidative etching.

The alkaline attack of polycarbonate is described by the reaction as follows:

$$\sim \mathrm{O-CO-O-C_6H_4-C(CH_3)_2-C_6H_4} \sim + \, 4\mathrm{OH^-} \rightarrow$$
$$\mathrm{CO_3^{2-}} + 2\,{}^{-}\mathrm{O-C_6H_4-C(CH_3)_2-C_6H_4-O^-} + 2\mathrm{H_2O}.$$

The carbonate ion and the bisphenol A anion are the etch products accumulating in the etching bath. The creation of phenolate end groups should increase the hydrophilicity of the plastic and thus draw in more aqueous etchant. The track region probably has an increased concentration of phenolic groups due to the radiolytic products therein. Increased formation of phenolates in the tracks provides an additional mechanism for enhanced etchant penetration into these regions. It should be kept in mind that the phenolic groups react much more rapidly with hydroxide than the carbonate groups. The differences in penetration behavior of neutral and alkaline solutions as well as the retention of etchant by the plastic may be explained in part by the above phenomena.

The alkaline etching of polyethyleneterephthalate is the sequence of cleavages of ester bonds:

$$\sim \mathrm{O-CH_2-CH_2-O-CO-C_6H_4-CO} \sim + \mathrm{OH^-} \rightarrow$$
$$\sim \mathrm{O-CH_2-CH_2-OH} + {}^{-}\mathrm{O-CO-C_6H_4-CO} \sim .$$

Final products are the ethylene glycol and the anion of terephthalic acid. Sodium or potassium salts of terephthalic acid have a low solubility in alkaline etchants. The solubility decrease rapidly with increasing alkali concentration. The terephthalates precipitate when a sufficient concentration of the anion is accumulated in the etch solution.

Oxidative etching generally involves several different reactions occurring simultaneously, which makes it difficult to compose an authentic equation. In most cases only a rough representation can be suggested, e.g., for polyolefins (PE, PP) undergoing the treatment with acidic solutions of Cr(VI) the following equation may be used [42]:

$$\mathrm{C}_n\mathrm{H}_{2n+2} + (n + 1/3)\mathrm{Cr_2O_7^{2-}} + 8(n + 1/3)\mathrm{H^+} \rightarrow$$
$$n\mathrm{CO_2} + (5n + 7/3)\mathrm{H_2O} + 2(n + 1/3)\mathrm{Cr^{3+}}.$$

This equation implies that the polymer is completely oxidized so that the final products are carbon dioxide and water. What actually happens is that the oxidation is incomplete and the process is quite intricate. Some other products – such as acetic acid – are formed during etching.

Due to the highly ionic character of the siloxane bond, silicones are sensitive to alkaline attack. This gives rise to depolarization, and consequently to reactions of the scheme [43]:

$$\equiv\mathrm{SiOSi}\equiv + \mathrm{NaOH} \rightarrow \mathrm{SiOH} + \mathrm{NaOSi}\equiv .$$

Hence treatment with alkalines readjusts the spectrum of molecular sizes of the polysiloxanes. In general, it has been found that the more highly substituted the silicon atoms are in a polymer, the more resistant is the siloxane against the alkaline attack [43]. This implies that, due to the possibility of irradiation-induced chain scissioning, some enhanced etchability of irradiated silicone rubber is found. Indeed, individual as well as overlapping tracks could be etched.

To change the surface activity of etchants, small amounts of a polar organic solvent can be added. It is common practice to use methanol or ethanol, which speed up the chemical attack considerably [1]. The role of added alcohol cannot be reduced to the decrease in the interface energy. First, apparently small molecules of alcohol diffuse into the polymer bulk and weaken the binding between the tightly packed chains. Secondly, alcohols may break ester groups in the polymer by the alcoholysis mechanism. Apart from alcohols, surfactants are often included in etch recipes[4] [44]. They improve the wettability and etching homogeneity. In contrast to alcohols, the surfactants normally reduce the bulk etch rate of polymers. The adsorbed layer of surfactant molecules partially protects the polymer surface from chemical attack. The surfactant also produces a strong effect on the shape of small pores. The pore shape conditioned by a surfactant will be considered below in more detail. The drawback of both agents – alcohols and surfactants – is the embrittlement that appears to a greater or lesser extent depending on the type of the surface-active substance and the polymer.

Table 4.1 lists recommended etching conditions for a number of polymers. Already in the 1960s the pioneers in the field of nuclear tracks found appropriate etching conditions for a wide variety of materials [1]. The role of M. Monnin and coworkers in the development of efficient etchants for polymers was also remarkable [58]. Note that no etchant has been found for ion tracks in polytetrafluoroethylene and other fully fluorinated plastics – in spite of numerous efforts made in different laboratories.

The recommended etching times are chosen so that fission tracks are large enough to be visible in the optical microscope. However, not all the polymers are thought to be suitable for use as track detectors. The track-recording plastics can serve as matrices for the production of micro- or nanostructured materials. Depending on the required size of structural units, the etching time can be considerably shorter or longer. Other etching conditions – temperature and concentration of chemicals – are also variables that allow one to control the attack rates of both the pristine and radiation-damaged polymers.

[4] Such surfactants are, e.g., nonyl benzene deca(ethylene glycol), Na-p-toluene sulfonate, or Na-dodecyl-benzene sulfonate. These surfactants have a hydrophobic (e.g., CH_3–) and a hydrophilic side (e.g., –SO_3 Na) [45].

Table 4.1. Recommended etching conditions for some polymers

Polymer	Etchant, temperature, time	Ref.
Cellulose acetate	1 ml 15% NaClO + 2 ml 6.25 *M* NaOH, 40°C, 1 h	[1]
Cellulose acetate butyrate	6.25 *M* NaOH, 70°C, 12 min	[1]
Cellulose nitrate	6.25 *M* NaOH, 23°C, 2–4 h	[1]
Cellulose propionate	28% KOH, 60°C, 100 min	[1]
Cellulose triacetate	1 ml 15% NaClO + 2 ml 6.25 *M* NaOH, 40°C, 1 h	[1]
Poly (allyl dyglicol carbonate) (PADC, CR-39)	6.25 *M* NaOH, 75°C, hours	[33]
Polyamide	25% $KMnO_4$, 100°C, 1.5 h	[1]
Polyarylate (product of polycondensation of phenolphthaleine and terephthalic acid)	4 *M* NaOH, 50°C, 1 h	[46]
Polybutylene terephthalate (PBT)	1 ml 6.25 *M* NaOH + 1 ml C_2H_5OH, 23°C, 24 h	[1]
Polycarbonate (poly diphenylol propane carbonate) (PC)	6.25 *M* NaOH, 50°C, 20 min 6.25 *M* NaOH + 0.4% Benax[5] 70°C, 20 min	[1] [1]
Polydimethylsiloxane (PDMSiO)	25 *M* NaOH, 115°C, 3 min	[1]
Polyetheretherketone (PEEK)	50 ml H_2SO_4 + 50 ml H_2O + 7 g $KMnO_4$, 70°C, 150 min	[47]
Polyethylene (PE)	10 g $K_2Cr_2O_7$ + 35 ml 30% H_2SO_4, 85°C, 30 min	[1]
Polyethylene naphthalate (PEN)	3–5 *M* NaOH, 45–65°C, hours	[48]
Polyethylene terephthalate (PET)	6.25 *M* NaOH, 70°C, 10 min	[1]
Polyimide (poly[N,N”-(n’,n-oxyddiphenylene)pyromellitimide]) (PI)	25% $KMnO_4$, 100°C, 8 h NaClO (10% Cl), 70°C, hours 30% H_2O_2, 100°C, hours	[1,49] [50] [51]
Polymethyl methacrylate (PMMA)	sat. $KMnO_4$, 85°C, 8 min	[1]

[5] Benax and Dowfax are surfactants produced by Dow Chemicals. The latter surfactant consists essentially of disodium salt of dodecyl (sulfophenoxy) benzenesulfonic acid.

Polyoxymethylene	5% $KMnO_4$, 60°C, 10 h	[1]
Polyphenoxide	25% $KMnO_4$, 100°C, 4 min	[1]
Polyphenylene oxide	sat. $KMnO_4$, 93°C, 24 h	[1]
Polypropylene (PP)	100 ml 40% H_2SO_4 + 25 g CrO_3, 80°C, 1 h	[52]
Polystyrene (PS)	sat. $KMnO_4$, 85°C, 2.5 h	[1]
Polysulfone (PS)	100 ml 40% H_2SO_4 + 30 g CrO_3, 50°C, hours	[53]
Polyvinylacetate (PVA)	6.25 *M* NaOH, 23°C, 400 h	[1]
Polyvinylaceto-chloride	sat. $KMnO_4$, 100°C, 30 min	[1]
Polyvinylchloride (PVC)	sat. $KMnO_4$, 85°C, 2.5 h	[1]
Polyvinylidene chloride	25% $KMnO_4$, 55°C, 2 h	[1]
Polyvinylidene flouride	5–12 *M* NaOH, 65–85°C, ≤ 100 h	[54]
	0.1 M $KMnO_4$ in 6 *M* KOH, 70°C, days	[50]
	10 M KOH + 0.25 *M* $KMnO_4$ + 0.5 NKIO_4, 90°C, 1 min	[55]
Other fluorinated polymers:		
Copolymer of ethylene and tetrafluoroethylene	12 *M* NaOH, 125°C, 250 h	[56]
Copolymer of vinylidene fluoride and tetrafluoroethylene, copolymer of vinylidene fluoride and haxafluoropropylene, copolymer of vinylidene fluoride and trifluorochloroethylene	100 ml 13% NaOH + 20 g $KMnO_4$	[57]
Polytetrafluoroethylene, polyhexafluoropropylene and other fully fluorinated polymers	Not found	
Silicone rubber	3 N NaOH, R.T., 10 h	[11]
Silicone-polycarbonate copolymer	6.25 *M* NaOH, 50°C, 20 min	[1]
Silicone-cellulose copolymer	8 *M* NaOH + 0.1% Dowfax[5] 80°C, 3 h	[1]

4.2 Etching Threshold and Track-Etch Sensitivity

4.2.1 Basic Definitions

An ion track can be revealed if the etch rate along the track, V_T, is larger than the bulk etch rate V_B. The bulk etch rate is the rate of dissolution of the non-disturbed material outside the track. The track-to-bulk etch ratio, $V = V_T/V_B$, depends on the damage density in a narrow zone around the ion path. In turn, the damage density is roughly proportional to the deposited energy density along the track. As a first approximation, the stopping power can be taken as a measure of the damage density. However, numerous experiments have shown that the V vs. $\mathrm{d}E/\mathrm{d}x$ curve, measured for a certain ion, is not a single-valued function. Faster ions produce tracks of lower etchability than slow ions of the same $\mathrm{d}E/\mathrm{d}x$. The maximum damage density is observed at a lower energy than would be inferred from the $\mathrm{d}E/\mathrm{d}x(E)$ function. This is due to the fact that the fast ions yield a higher fraction of fast electrons that leave the track core and do not contribute to the damage.

Several approaches have been developed in an effort to find a criterion for predicting the track etchability [1]. The track-forming capability has been related to the primary ionization [1], the energy deposited by secondary electrons within a narrow cylinder of a certain critical radius [59], the energy deposited in collisions with the transferred energy below a certain limit ω (REL_ω – "restricted energy loss criterion" [60]), and the critical dose due to the energy deposited in primary and secondary collisions [61]. Although the secondary processes of energy transfer are more important for polymers than for inorganic solids, the criterion of primary ionization remains the most applicable one for minerals and for plastics as well. The primary ionization density J can be calculated from [62]

$$J = \frac{Z_{\mathrm{eff}}^2 C_1 C_2}{I_0 \beta^2}[\ln(W_{\mathrm{max}}/I_0] - \beta^2 - \delta + K], \tag{4.8}$$

where $C_1 = 2\pi n_e e^4/mc^2$; n_e is the number of electrons in $1\,\mathrm{cm}^3$; m is the electron mass; $W_{\mathrm{max}} = 2mc^2\beta^2\gamma^2$; $\gamma^2 = 1 - \beta^2$; β is the ion velocity in units of the velocity of light c; δ is a polarization correction at relativistic velocities; K is a recorder-specific constant; C_2 and I_0 are the fraction and the ionization potential of the most loosely bound electrons of the recording material, respectively; and Z_{eff} is the effective charge of the ion.

The following relationship between the primary ionization density and the track etch rate – known as the "response function" or "sensitivity function" – was suggested in [63]:

$$V_T/V_B = 1 + aJ^b. \tag{4.9}$$

The exponent b was found to be equal to 2.6 for alpha tracks in cellulose nitrate [63]. For polycarbonate and polyethylene terephthalate $b \approx 3$ [64]. The general form of (4.9) is

$$V_T/V_B = 1 + a\,X^b, \tag{4.10}$$

where X can be replaced by J, REL or radius-restricted energy loss L_r [23,64]. Since the etching response is governed by a power-law function – at least in the range of relatively low damage densities – there is an abrupt change in the etched-track observation probability within a narrow interval of the ionization (damage) density. The minimum primary ionization (or REL, L_r) required for the formation of an etchable track is called the "registration threshold". On the one hand, it is clear that the threshold cannot be defined in a rigorous sense because (4.10) is a continuous function. On the other hand, for practical purposes it is convenient to have such a characteristic of a track-recording medium. One can describe the threshold as the transition from discontinuous, inhomogeneous etching (where the track diameters show a wide variation in sizes) to continuous, homogeneous etching with narrow etch-track diameter distributions [65].

Data on the linear ionization density and track-etch threshold of various materials, including some polymers, are given in [1]. Table 4.2 lists some examples of polymers having different thresholds. The thresholds are given in terms of the lightest particles that can be registered, and their energies. Such a presentation does not depend on the algorithm used to calculate the ionization density along the track. The thresholds are also given as $(\mathrm{d}E/\mathrm{d}x)_{\mathrm{thr}}$, which is reasonable as in all cases the projectiles have energies below or around 1 MeV/amu. (Otherwise $\mathrm{REL}_\omega (= \mathrm{REL}_{200\,\mathrm{eV}})$ would be a more adequate characteristic value.) Thresholds presented by different authors are

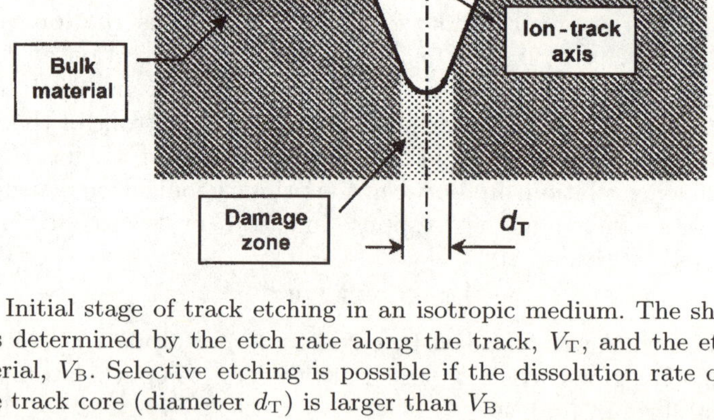

Fig. 4.6. Initial stage of track etching in an isotropic medium. The shape of the etch pit is determined by the etch rate along the track, V_T, and the etch rate of bulk material, V_B. Selective etching is possible if the dissolution rate of material within the track core (diameter d_T) is larger than V_B

Table 4.2. Track-etch thresholds in several polymers, for impact of ions with $E \sim 1\,\mathrm{MeV/a.u.}$ or above. The $\mathrm{d}E/\mathrm{d}x$ values are theoretical estimations for the reported systems

Polymer	Registered particle and its energy	$\mathrm{d}E/\mathrm{d}x$ [eV/nm]	Reference
CR-39	^{1}H, >1 MeV	>30	[33]
Cellulose nitrate	^{1}H, 0.55 MeV	50	[1]
Polycarbonate	^{4}He, 0.3 MeV	250	[1]
Polyethylene terephthalate	^{16}O, 36 MeV	900	[1]
Polyimide	^{16}O, 36 MeV	900	[1]
Polyvinylidene fluoride	^{11}B ions, 1.2 MeV	950	[57]

often inconsistent with each other. For example, Trautmann et al. [65] give a threshold of 4.5 keV/nm for polyimide etching, whereas the value calculated for the system given in Table 4.2 amounts to 0.9 keV/nm. These differences may arise from different crystallinity, preparation techniques, fillers or additives, different storage conditions, etc.

Note that there is a wide variety of track-etch responses for identical formulations with different trade names and even for samples of different thicknesses purchased from the same manufacturer. The data presented in Table 4.2 only give some idea about the range of thresholds for polymers having different sensitivities.

4.2.2 Sensitivity Depends on the Polymer Structure

The sensitivity of a polymer to ionizing heavy particles depends on its molecular and supramolecular structure. In general, polymers may undergo chain scission and cross-linking under irradiation. Since both the dissolution rate and the etching rate are decreasing functions of molecular mass one can expect that polymers that predominantly undergo chain scission will record tracks more easily. The correlation between the sensitivity of polymeric track detectors (in terms of the minimum value of J for which an etchable track is produced) and the yields of chain scissions $G(S)$ and cross-links $G(X)$ was studied in [66]. Values of $G(S)$ and $G(X)$ have been measured in experiments with low-LET radiation, i.e., high-energy electrons and gamma rays. The track-etch sensitivity was found to correlate strongly with $G(S)$ and not at all with $G(X)$. Thus, typical radiation-degradable materials having "weak" chemical bonds in the main chain (such as ether and ester groups or quaternary carbon atoms) are expected to be good track recorders. On the other hand, the absence of a clear correlation between $G(X)$ and the registration threshold does not mean that cross-linking has no effect on the track etching. Links between macromolecules in tracks should inevitably reduce the chain mobility thus slowing down the etching reaction. Thus the slope of the response function may depend on the cross-linking yield.

Cellulose nitrate and CR-39 are the most sensitive representatives of the family of polymeric track detectors. The monomer used for the synthesis of CR-39 is the oxydi-2, 1-ethanediyl di-2-propenyl diester of carbonic acid [33]. Another chemical name of the monomer is diethylene glycol bis allyl carbonate. CR-39 is a cross-linked polymer containing ester and ether bonds as well as tertiary carbons in a high concentration that endows its structure with a high sensitivity to ionizing radiation. In the case of cellulose nitrate, glucoside and nitrate groups serve as easily breakable structure elements.

Copolymers of CR-39 with various radiation-sensitive monomers were prepared and tested in [67]. The copolymer containing 3% of N-isopropylacrylamide was found to be significantly more sensitive than pure CR-39, especially in the range of the lowest REL values. The new copolymer can record protons with energies up to 20 MeV ($\mathrm{REL}_{200\,\mathrm{eV}} = 2\,\mathrm{keV/mm}$). Small quantities of antioxidants in CR-39 improve the track registration properties of the polymer [68]. Similarly, phenolic antioxidants increase the track-to-bulk etch ratio in isotactic polypropylene. Apparently, the antioxidant molecules act as scavengers for active radiolysis intermediates (radicals, excitons). Capture of the mobile radiolysis products by the scavenger fixes bond cleavages and prevents the repair of broken macromolecules [69].

Aromatic polymers are known to have a high resistance to low-LET ionizing radiation due to the protecting effect of the benzene ring. However, the protection weakens at high LET, which makes some aromatic polymers such as polycarbonate or polyethylene terephthalate quite sensitive to energetic heavy particles.

The registration threshold for amorphous polymers is notably higher than that for the same polymers in their crystalline state [70]. There are at least two reasons for the better recording properties of crystalline polymers. First, the etchant distinguishes easier between the ordered crystalline phase and the disordered track area. Secondly, processes leading to restoration of the initial polymer structure occur in the amorphous matrix because of higher mobility of macromolecule segments and radiolysis products.

4.2.3 Sensitivity Depends on the Etching Conditions

The temperature, etchant composition and concentration of chemicals are variables that allow one to control the etch rate and to change the ratio between V_{B} and V_{T}. Equations (4.5) and (4.6) explain why the track-to-bulk etch rate ratio depends on the etching conditions. The concentration of active sites, $C_{\mathrm{m,s}}$, and the reaction rate constant, k_{s}, take different values on the surface of a pristine polymer and in its damaged zone. Therefore, the relative contributions of the diffusional term, $\Delta x/D_{\mathrm{et}}$, and the chemical reaction term, $1/k_{\mathrm{s}}C_{\mathrm{m,s}}$, are different inside and outside the track. The diffusion coefficient D_{et} is generally a function of the concentration. Moreover, the etched track is a confined space in which the diffusion coefficient is smaller than in the bulk solution. This means that, above a certain etchant concentration, the track etch rate can be limited by diffusion, whereas the bulk etch rate is still determined by the term $1/k_s C_{m,s}$.

The constant k_s characterizes a complex interaction between etchant and various species formed in the ion track. The interaction depends on specific properties of particles participating in the reaction, e.g., ion size, ion mobilities, etc. This results in different values of V_T/V_B for different etchants used for the same particular substance. It is found empirically that a maximum in V_T/V_B occurred for cellulose acetate, polycarbonate and CR-39 at 6M concentration of alkali [1, 71]. Even the type of cation in the alkali plays a role: the etching selectivity in CR-39 is higher for NaOH than for KOH [71]. Replacement of univalent cations by divalent cations reduces significantly the track-to-bulk etch rate ratio in polyethylene terephthalate [7]. As was mentioned above a poor choice of the etch agent can lead to a nonselective process ($V_T/V_B = 1$) or to a very high etching threshold. The temperature dependence of the reaction rate constant typically obeys the Arrhenius law:

$$k_s = k_s^0 e^{-E/RT} . \tag{4.11}$$

Here k_s^0 includes a frequency factor and a geometrical factor, E is the activation energy, R is the universal gas constant, and T is the temperature.

Accordingly, experimentally measured V_T and V_B are normally exponential functions of the inverse temperature, with the activation energies E_T and E_B, respectively. If $E_T = E_B$, the sensitivity does not change with increasing temperature, and vice versa, if $E_T \neq E_B$, the track-to-bulk etch rate ratio can be varied by changing the etchant temperature. For cellulose nitrate $E_T < E_B$ [72], therefore the sensitivity is higher at lower etching temperatures. Most likely, this is conditioned by the diffusion-limited kinetics, which, in turn, is related to building up a layer of etch products in ion tracks and on the polymer surface. Different types of polycarbonate show variations of the temperature dependence of the track etch rate [73].

For instance, $E_T < E_B$ was observed in [74] whereas the opposite was found in [75]. Thus, no definite recommendation for the optimal temperature can be given a priori for a given polycarbonate sample.

For polyethylene terephthalate an increase in the etching temperature typically improves the track-to-bulk etch ratio [27]. In this polymer, the difference between E_T and E_B depends on the stopping power. The heavier the ion, the higher is the activation energy for track etching. Figure 4.7 illustrates the temperature dependences of the bulk and track etch rates for PET at different alkali molarities. For 6 M NaOH, the slope of the $\log V_T$ vs. $1/T$ curve is distinctly greater than that for the $\log V_B(1/T)$ function. The activation energies for the bulk etching and for the etching of Xe tracks are equal to 69 and 130 kJ/mol, respectively. Note that at a higher alkali concentration (9 M) the apparent value of E_T decreases with increasing temperature. At

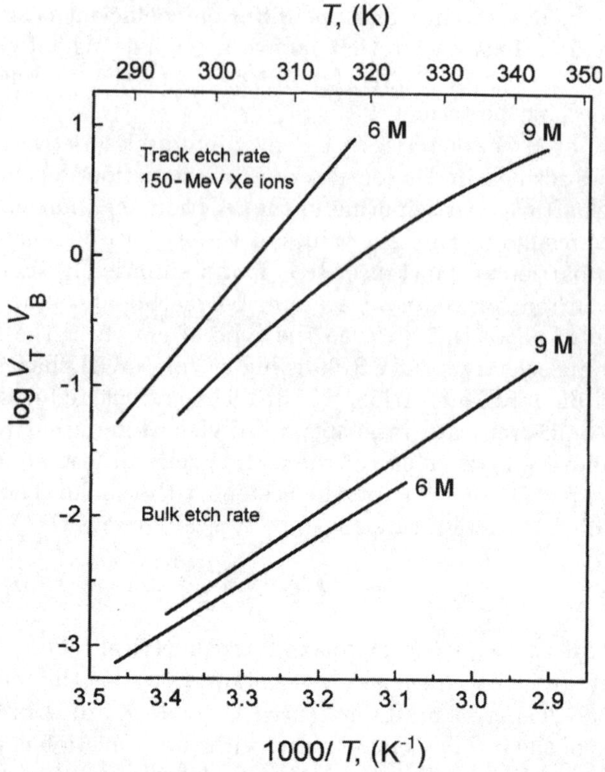

Fig. 4.7. Arrhenius plot for the bulk etch rate of polyethylene terephthalate and the track etch rate of Xe ions at two different NaOH concentrations (6 M and 9 M)

high etchant temperature and concentration, local buildup of etch products in the tracks sets a limit on the etch rate. The solubility of the etch products (sodium terephthalate) decreases drastically at high NaOH concentrations. Precipitation of the etch products within an etch pit results in the change of the apparent activation energy of track etching, as is seen from changing the slope of the $\log V_{\mathrm{T}}$ vs. $1/T$ curve.

4.2.4 Track-Etch Response at High Stopping Power

Whereas the sensitivity of polymers to protons and alphas is essential for particle track detectors, the track-etch response at high $\mathrm{d}E/\mathrm{d}x$ is of special importance for the creation of micro- or nanopores with high aspect ratio. Equation (4.10) does not properly describe the response of a polymer in the MeV/μm range of the ion's energy losses. At high $\mathrm{d}E/\mathrm{d}x$ the response function is never as sharp as in the vicinity of the registration threshold.

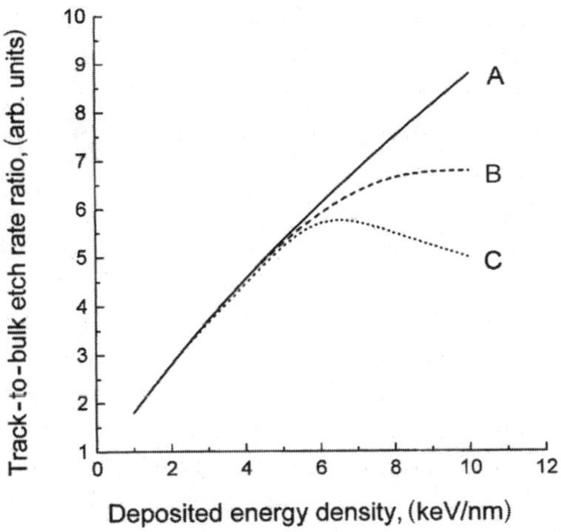

Fig. 4.8. Typical shapes of response functions at high $\mathrm{d}E/\mathrm{d}x$ (qualitative representation)

Possible shapes of the V vs. $\mathrm{d}E/\mathrm{d}x$ curves are shown in Fig. 4.8. The monotone curve A – deviating from the straight line to a greater or lesser degree – was measured for polycarbonate and cellulose acetate in [76]. Another typical shape of response function is represented by curve B that approaches saturation at high values of deposited energy density. Such a behavior was observed for ion tracks in cellulose nitrate [76], polycarbonate [73, 77], and polyethylene terephthalate [53]. The saturation may be attributed to two different reasons. In the case of cellulose nitrate, it is believed that the track etch rate is limited by diffusion. For polycarbonate the plateau was observed at two different etch temperatures [73], i.e., at considerably different absolute rates of the process. Once the absolute rate of etching does not influence the shape of the V vs. $\mathrm{d}E/\mathrm{d}x$ dependence, one has to conclude that the plateau is caused by a reason other than the diffusion limitation. Sometimes this phenomenon is considered as saturation due to the "total damage" of the initial polymer matrix. However, the term "total damage" can hardly be applied to ion tracks in polymers. The material in tracks is not decomposed into separate atoms. The saturation in etching response is rather due to a competition between degradation of initial macromolecules and synthesis of new chemical structures in the tracks. The new chemical structures are depleted of H atoms because hydrogen forms volatile compounds and leaves the matrix. This leads to the production of unsaturated carbon–carbon bonds (which increases etchability)and to the formation of cross-links (which decreases etchability). As the dissipated energy density becomes greater, the polymer structure undergoes a deep rearrangement in the track core [78]. The carbon atoms may unite into stable compounds that slow down the etch rate. These processes manifest themselves as the "decrease" of damaging power at high $\mathrm{d}E/\mathrm{d}x$. As a result, the response curve of type C (see Fig. 4.8) can be obtained. In polypropylene, the maximum track etch rate was observed at $\mathrm{d}E/\mathrm{d}x \approx 8\,\mathrm{keV/nm}$ [53], i.e., xenon ions provided a higher aspect ratio than the heavier Au and U ions.

Note that Fig. 4.8 is only a qualitative representation of the situation at the deposited energy densities exceeding the track-etch threshold by 2–3 orders of magnitude. Primary ionization or restricted energy loss can be taken as measures for the deposited energy density in ion tracks. If the ions' velocities do not differ too much, the total stopping power $\mathrm{d}E/\mathrm{d}x$ can also be plotted on the abscissa.

The maximum values of $V_\mathrm{T}/V_\mathrm{B}$ depend strongly on the track-recording plastic. For some of them (polyethylene, polypropylene) the etch-rate ratio does not exceed a few tens; in the case of polyethylene terephthalate and especially polycarbonate this quantity may be as large as several thousands. The maximum $V_\mathrm{T}/V_\mathrm{B}$ can be achieved by irradiating the plastics with sufficiently heavy ions (Kr, Xe) and performing sensitization of tracks before etching.

The ratio between the rate of penetration of etchant into the latent track and the rate of propagation of the etching front outside the latent track, $V_\mathrm{T}/V_\mathrm{B}$, is the most important parameter for all possible applications, as it determines the shape of the etched track. As this ratio depends on $\mathrm{d}E/\mathrm{d}x$, and taking into account that different polymers show different changes of V_T with increasing $\mathrm{d}E/\mathrm{d}x$, it is difficult to illustrate the subject by a simple set of numbers. Therefore we only present some representative numbers to give a feel for their order of magnitude: observed ratios amounts to less than 2.3 for PSU [53], 21 for PP [53], 127 for PVDF [55], and slightly less than 1000 for PET [53].

4.2.5 The Realization of Ion-Track Etching

There are several ways in which controlled ion-track etching can be realized. One approach is to combine the etching process with conductometry. Here a chosen electric voltage is applied across the track membrane during etching. The etching process is interrupted by a stopping medium after reaching a preset value of the electric current through the membrane. In the case of etching from one side only (to obtain conical pores) this stopping medium can be, e.g., a neutral solution that dilutes, or an acidic solution that neutralizes the etchant at the moment of breakthrough. Thus the measured current results from a superposition of electromigration, diffusion, osmotic flux and neutralization reactions. It was shown that, for the case of etching of an irradiated PET membrane with NaOH on the one side, and KCl/HCOOH solution on the other side of an etching vessel, electromigration predominates over diffusion already at voltages above 0.1 V, hence electrostopping is much more efficient than chemical stopping already at quite small applied voltages [79]. For more details of conductometry see Sect. 4.5.

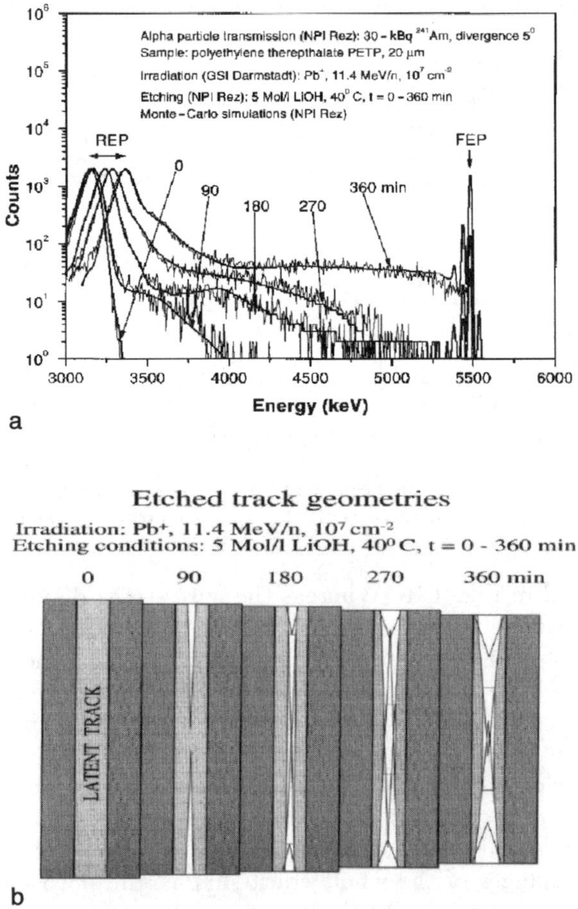

Fig. 4.9. (**a**) ITS spectra during different stages of track etching; (**b**) reconstructed track shapes. In the reconstruction, the track has been approximated by a superposition of cones of different shapes. Please note also the thinning of the foil due to bulk etching [80]

Other possibilities to control ion-track etching are to perform etching and SEM or ITS iteratively. For example, Fig. 4.9a shows some ITS spectra of 11.4-MeV/amu Pb ion tracks in PET, as measured at different stages of track etching, and Fig. 4.9b illustrates the track shapes as reconstructed from the spectra by means of a Monte-Carlo program. It is seen that the general etch track shape is usually hyperboloidal, which has been approximated here by the superposition of a number of conical structures.

Etched tracks in thin foils can be usually described by two parameters, the outer (at the foil surface) and the inner (at the position where the track is narrowest) track diameters. The evolution of these parameters with etching

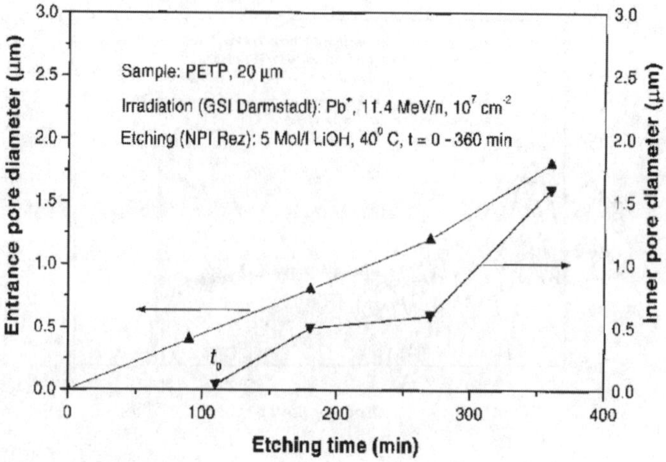

Fig. 4.10. The development of inner and outer track diameters of tracks during the etching process. This figure refers to the tracks characterized in Fig. 4.9 [80]

time is illustrated in Fig. 4.10. Whereas the outer track diameter shows nice proportionality with etchant time, the inner one becomes nonzero only when the foil is perforated. The longer the etching time, the more the inner track diameter approaches the outer one, i.e., the more cylindrical the tracks become.

Please note in this connection that different analytical techniques are sensitive to different track parameters. Whereas the conductometric approach records the development of the effective pore diameter (which is equal to the square root of the product of the inner and outer diameters of a conical pore), and SEM records the one of the outer pore radius, ITS gives information about the changes of the whole track shape. All approaches except for SEM neglect the scattering δ_r of the pore sizes that increase with increasing pore radius r, due to statistical variations of the etching process and the experimental conditions.

4.3 Environmental Effects and Track Sensitization

Attempts to improve the reproducibility of etching, on the one hand, and search for the methods of track enhancement, on the other hand, have shown that a number of environmental factors may alter the latent tracks. This knowledge leads to caution in the choice of laboratory procedures and new applications. A complete analysis of environmental effects on track etching is not our aim in this section. We will give an outline of the most important aspects from the practical point of view.

Radiolytic products in heavy-ion tracks are chemically active species undergoing postirradiation reactions, such as recombination, disproportionation, oxidation, photo-oxidation, etc. [60]. Some of the reactions are very

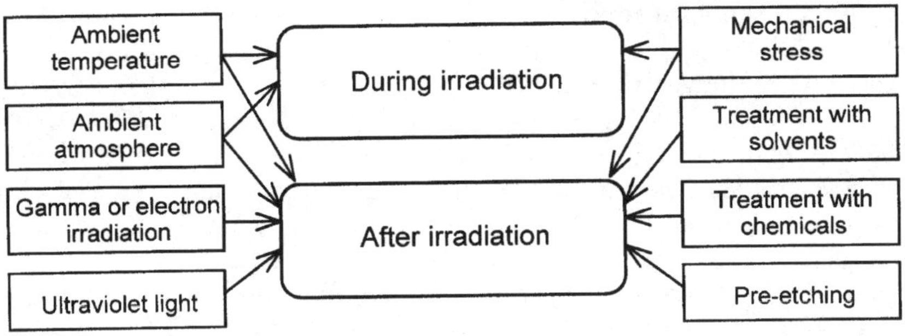

Fig. 4.11. Simplified classification of environmental effects on track etching

fast; for this reason the environmental conditions play a crucial role not only after irradiation but also during the irradiation (Fig. 4.11). Temperature and ambient atmosphere are the most important factors for latent-track formation in plastics. The presence or absence of oxygen affects the radiation effects in organic materials and formation of etchable tracks as well. Apart from oxygen, other gaseous agents such as humidity or NO may increase the sensitivity [25, 81]. Undesirable desensitization of tracks can arise from irradiation in vacuum. Depending on the duration of stay of polymeric samples in a pumped irradiation chamber, the track etch rate can be altered to a greater or lesser degree. Thus individual tracks in the samples exposed to different ion-track densities (at a fixed ion flux) may have different etching characteristics.

The track-etch response sensitivity of polymers (PC, CR-39, PET, PP) generally decreases with increasing temperature during irradiation [82]. However, the track etch rate vs. registration temperature dependence can be substantially nonmonotonic [83]. The most pronounced changes in the track-etch response occur at temperatures around relaxation transitions [84]. A considerable effect can be observed even at temperatures close to room temperature. The correlation between the relaxation transitions and the alterations of V_{T} is the evidence for the strong influence of molecular and radiolysis intermediates' mobilities on the track formation in polymers. The balance between chain scissions, repair of cleaved bonds, cross-linking and oxidation processes changes with changing temperature during irradiation and after irradiation.

Heating of the ion-irradiated polymers leads to fading of the latent tracks. At a certain temperature the tracks disappear completely and cannot be developed any longer by etching. However, there are some polymers showing an "anomalous" behavior. An example is polypropylene in which a moderate heating enhances the particle tracks, presumably due to thermo-oxidation [69].

The storage of some ion-irradiated polymers (such as PC or PET) in air leads to a significant increase in the track etch rate [27, 28, 68, 85]. The acceleration of this process with the help of exposure to ultraviolet (UV) light is used in a technological process of production of polyethylene terephthalate track membranes [86]. Unirradiated PC and PET are optically transparent at wavelengths longer than 275 and 313 nm, respectively. Being irradiated with ionizing particles, these polymers absorb electromagnetic radiation at $\sim$ 305 nm (PC) and 310–340 nm (PET), respectively. Exposure to UV with the wavelength that corresponds to the absorption maxima is an elegant way of track sensitization. Photo-oxidation occurs only along the tracks and not in the bulk of the virgin polymers [87]. Untreated Kr or Xe ion tracks in PC and PET have an etch rate ratio of about 10^3 and 10^2, respectively. UV irradiation in an appropriate wavelength range increases V_T/V_B by a factor of 10 in polycarbonate and 3–5 in polyethylene terephthalate [28, 74].

Low-LET irradiation such as gamma-rays or high-energy electrons can also increase the track etchability. For instance, the beneficial effect of gamma rays on ion tracks was observed for PET [88], polyimide [89], and PADC [90]. The mechanism of the etching selectivity increase is probably associated with the migration of short-lived radiolytic products. The ion tracks serve as sinks for the mobile species generated by the low-LET radiation throughout the volume of the polymer. Trapping of the reactive species increases the defect concentration in the tracks and thus improves the etching selectivity.

However, there are also cases where both track and bulk etching are enhanced by preirradiation. In these cases, the gain in etching response (ratio of track/bulk etching) may be largely compensated by an increase in surface roughness. As an example, see Fig. 4.12. Anyhow, even the small gain in etching response observed there (from 28.5 to 31) already enables an increase of the etch-track diameter by a factor 4 under identical etching conditions [90]. Typical values for V_B of PI by 6.25 mole/l NaOH range from 2.2 µm/h at 50°C

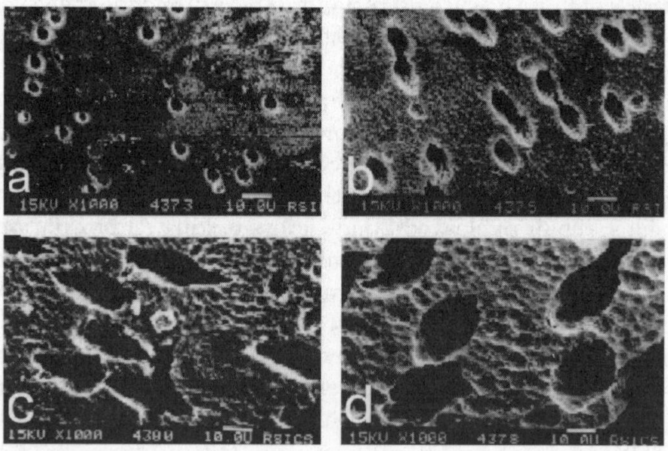

Fig. 4.12. Influence of preirradiation of PADC with 2-MeV electrons at doses increasing stepwise from zero (**a**) up to 230 kGy (**d**), on the etching of 140-MeV ^{28}Si ion tracks [90]

to 6.9 μm/h at 65°C. Irradiation of PI with 2-MeV electrons up to 23 kGy dose enhances the etching speed to 3.6–7.6 μm/h, respectively. Similar data for PET, PC and some other polymers irradiated by electrons are presented in [75]. For the sake of completeness it should be noted at this point that it was proposed to use such electron-sensitization processes as a means for electron dosimetry.

A peculiar way of track sensitization is to deposit a thin fullerene layer onto the surface of a polymer foil before ion irradiation, by which the near-surface track etching is slightly enhanced. The underlying mechanism is not yet fully understood; it might be the consequence of direct energy transfer during the ion-induced C_{60} destruction, or polymer sputtering by energetic destruction products, or the chemical action of C_{60} destruction products onto the polymer [91].

There are some other methods of track sensitization. Heavy-ion tracks in PET can be effectively sensitized by treatment with organic solvents such as dimethylformamide, dioxane and others at 50–70°C [92]. For PC, soaking in methanol and eventually trifluoroacetic acid appears to be beneficial. For other polymers, this type of sensitization has not been achieved. In contrast to the photo-oxidative treatment, the solvents do not cause chemical degradation. Only free volume of molecular size is generated by the solvent-induced crystallization and extraction of degraded polymer products. The method makes it possible to increase the track etch rate by 1–2 orders of magnitude and is especially suitable for sensitization of tracks of relatively light ions (Si, Cl). This method is quite useful for producing very long tracks with cylindrical shapes in thick polymer foils, as at lower etching speed their shapes would be more conical.

Pre-etching can also be considered as a kind of chemical sensitization if some components of etchant diffuse into latent ion tracks before the real etching starts. Soaking of ion-irradiated CN in alkali at a low temperature has a considerable influence on the subsequent track etching at a higher temperature [93]. Similarly, track etching of polycarbonate proceeds at a higher rate after the polymer was subjected to pre-etching, rinsing and storage for several days [94]. An interesting effect is the increase in the track etch rate caused by postirradiation treatment of CR-39 with CO_2 [95,96]. Carbon dioxide seems to be passive with regard to possible chemical reactions in ion tracks. Perhaps the sensitization with CO_2 follows the same mechanism that has been suggested for the sensitization with water-soluble gases (HCl, SO_2) [97]. Such acidic species, being absorbed in tiny voids along the damage trails, easily dissolve in alkaline etchant thus promoting penetration of the etchant into the tracks.

This brief description of the environmental effects is aimed at demonstrating the fact that the ion-irradiated polymer is a metastable system that, sometimes, is extremely sensitive to ambient conditions. Reproducible etching is possible only if one takes precautions to minimize the uncontrolled influence of environment on the ion tracks during and after their formation.

Controlled environmental effects, instead, can be used for desirable alterations of track-etching properties.

4.4 Shapes of Etched Tracks

4.4.1 Modeling the Track Shape on the Microscopic Scale

On the micrometer scale the two etch rates, V_{B} and V_{T}, are sufficient to describe the shape of an etch "pit". The formation of an etch cone around the ion path can be easily understood when we apply Huygens' principle to the track-etching process. The process can be considered as a superposition of elementary time-delayed waves emitted from the moving end point of the etched track as it proceeds into the depth of the polymer bulk. The waves merge to an envelope that corresponds to the actual shape of the etched track. A satisfactorily quantitative description of the track shape can be accomplished only with the knowledge of the complete form of the V_{T} vs. ion-range function. For increasing, constant, and decreasing track etch rates, the shape of the etch pit is concave, conical, or convex, respectively. Situations when V_{T} may be approximated as a constant are generally restricted to small layer removals and to track portions of higher-energy ions where the damage density is only a slowly varying function of the range.

A systematization for the variety of etch-track kinetics is shown in Fig. 4.13. Distinctions are made between isotropic and anisotropic solids, constant and varying track etch rates, thick and thin track-recording samples [98]. By definition, the thin sample represents the case when during etching the foil is perforated through completely. Most polymers are isotropic with

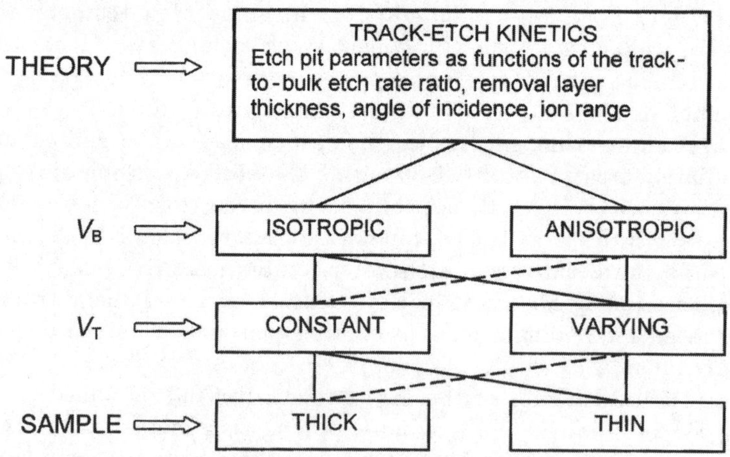

Fig. 4.13. Etch-track kinetics classification according to the etching and detector properties (after Somogyi [98])

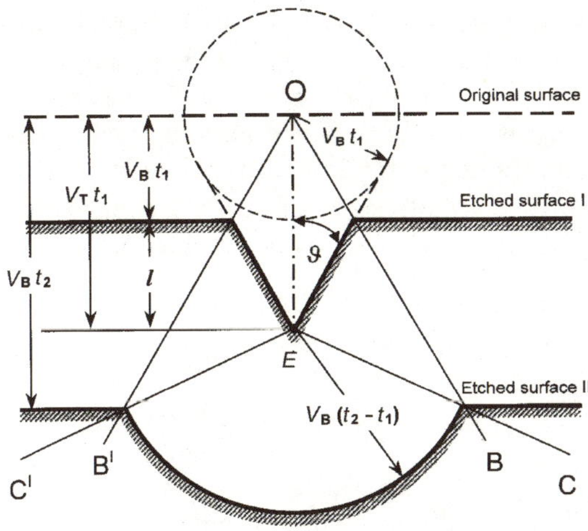

Fig. 4.14. Etched-track geometry in an isotropic polymer at a normal incidence of ion

regard to etching, however, thin films, which have been oriented (stretched) during the production process, may show a pronounced anisotropy. An exhaustive description of various track shapes is presented in books [1, 2] and some comprehensive papers [98–100].

Let us show here a simple demonstrative model of track evolution in the case of normal incidence and constant V_T. The track shown in Fig. 4.14 is formed by an ion that crossed the original surface at point O and stopped at point E. When the chemical dissolution proceeds with the velocity V_T along the track axis and with the velocity V_B in other directions in the medium, mathematically the track shape can be considered as a "normal cone". The surface of this cone is generated by the rotation of a straight line given by the equation

$$y = -x \tan\vartheta + L \tan\vartheta \tag{4.12}$$

around the x-axis that coincides with the track axis; the origin of the coordinate system is associated with the point of ion incidence. L and $\tan\vartheta$ can be defined in terms of the removal layer thickness H and the track-to-bulk etch rate ratio V:

$$\tan\vartheta = (V^2 - 1)^{-1/2}, L = VH\,. \tag{4.13}$$

After a layer removal $H = V_B t_1 = R_o/V$, where R_o is the ion range (length OE in Fig. 4.14) and $t_1 = R_o/V_T$, there is no longer preferential etching along the axis of the track. The end of the track cone in this phase becomes spherical, i.e., the track is composed of a conical and a spherical portion. It is a "partial cone and sphere" or "transition" phase of etch-pit formation ($t_1 <$ etching time $< t_2$) during which the circular contour of the track on the sample surface continues moving along the cone generated by rotation of line OB. The development of the contour of the spherical part of the track is determined by the conical surface generated by the line EC. The radius of the spherical part is equal to $V_B(t - t_1)$. Finally, at the moment $t \geq t_2$, the conical track portion is entirely removed and the etch pit becomes spherical. All intermediate stages of the track evolution are shown in Fig. 4.17.

Directly measurable quantities – the track diameter on the surface D and the visible track length l – are the result of the competition between the two etch rates and can be represented in terms of these parameters:

$$l = (V_T - V_B)t, \tag{4.14}$$

$$D = 2V_B t[(V - 1)/(V + 1)]^{1/2}. \tag{4.15}$$

Equation (4.14) is valid for the conical phase, $t < t_1$, whereas (4.15) applies for both the conical and transition phases, $t < t_2$.

More complex geometrical relations have been derived for the case when the track is tilted to the surface. The intersection of the conical track with the etched surface is an ellipse, with the track position displaced from the center. The interested reader can find a detailed description of the etch-pit geometries in the above-mentioned publications. "Catalogues" of typical track shapes are given in [2] and [100].

Note that at a certain incidence angle the tracks cannot be developed, which results from the finite value of the etch-rate ratio V. When the angle of incidence α is smaller than $\arcsin(V_B/V_T)$, the surface advances at a greater rate than the normal component of V_T. Figure 4.15 illustrates the situation. The critical registration angle α_0 is defined by

$$\sin \alpha_0 = H/L = V_B/V_T. \tag{4.16}$$

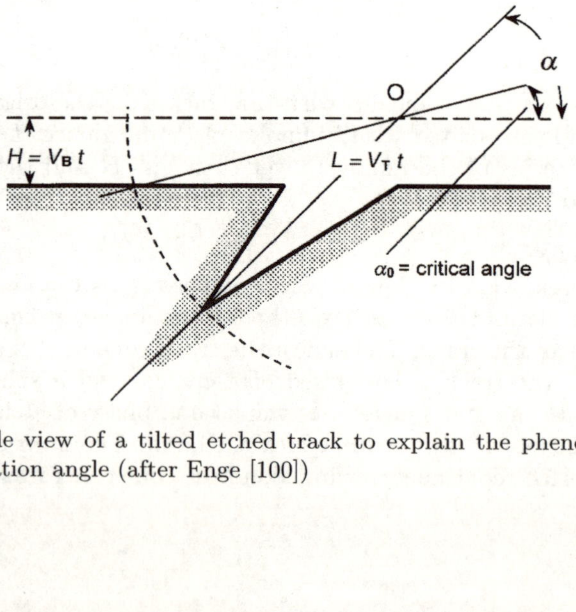

Fig. 4.15. Side view of a tilted etched track to explain the phenomenon of the critical registration angle (after Enge [100])

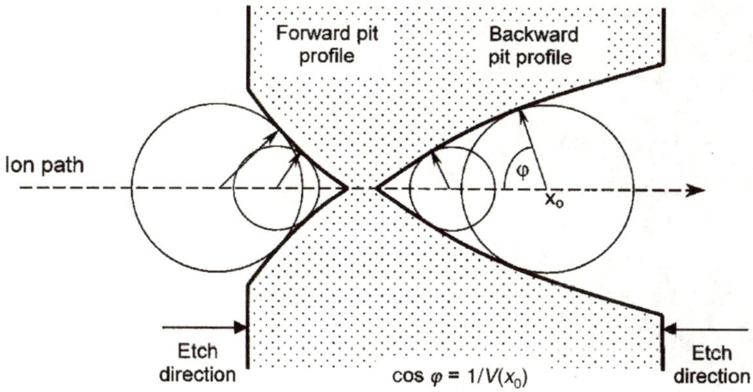

Fig. 4.16. Formation of forward and backward etch pits in thin isotropic foil etched from both sides

For all incident angles $\alpha \leq \alpha_0$, there will be no etch pits. For an array of latent tracks intersecting the polymer surface at different angles, the fraction that is etched is called *etching efficiency*. It can be easily shown that for the irradiation in "2π-geometry" the etching efficiency η is directly related to the track-to-bulk etch-rate ratio [1]:

$$\eta = 1 - \sin\vartheta. \tag{4.17}$$

This is, for example, the case for exposing a surface to a thin infinite source of fission fragments parallel to the surface. For fission fragments, the etched cone angle in most polymers is a few degrees and thus the etching efficiency is 90–95%. The etching efficiency has to do with purely geometrical requirements under specified etching conditions. In contrast to the *registration efficiency*, the etching efficiency is not related to the registration threshold; it deals with the particles producing etchable damage, i.e., above the registration threshold.

The case of the varying track etch rate is shown is Fig. 4.16. The etched track in a thin isotropic sample is assumed to have been created by an ion with increasing $\mathrm{d}E/\mathrm{d}x$ along its trajectory. The drawing illustrates the above-mentioned Huygens' principle. The description of the evolution of an etch pit wall is equivalent to the description of the evolution of a head wave produced by an object moving at a variable speed $V_\mathrm{T}(x)$ in a medium having a wave propagation speed V_B. The forward pit profile is concave due to increasing V_T in the direction of the ion travel; whereas the backward etch pit profile is convex.

4.4.2 Examples of Etched Tracks in Polymers

A series of photomicrographs in Fig. 4.17 shows the main features of the etched track evolution. A tilted track of a ^{4}He ion in CR-39 was treated in

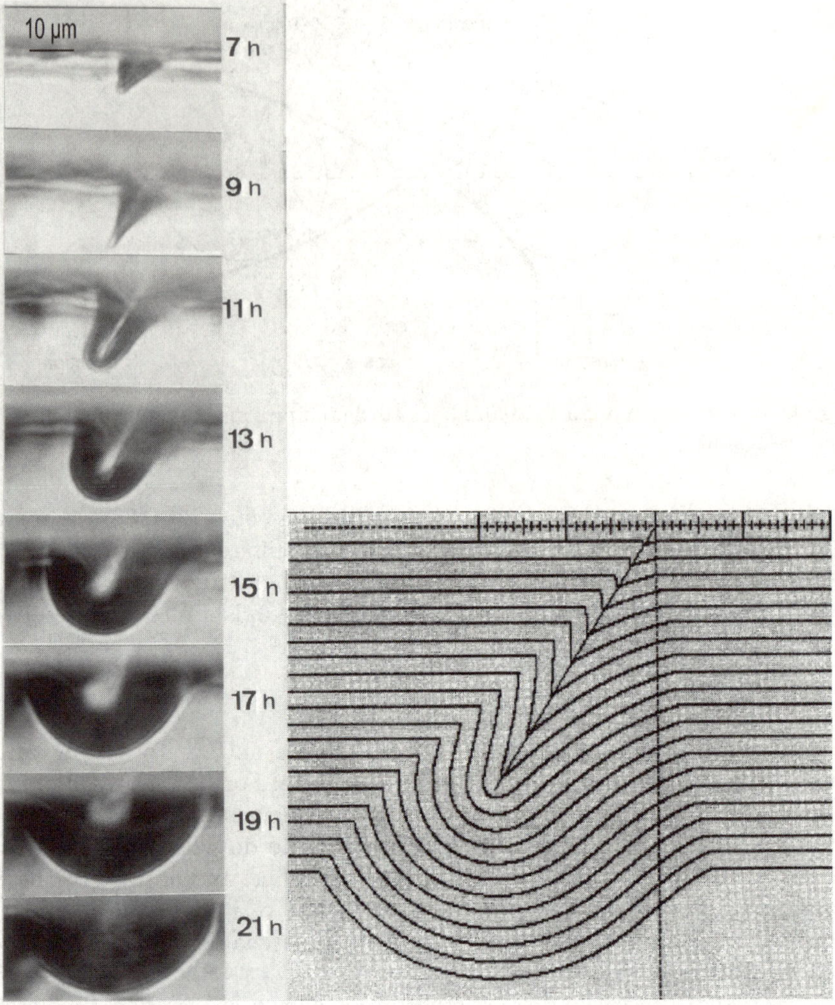

Fig. 4.17. Experimental and simulated etched–track profiles [101]. Ion: ^{4}He with an energy of 5.5 MeV. Polymer: CR-39. Etching conditions: 7 *M* NaOH, 70°C: Angle of incidence: 30° (courtesy of M. Fromm)

7 *M* NaOH at 70°C step-by-step. As the etching time increases, the track assumes the form of the short cone, concave cone, rounded cone and, finally, the spherical pit. The sequential phases of the track-profile evolution, obtained with the use of the theoretical simulation, are also presented. Note that after the first few hours of etching the track is faintly visible. This is because of the low ionization rate of the ion at the entrance into the polymer.

Shapes of etch pits obtained at different etch-rate ratios are presented in Fig. 4.18. Foils of PET and PC were irradiated with a beam of krypton ions

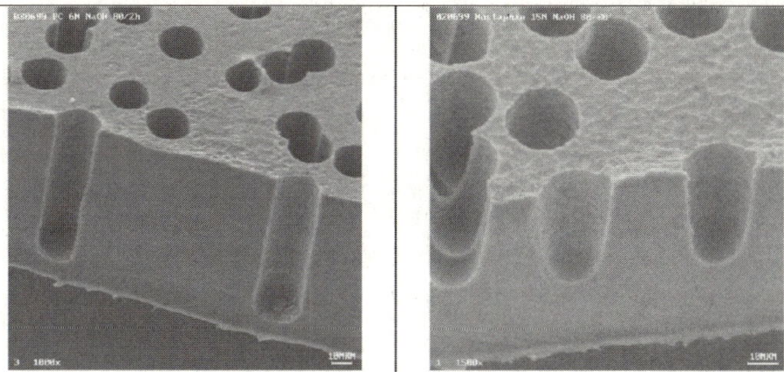

a b

Fig. 4.18. Polycarbonate foil Pokalon irradiated with 640-MeV Kr ions and etched in 6 mole/l NaOH at 80°C for 2 h (**a**). Polyethylene terephthalate foil Hostaphan irradiated with 250-MeV Kr ions and etched in 15 M NaOH at 80°C for 30 min (**b**). The samples were cut to show profiles of etched tracks. Initial thickness of the foils is 100 µm. Scale bars: 10 µm. Courtesy of V.V. Shirkova, Flerov Laboratory JINR, Dubna (unpublished)

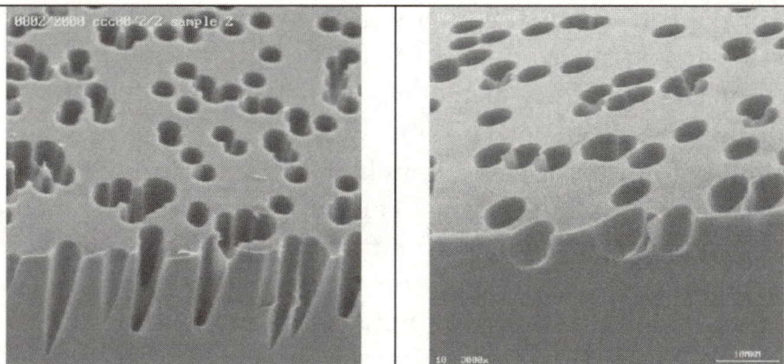

a b

Fig. 4.19. Polyethylene terephthalate foil Hostaphan irradiated with Kr ions and etched in polishing etchants: mixture of ethanol and water (50 : 50, v/v) saturated with NaOH (**a**); isopropanol saturated with NaOH (**b**). Scale bar: 10 µm. SEM photos were taken by O.L. Orelovitch, Flerov Laboratory JINR, Dubna

perpendicular to the surface. Etching of PET was performed in a highly concentrated alkaline solution that provides a low etching selectivity. The etched tracks have appreciable conical parts. In the case of PC the etchant used (6 M NaOH) is quite selective. Due to the high V_T/V_B cylindrical channels with rounded ends are obtained. In both cases the etching time t corresponds to the transition phase of the track evolution ($t_1 < t < t_2$).

Conical and almost hemispherical etched tracks are presented in Fig. 4.19. Note that the surface of PET in Fig. 4.19 is markedly smoother than that in Fig. 4.18b. This is because of the polishing effect of the low-selective etchants on the basis of alcohols.

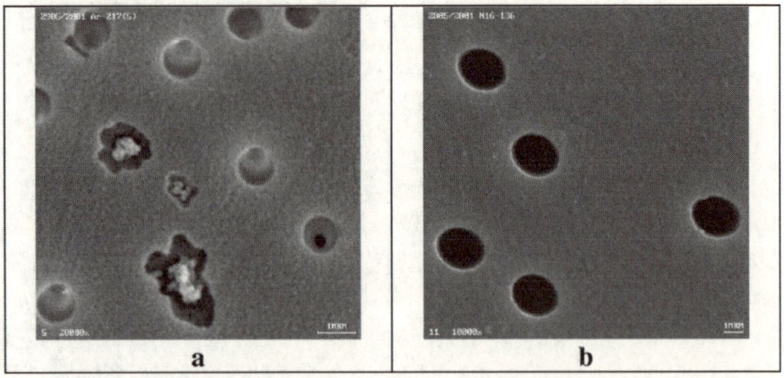

Fig. 4.20. (**a**) Biaxially oriented PET film with a thickness of 10 μm, irradiated with 225-MeV Ar ions and etched in C_2H_5OH + H_2O + NaOH. (**b**) Biaxially oriented polypropylene film Torayfan with a thickness of 10 μm, irradiated with 125-MeV Xe ions and etched in H_2O + CrO_3. Scale bars: 1 μm

Commercially available microporous foils contain, in general, tracks of cylindrical or conical shape. There are, however, also reports of cigar-shaped pores in fission-fragment irradiated PC foils [102], probably arising from heterogeneities in the cast films (specifically in the case of PC), and from the action of surfactants, see Sect. 4.5 and Fig. 4.26.

An example of anisotropic track etching is shown in Fig. 4.20b. Typically the ion tracks in biaxially stretched polypropylene film have elliptical contours. The ratio between the major and minor axes of the ellipse ranges from 1.1 to 1.6 depending on the type of film. The direction of the minor axis coincides with the direction of longitudinal stretching. In fact, the track-etching process is described by three mutually perpendicular etch rates. The least of them is the "surface" etch rate, i.e., the etch rate determining the decrease in the film thickness.

4.4.3 Track Etching in the Case of Insoluble Fillers or Etching Products

The PET foils shown in Figs. 4.18 and 4.19 do not contain any additives in the form of small particles. Homogeneity of the track-recording polymer is a necessary condition for uniform etching. Commercially available thin PET films are often produced of a resin containing fillers. The fillers are fine powders of inorganic substances such as TiO_2, SiO_2, kaoline, etc. This kind of inhomogeneity has a strong effect on the quality of the surface and tracks after etching. When the etching front reaches a filler particle, the particle is etched out from the matrix. The irregularities caused by the presence of the filler are demonstrated in Fig. 4.20a. One can see nonuniform caverns with the titanium dioxide aggregates inside.

In the case of a pristine or ion-irradiated polymer that contains insoluble additives and fillers, the average surface recession of that polymer is a function of the etching time. Corresponding experiments have been made with, e.g., commercial silicone rubber. Whereas in ion-irradiated silicone rubber marked surface recession was recorded immediately, pristine samples show a strongly retarded surface dissolution. Differentiation of the time-dependent removed layer thicknesses yields the time-dependent etching speeds $V_{\text{etch}}(t)$ that, at least in the case of silicone rubber, follow in both cases a power-law relationship:

$$V_{\text{etch}}(t) \sim t^{\alpha} , \tag{4.18}$$

with $\alpha \approx -(0.5 \pm 0.05)$ for the pristine, and $\alpha \approx -(0.63 \pm 0.05)$ for the irradiated case [11]. These correlations can be understood to be results typical for etching of materials with insoluble fillers or reaction products. For simplicity, let us assume that a volume fraction $F_{\text{insol}} = U_{\text{insol}}/U_{\text{tot}}$ of the removed material is insoluble – this may be due to the fillers, or due to insoluble etching products, and that the presence of insoluble objects exposed on the sample surface reduces the etchant attack by hindering the free access of the etchant to the sample surface:

$$U_{\text{etch}}(t) = V_{\text{etch}}(t=0)^{*} c^{\#} / U_{\text{insol}}(t) , \tag{4.19}$$

with U_{insol} being the volume of insoluble objects enriching in the surface-near region per unit area within the exposure time t, and $c^{\#}$ being a proportionality factor.

Then the totally etched volume U_{tot} can be calculated from the etching speed as:

$$U_{\text{tot}}(t) = \int_0^t V_{\text{etch}}(t)\mathrm{d}t. \tag{4.20}$$

Expressing the total volume U_{tot} by U_{insol}, and inserting U_{insol} into (4.19), one obtains an integral equation of the type:

$$V_{\text{etch}}(t) = c^{*} / \left(\int_0^t V_{\text{etch}}(t)\mathrm{d}t \right) , \tag{4.21}$$

with $c^{*} = V_{\text{etch}}(t=0)^{*} c^{\#} / F_{\text{insol}}$. Let us first treat the etching of unirradiated matter, for which we assume that $F_{\text{insol}} = \text{const}$. Then the solution of (4.21) is: $V_{\text{etch}}(t) = \gamma/\sqrt{t}$, with γ being a system-dependent proportionality factor. This corresponds to (4.18) with $\alpha \approx -1/2$.

Now let us regard the etching of the irradiated zone. In order to describe the experimentally observed $t^{-2/3}$ dependence of the etching speed in this case, let us assume that, due to the greater etching speed, part of the insoluble objects exposed at the sample surface can be floated away during the etching process into the liquid together with the soluble products. Therefore, let F_{insol} describe now only the insoluble fraction that remains on the sample surface. Then F_{insol} is no longer constant, but depends on V_{etch} itself.

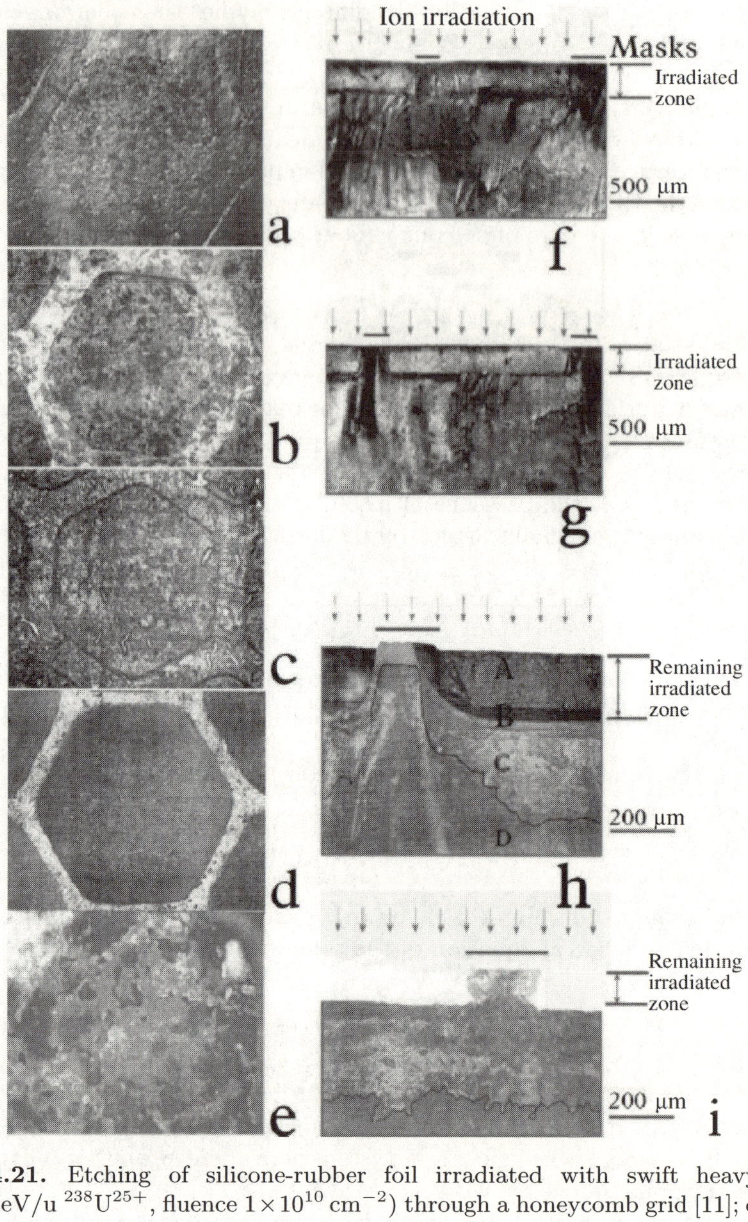

Fig. 4.21. Etching of silicone-rubber foil irradiated with swift heavy ions (11.4 MeV/u $^{238}U^{25+}$, fluence 1×10^{10} cm^{-2}) through a honeycomb grid [11]; optical microscope images. The *darker areas* are regions of stronger, and the *brighter areas* regions of weaker etching. The white color stems from insoluble fillers that have become exposed to the surface by etching, and that are apparently better removed, the higher the etching speed. Etching times: 10 s (*a*), 5 min (*f*) 15 min (*b*,*g*), 1 day (*c*,*h*), 3 days (*d*), 2 months (*e*,*i*)

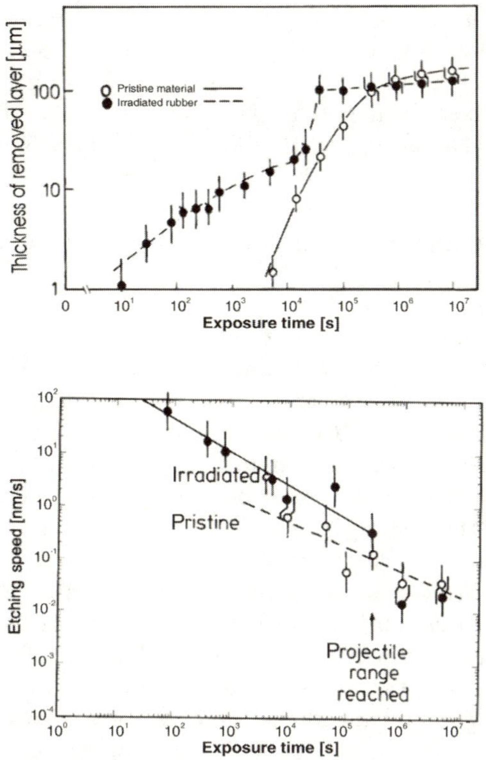

Fig. 4.22. (**a**) Thickness of silicone-rubber layer, and (**b**) etching speed of silicone rubber, as a function of exposure time to etchant solution. (*b*) is derived by differentiation of (*a*). The irradiation conditions were chosen so that the polymer foil was irradiated at a fluence that roughly corresponds to the onset of track overlapping [11]

The functional dependence of $F_{\text{insol}}(V_{\text{etch}})$ must be such that for high V_{etch}, a major fraction of the insoluble products will be lost by floating, i.e., $F_{\text{insol}} \to 1$. For very low V_{etch}, one should expect that nothing of the solid objects will leave the surface, hence F_{insol} should correspond in this case to the true ratio of soluble-to-insoluble etching products: $0 \geq F_{\text{insol}} \geq 1$. A possible correlation that describes the above-discussed tendency would be: $F_{\text{insol}} \sim V_{\text{etch}}^{-1/2}$. This case would also signify that: $F_{\text{insol}} \sim t^{1/3}$. Inserting this into (4.21), one obtains $V_{\text{etch}} \sim t^{-2/3}$, which corresponds to the correlation measured for the irradiated sample.

The above assumption that $F_{\text{insol}} = \text{const.}$ for pristine, and $F_{\text{insol}} \sim V_{\text{etch}}^{-1/2}$ for irradiated silicone rubber is supported by the experiment, see Fig 4.21. Apparently the overall fraction of insoluble objects that remain adhering to the sample surface decreases with increasing etching speed, as their detachment is facilitated by more rapid changes of the surface topology during etching.

This fraction increases strongly during the onset of the etching process, but grows slower at a later stage of etching. For the case of SHI-irradiated silicone rubber cited here (Fig. 4.22) [11], the fraction of insoluble objects remaining on the sample surface during etching is about 20%. The remaining insoluble fraction is found sedimenting on the bottom of the etchant solution vessel.

4.5 Etched-Track Profiles in the Nanometer Range

Conductometric studies. So far we have ignored the fact that the damage trail along the ion trajectory has a nonzero diameter. To go into more detail, we have to consider peculiarities of etching of the damage zone within which the properties of polymer are altered by ionization/excitation cascades and subsequent chemical reactions. Observations made using transmission electron microscopy have shown that the diameter of the selectively etched cylinder around the ion path is of the order of 10 nm. More detailed and quantitative evaluation of the etching kinetics within the damage zone was obtained from conductometric measurements performed on thin foils. The technique was suggested by Walker and coworkers [1] and used by Bean et al. [103] to study the process of fission-track evolution in thin samples of mica and polycarbonate. The basic principles of the method are as follows. The electrical conductivity across the sample is monitored while transverse particle tracks are being etched. The etchant in the growing pores serves as the conducting electrolyte. The relative simplicity and high accuracy of electrical measurements make it possible to monitor the pore growth with high resolution. Both single-track and multi-track samples can be used for measurements. The effective pore radius r_{eff} as a function of time t is calculated from

$$G(t) = N\pi k(r_{\mathrm{eff}})^2/l, \tag{4.22}$$

where $G(t)$ is the electrical conductance, N is the number of tracks, k is the specific conductivity of the solution at the etching temperature, and l is the thickness of the sample (here we assume that all the tracks are perpendicular to the sample surface). It is clear that (4.22) is valid if the etched tracks are cylindrical in shape, e.g., V is high and l is small. A criterion can be suggested to define the applicability of the conductometric method for the measurements of transversal dimensions of the radiation-damaged zone in an ion track: the quantity $V_{\mathrm{B}}l/2V_{\mathrm{T}}$ must be smaller than the radius r_{T} of the zone where the polymer is significantly disturbed:

$$V_{\mathrm{B}}l/2V_{\mathrm{T}} > r_{\mathrm{T}}. \tag{4.23}$$

Under such a condition, the openings on both ends of the etched track are small enough at the moment when the etchant penetrates the whole length of the track. In other words, the highly selective etching produces in the thin foil an extremely narrow channel that is localized entirely within the damage zone. Its further growth during etching reflects the local response of the polymer at a certain distance from the ion trajectory. This distance is equal to the measured effective pore radius. The derivative of the effective pore radius is the radial etching rate V_r. Thus, from the measured r_{eff} vs. t curve one can derive a function $V_r(r)$ that shows what happens to the polymer at small radii within the track.

There are some factors that may interfere with the application of (4.22). For the multitrack samples, the nonsimultaneous perforation of individual tracks causes a problem. Surface conductivity, heating of the etchant in pores, electrolyte exhaustion due to chemical reaction and accumulation of etch products can also be sources of error. These complications were theoretically analyzed in [103,104]. In any particular case one can check if the above factors influence the etching kinetics by comparing (1) functions $r_{eff}(t)$ measured on single and many tracks; (2) etching curves obtained at significantly different absolute rates of the process; (3) etching curves obtained using concentrated and diluted etchants; (4) values of r_{eff} measured at different pH. The analysis of the results obtained under different conditions makes it possible to exclude possible sources of errors.

Another difficulty is the low track-to-bulk etch ratio in the majority of all plastics. Only a few types of polymeric films have been found to fulfill the applicability criterion for the conductometric studies of etching kinetics on the nanometer scale. Figure 4.23 presents results of successful measurements carried out with ion tracks in thin PC and PET foils.

The curve A in Fig. 4.23 is a monotonic function with a very fast rise just after the breakthrough. The effective pore radius reaches about 5 nm in 1 min after the moment of perforation. Further growth of the pore occurs at a rate falling by a factor of ten in 1 nm [1]. The decreasing radial etch rate is believed to reflect the decreasing density of damage around the particle path. The highly damaged track core manifests itself on the r_{eff} vs. t curves as the step-like rise of the pore diameter followed by a phase of much slower etching. The diameter of the track core is only a few nanometers. The etchant penetrates along the track core at the rate that is almost 4 orders of magnitude higher than V_B. The track core is surrounded by the track halo ("penumbra") in which the etch rate slowly decreases to a constant value at large radii.

In contrast to curve A, curve B – measured also for an ion track in polycarbonate – is not monotonic. Curve B has a linear segment at radii between 6 and 10 nm. Though it does not seem remarkable, this qualitative difference between curves A and B is significant and representative of the performed series of experiments [105]. Based on this result, it has been suggested that the damage of the polymer is not degradation only. The mutually opposing effects of chain breaking and cross-linking might lead to a flat step in the radial etch rate in the halo.

For ion tracks in PET, the susceptibility of material to chemical attack in the core and that in the halo are surprisingly different. There is a distinct minimum of the radial etch rate at radii of 5–10 nm (curve C in Fig. 4.23).

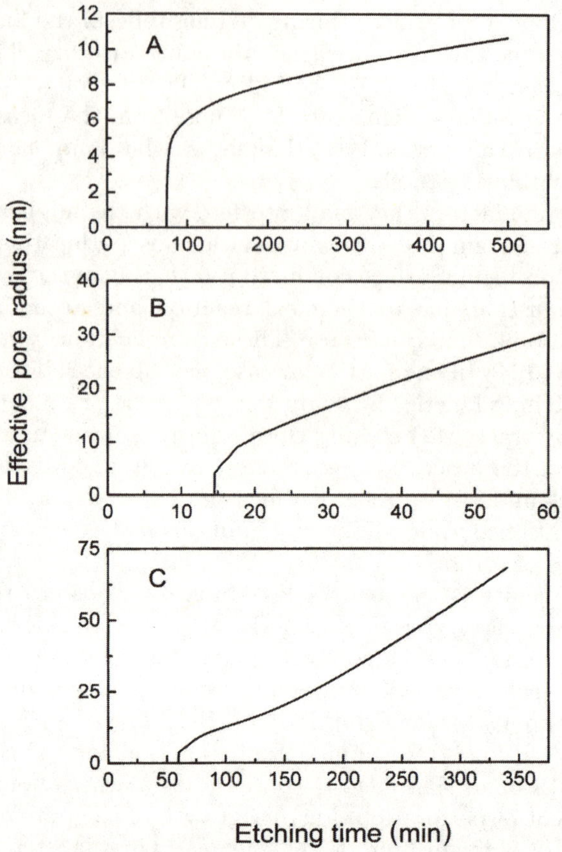

Fig. 4.23. Effective pore radius as a function of etching time. (**a**) PC foil with a single fission track etched in 3.1 M NaOH at 7.2°C [1]. (**b**) PC foil with a single fission track etched in 6 M NaOH at 24.5°C [105]. (**c**) PET foil with a single gold ion track etched in 0.1 M NaOH at 80°C

Curves of such a shape were measured for ions of different masses and energies [53,106,107]. In contrast to the situation with tracks in PC, here the local etch rate in the halo increases with increasing radius. The higher resistance of the material to chemical attack is assumed to originate from cross-linking that prevails over chain scission around the track core. The radial distribution of the local etch rate in a track of a uranium ion is shown in Fig. 4.24. The local etch rate V_r at $r = 0$ is the measured quantity V_T. The profile of V_r at $r > 5\,\mathrm{nm}$ is determined from conductometric data presented in [107]. At $r \approx 100\,\mathrm{nm}$ the local etch rate approaches a constant value V_o (which is equal to V_B for the isotropic material).

It is remarkable that the radial extent of the halo reaches up to almost 100 nm. A reduction of the etch rate in the track halo was observed also

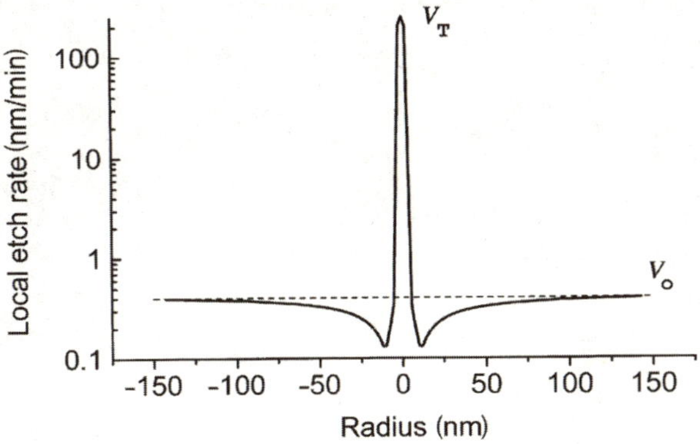

Fig. 4.24. Profile of the local etch rate in tracks of uranium ions with energies of some MeV/u. Material: polyethylene terephthalate

for polymers other than PET and, therefore, seems to be a quite common phenomenon [106]. The nature of modifications of material in this zone is still under discussion. There is no strong evidence that the cross-linking is the only reason for the reduction of the etch rate. Another possible mechanism can be associated with the compaction of the polymer, i.e., with the decrease in the free volume caused, in turn, by a pressure pulse propagating from the track core.

Replica studies. Regardless of the mechanisms responsible for the change in etchability, the submicroscopic evaluation clearly shows that the ion track cannot be restricted to a straight line, i.e., the trail of a zero radius. This is especially true for ions of high masses. The conductometric measurements demonstrated that the extent of the primary damage in gold or uranium tracks approaches the micrometer size. For lighter ions another experimental technique has been applied to evaluate the shape of tracks at the early stage of development. The technique is based on the electron microscopic analysis of replicas with the resolution of ≤ 1 nm [108]. The replicas are folded in such a way that the "protuberances" generated by particle tracks appeared clearly contrasted on the edge of the replica fold.

The front profile of a slightly etched track is directly imaged in the transmission electron microscope. Shapes of etched pits produced by alpha particles and some light ions (^{7}Li, ^{12}C, ^{16}O) in polycarbonate Makrofol E were presented in a series of publications [109, 110]. The variational principle was applied to obtain the general equations for chemical etching inside and outside the damage zone. Based on the experimentally measured etched profiles, Mazzei and coworkers reconstructed the local etch rate as a function of the radial distance. The evolution of the etched track was described by an array of trajectories representing the movement of points on the etched surface in

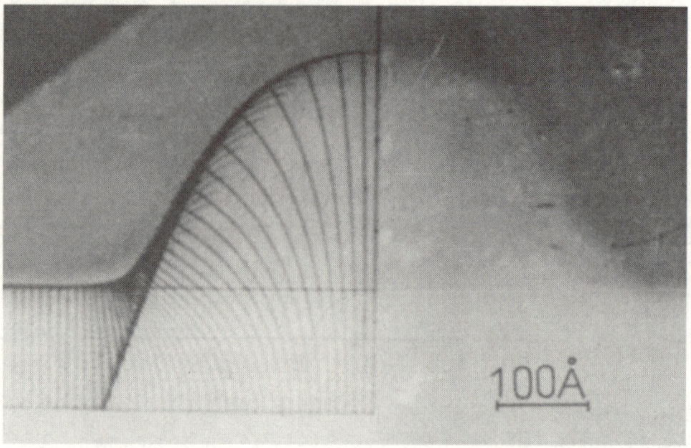

Fig. 4.25. Observed and theoretically constructed profiles of a 2-MeV ^{16}O ion track in Makrofol E after short etching (courtesy of O. Bernaola). For details see [112]

time. An example of the experimental etched track profile together with the theoretically constructed profile is shown in Fig. 4.25.

Surprisingly, the track diameters measured by this means in polycarbonate did not correlate with the size of physical damage produced by secondary electrons [111]. The tracks of alphas were too large, which can be caused by the fact that the etching agent and product transportation need a certain minimum track radius (~ 2.5 nm). Postirradiation rearrangement of the damage due to transport and diffusion processes would also play a large part in broadening the preferentially etched zone of a track. At the same time, for the fission-fragment tracks the size of the preferentially etched core was smaller than expected. Again, this can be further evidence of the cross-linking taking place outside the track core at quite high $\mathrm{d}E/\mathrm{d}x$.

Surfactant-controlled shape of etched tracks. We have already mentioned that the size of attacking molecules and etching products may have an effect on the submicroscopic etching kinetics. This results from the fact that the radius of a latent track is close to molecular dimensions. Here we describe another phenomenon of this sort – the effect of surfactant in the etch solution.

First consider the influence of a surfactant on the chemical reaction at the solid/liquid interface. Large-molecule surfactants are composed of a hydrophobic part, normally an alkyl radical having 8–12 carbon atoms, and a hydrophilic part, that is a polar moiety (ionic or nonionic). The surfactant molecules orient themselves with their hydrophobic part towards the surface and the hydrophilic part towards the aqueous solution. This is shown in Fig. 4.26a where the hydrophilic "heads" of the molecules are depicted as circles. The formed surfactant layer is quasisolid and partially protects the surface from the etching agent. However, the adsorbed layer not only consists of surfactant molecules but also contains solvent (water) molecules. Transport of water and ionic solutes through the barrier formed by the surfactant

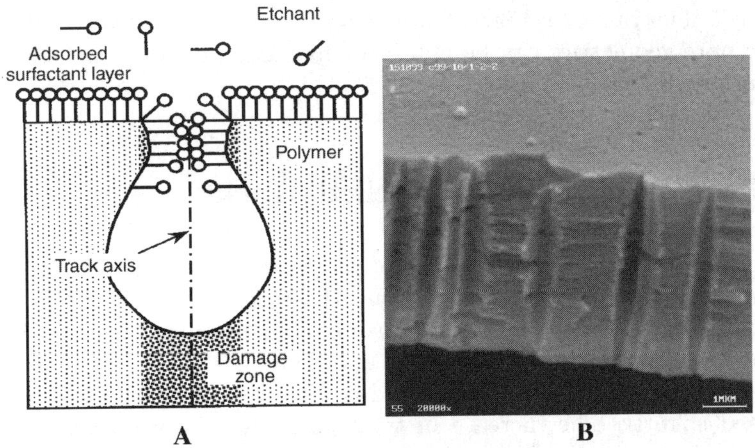

Fig. 4.26. Geometry of etched ion track in the presence of surfactant (**a**); SEM photo of cigar-like pores produced in thin PET film (**b**) [114]

is probably similar to that through lipid bilayers. Normally the diffusion of H^+ and OH^- ions across such molecular layers is significantly faster that the diffusion of other ions.

The thickness of the adsorbed layer h is of the order of the fully extended length l_S of the surfactant molecules. For most commonly used surfactants l_S is 2–4 nm. The formation of the adsorbed layer leads to a reduction in the bulk etch rate of the surface of a solid that undergoes etching. Typically the polymer/solution interface advances 2–3 times slower if a surfactant is added to the etchant [112, 113].

Consider now the development of an ion track at the beginning of etching. Diffusion of etchant molecules across the surfactant layer initiates the formation of a small hole at the entrance of the ion track. When the hole attains a diameter of a few nanometers, the surfactant molecules penetrate into the developing pore and block its entrance (Fig. 4.26a). Because the track core diameter d_T is smaller than 2 h, further diffusion of the surfactant molecules into the growing pore is hindered. The adsorbed surfactant layer is not permeable to large molecules. At this stage the transport of surfactant molecules into the pore can proceed only by lateral propagation on the pore internal wall surface. This process is slower than the diffusion in the bulk solution. Therefore, the volume inside the pore remains free of surfactant molecules. However, the diffusion of hydroxide ions into the pore or etch products out of the pore, is not hindered due to their relatively small size. A hydrophilic channel formed by the surfactant molecules provides transport of alkali along the pore axis. As a result, the inside diameter of the pore grows at a higher etch rate than the pore entrance. After some time "bottleneck" or "cigar-like" pore channels are formed. The shape of pore channels produced in this manner is illustrated by Fig. 4.26b. Using the surfactant-controlled etching, diverse pore geometries can be obtained that may provide benefit in various applications [113].

4.6 Thermal Stability of Etched Tracks

Ion transmission spectrometry was applied to study the changes of ion-track shapes upon thermal treatment. For this, the above-characterized tracks were systematically annealed and their transmission determined by ITS. The radii of etched tracks in PET first decreased and then increased upon annealing. By contrast, etched tracks in PI remained stable up to $\sim 500°C$ and thereafter increased slightly. The increase in track radii may be ascribed to the loss of polymeric matter at that temperature. The transiently slight decrease in track radius observed for PET may stem from thermal polymer relaxation. By contrast to these results, tracks stabilized by a thin metallic tubule (as prepared by chemical deposition of, e.g., Ag) maintain stability far above the polymer's decomposition temperatures [114, 115].

4.7 Techniques to Improve the Visibility of Very Small Etched Tracks

Apart from the above-mentioned ways to enlarge etched tracks for better visibility by suitable choice of highly sensitive polymers, optimum etching conditions, and the addition of sensitizers, one can exploit electrical discharges for further enlargement of the etched track zones. The idea is that a very high field strength emerges from the tip of a small electrolyte-filled etch pit at which a voltage is applied. If the field exceeds a critical value, electrical breakthrough occurs that leads to tree-like long-ranging cracks. By enlarging these cracks by etching the total etched volume they can be magnified considerably [116, 117], Fig. 4.27. The shape of the etch pits reflects the random distribution of the water trees emerging upon electrical breakthrough (for more details, see, e.g., [118]). As thus even tiny etch pits of low-energy light ions can be made detectable, this technique has become an important tool in dosimetry.

Another approach to obtain amplified images of track distributions is to enable a voltage breakthrough through a stack of both the microporous etched foil and another thin Al-coated polymer foil. Due to the delivered heat during electrical breakthrough the Al-coating melts and/or is oxidized so that a mm-sized transparent spot appears on the Al-coated foil.

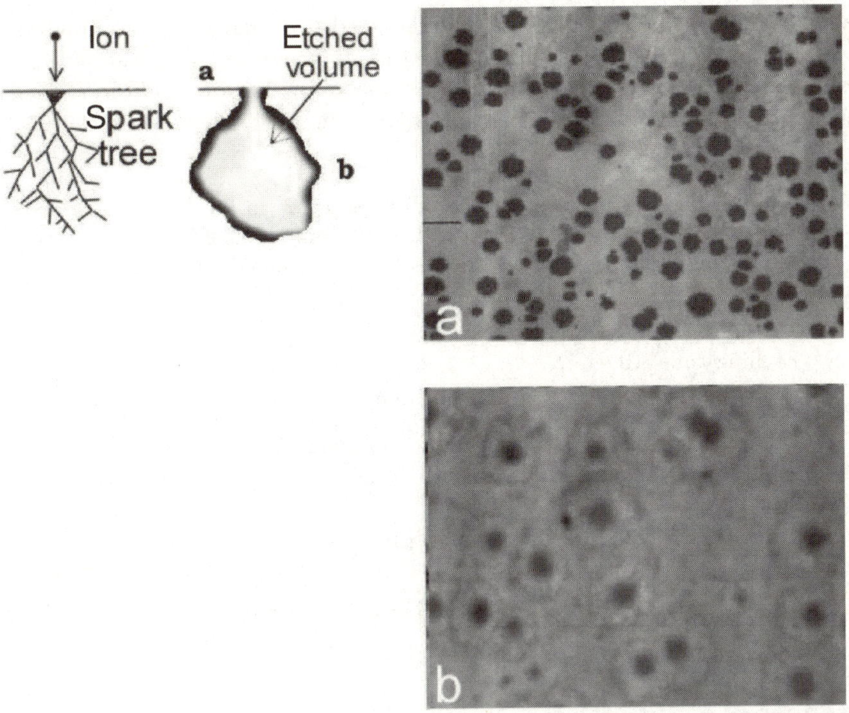

Fig. 4.27. Example of tracks of 100 keV H in CR39 that have been enlarged by spark breakthrough during etching. Image size: (**a**) 160 μm × 200 μm; (**b**) 40 μm × 50 μm. (*a*) is focused on the sample surface, and (*b*) is focused on the etched spark tree regime (Samples obtained from Dörschel B and Hermsdorf D, TU Dresden)

4.8 Cases of Inverse Etching

A precondition for the etch-track formation is that $V_t \gg V_b$, with V_t being the track etch rate and V_b the bulk etch rate. For $V_t \approx V_b$; $V_t > V_b$ only shallow etch pits emerge, and for $V_t \leq V_s$, no surface depression is developed at all. In this case the pristine matter is removed faster than the track material, so that the irradiated zones should protrude from the neighborhood after prolonged etching – provided that they exhibit sufficient rigidity. Otherwise they will collapse and form a spaghetti-like surface topology.

The underlying reason for the reduced track etchability is the dominance of cross-linking over chain scissioning upon swift heavy-ion impact. Materials that show this behavior are, e.g., polysilanes. In fact, ITS tests on irradiated polysilanes did not reveal any pore formation even after prolonged etching[6]

[6] Ion-track studies in polysilanes are complicated due to the high brittleness of polysilane foils, the appearance of occasional giant-size etch pits (of up to 30 μm radius in pristine PMPySi, and of up to 140 μm size in pristine PDHSi), and due

with 5 mole/l NaOH.[7] When ion-irradiated polysilanes are completely dissolved in solvents such as toluene, needle- or spaghetti-like residues show up (see Fig. 4.29) that are thought to be the remnants of single ion tracks, as their areal density roughly corresponds to the ion fluence.

Table 4.3. The emergence of visible "dark objects" in ion-irradiated polysilanes (11.4-MeV/nucleon Au or Pb ions at $1 \times 10^{10}\,\mathrm{cm}^{-2}$ fluence) in the optical microscope during etching with 3 mole/l NaOH at ambient temperature (own work, unpublished)

Etching time	PDHSi		PMPySi	
	Object size [μm]	Object density [cm^{-2}]	Object size [μm]	Object density [cm^{-2}]
0	–	0	–	0
70 min	–	0	–	0
4 h	0.3 ± 0.1	$(3 \pm 1)\times10^7$	–	0
2 d	0.5 ± 0.1	$(2 \pm 1)\times10^7$	–	0
6 d	1.0 ± 0.1	$(9.5 \pm 1)\times10^7$	1.0 ± 0.1	$(1.5 \pm 1)\times10^7$

Pristine as well as ion-irradiated polysilanes membranes are transparent under the optical microscope (Fig. 4.29 left). However, after etching additionally some tiny metallic-grey dark objects gradually emerge in the optical microscope (Fig. 4.28 right), the areal density of which increases with etch time, but is always much lower than the expected track density (Table 4.3). As consequently these spots cannot be identified with ion tracks, these objects are tentatively assigned to zones of multiple ion impact. It is assumed that only after multiple ion impact is the material's degassing and cross-linking density sufficiently high to enable an observable optical phase contrast after prolonged etching.

to the splitting-up of spin-coated PDHSi films into irregular stacks of loosely connected layers of different thickness upon etching, until finally large fibrous residues remain.

[7] Fink D, Herden V, Seki S (1999), unpublished work

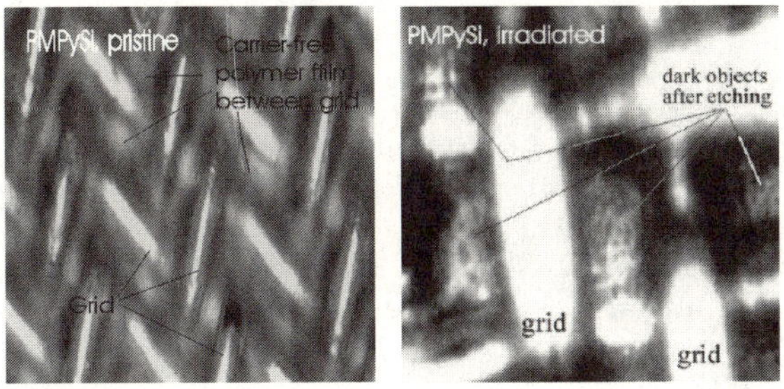

Fig. 4.28. Optical microscopic image of a self-carrying PMPSi film spun over a metallic grid for support. *Left side:* Unirradiated, etched. *Right side:* Swift heavy-ion irradiated at $1 \times 10^{10}\,\text{cm}^{-2}$, and subsequently etched. Whereas the pristine film remains featureless after etching, *dark spots* show up in the irradiated polysilane matrix, possibly stemming from the optical phase contrast of multiply overlapping tracks. *Left side:* 1 mm × 1 mm, *right side:* 0.3 mm × 0.3 mm[8]

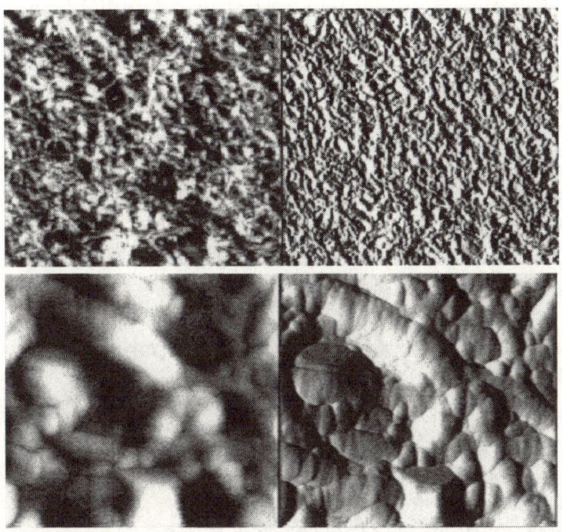
a

b

Fig. 4.29. AFM (*left*) and LFM (*right*) images of (**a**) swift heavy-ion-irradiated (3-GeV U) polysilane PMPySi[9] (spin coated onto a Cu backing) and (**b**) swift heavy-ion-irradiated (200-MeV Xe) polysilane PBMSi[10] (spin coated onto a Cu backing), after leaching in toluene, to eliminate the remaining soluble matter. As the SiC_xH_y remnants along the ion tracks do not have sufficient stiffness they appear to lie on the substrate like spaghetti. Note that the SiC_xH_y wires have some fine structure. Possibly this reflects the projectile ions' statistical energy transfer processes to the target. Scales: (**a**) 1 μm × 1 μm; (**b**) 50 nm × 50 nm

---

[8] Seki S, Fink D, Spohr R (1999), unpublished work

[9] Fink D, Seki S, Spohr R, Farenzena L, Papaleo RM (1999), unpublished work

[10] Herden V, Funk D, Richter A (1999), unpublished work

References

1. Fleischer RL, Price PB, Walker RM, Nuclear Tracks in Solids: Principles and Applications. University of California, Berkeley, 1975
2. Spohr R, Ion Tracks and Microtechnology. Principles and Applications. Vieweg Verlag, Braunschweig, 1990
3. Moiseev YuV, Zaikov GE, Chemical Stability of Polymers in Aggressive Media (in Russian). Khimia, Moscow, 1979
4. Schnell H, Chemistry and Physics of Polycarbonate. Interscience Publishers, New York, 1964
5. Born M, Volumen und Hydrationswärme der Ionen. Z Phys **1**, 45–48 (1920)
6. Markin VS, Chismadzhev YuA, Induced Ionic Transport (in Russian). Nauka, Moscow, 1974
7. Samoilova LI, Apel PYu, Etching of small pores in PETP by different alkalis. Radiat Meas **25**, 717–720 (1995)
8. Parsegian A, Energy of an ion crossing a low dielectric membrane: solutions to four relevant electrostatic problems. Nature **221**, 844–846 (1969)
9. Schulz A, Danziger M, Akapiev GN, Trofimov VV, Prokert K, The pore opening process of etching PETP films irradiated by multiply heavy ions and the three-phase-model of the etching process of the single ion track. In: Yu.Ts. Oganessian and R. Kalpakchieva (eds.), Heavy Ion Physics, World Scientific, Singapore, 1998, pp. 792–795
10. Fink D, Müller M, Capillaric penetration of etchant solution into swift heavy ion irradiated silicone rubber. Nucl Instrum Methods **B170**, 134–144 (2000)
11. Fink D, Müller M, Petrov A, Etching kinetics of swift heavy ion irradiated polymers with insoluble additives or reaction products. Presented at the 5th Intl. Symposium on "Swift Heavy Ions in Matter", May 22–25, 2002, Giardini Naxos, Italy
12. Fink D, Dwivedi KK, Müller M, Ghosh S, Hnatowicz V, Vacík J, Červena J, On the penetration of etchant into tracks in polycarbonate. Rad Meas **32**, 307–313 (2000)
13. Fink D, Ghosh S, Klett R, Dwivedi KK, Kobayashi Y, Hirata K, Vacík J, Hnatowicz V, Červena J, Chadderton LT, Transport processes during the incubation time of ion track etching in polymers. Nucl Instrum Methods **B146**, 486–490 (1998)
14. Fink D, Petrov A, Müller M, Hnatowicz V, Vacík J, Červena J, Marker penetration into high energy ion irradiated polymers. Surf Coat Technol **158–159**, 228–233 (2002)
15. Ghosh S, Klett R, Fink D, Dwivedi KK, Vacik JJ, Hnatowicz V, Červena J, On the penetration of aqueous solutions into some pristine and heavy-ion irradiated polymers. Radiat Phys Chem **55**, 271–284 (1999)
16. Fink D, Asmus T, Müller M, Apel PYu, Chadderton LT, Ion tracks in Mylar. Paper in preparation; to be published (2005)
17. Baur D, Apel PYu, Korchev YE, Müller C, Siwy Z, Spohr H, Spohr R, Surface gel in ion track etching – observations and consequences. Proc of the workshop on European Network on Ion Track Technology, Caen, France, 24–26 Feb. 2002
18. Shen M, Bever MB, Gradients in polymeric materials. J Mater Sci **7**, 741–746 (1972)
19. Fleischer RL, Price PB. Tracks of charged particles in high polymers. Science **140**, 1221–1222 (1963)

20. Fink D, Petrov A, Stolterfoht N, Wilhelm M, Richter A, Behar M, Farenzena L, Papaleo R, Hirata K, Kobayashi Y, Chadderton LT, Schulz A, Fahrner WR, Creation of nano-scale objects by swift heavy ion track manipulations. Proc. 2nd Int. Symp. On Material Chemistry in Nuclear environment, Tsukuba, March 13–15, 2002
21. Edmonds EA, Durrani SA, Relationships between thermoluminescence, radiation-induced electron spin resonance and track etchability of lexan polycarbonate. Nucl Tracks **3**, 3–11(1979)
22. Koul SL, Campbell ID, Chadderton LT, Langroo M, Fink D, Biersack JP, ESR and track-etch studies of irradiated polymers. Nucl Instrum Methods **B32**, 383–388 (1988)
23. Lück HB, Mechanism of particle track etching in polymeric nuclear track detectors. Nucl Instrum Methods **202**, 497–501 (1982)
24. Bernas A, Chambaudet A, Sur les dégâts chimiques créés par des faisceaux d'électrons ou d'ions lourds dans différents plastiques détecteurs de traces. Rad Effects **22**, 129–134 (1974) (in French)
25. Boyett RH, Johnson DR, Becker K, Some studies on the chemical damage mechanism along charged-particle tracks in polymers. Radiat Res **42**, 1–12 (1970)
26. Picq V, Ramillon JM, Balanzat E, Swift heavy ions on polymers: Hydrocarbon gas release. Nucl Instrum Methods **B146**, 496–503 (1998)
27. Benton EV, Henke RP, Sensitivity enhancement of Lexan nuclear track detector. Nucl Instr Methods **70**, 183–184 (1969)
28. Tretyakova SP, Apel PYu, Jolos LV, Mamonova TI, Shirkova VV, Study of registration properties of polyethyleneterephthalate. In: Francois H et al. (eds.), Solid State Nuclear Track Detectors. Pergamon Press, Oxford, 1980, pp. 283–287
29. Chambaudet A, Roncin JIR, UV and ESR studies on plastic track detectors irradiated by heavy ions. In: Fowler PH, Clapham VM (eds.), Solid State Nuclear Track Detectors. Pergamon Press, Oxford, pp. 15–21 (1982)
30. Balanzat E, Betz N, Bouffard S, Swift heavy ion modification of polymers. Nucl Instrum Methods **B105**, 46–54 (1995)
31. Steckenreiter T, Balanzat E, Fuess H, Trautmann C, Chemical modifications of PET induced by swift heavy ions. Nucl Instrum Methods **B131**, 159–166 (1997)
32. Popov AA, Rapoport NYa, Zaikov GE, Oxidation of Oriented and Strained Polymers, Khimia, Moscow, 1987 (in Russian)
33. Cartwright BG, Shirk EK, Price PB, A nuclear-track-recording polymer of unique sensitivity and resolution. Nucl Instrum Methods **153**, 457–460 (1978)
34. Fischer BE, Spohr R, Production and use of nuclear tracks: Imprinting structure on solids. Rev Mod Phys **55**, 907–948 (1983)
35. Rybnikař F, Selective etching of polyolefines. I. Isotactic polypropylene. J Appl Polym Sci **30**, 1949–1961 (1985)
36. Enge W, Grabish K, Beaujean R, Bartholomä KP, Etching behavior of a cellulose nitrate plastic detector under various etching conditions. Nucl Instrum Methods **115**, 263–270 (1974)
37. Enge W, Grabish K, Dallmeyer L, Bartholomä KP, Beaujean R, Etching behavior of the Lexan polycarbonate plastic detector. Nucl Instrum Methods **127**, 125–135 (1975)

38. Gruhn TA, Li WK, Benton EV, Cassou RM, Johnson CS, Etching mechanism and behaviour of polycarbonates in hydroxide solution: Lexan and CR-39. In: Francois H et al. (eds.), Solid State Nuclear Track Detectors. Pergamon Press, Oxford, 1980, pp. 291–301
39. Lück HB, On the use of polyethyleneterephthalate as solid state nuclear track detector: mechanism and kinetics of bulk etching. Nucl Instrum Methods **200**, 517–523 (1982)
40. Lück HB, On the use of polyethyleneterephthalate as solid state nuclear track detector: Mechanism and kinetics of bulk etching. Nucl Instrum Methods **213**, 507–511 (1983)
41. Somogyi G, Varnagy M, Medveczky L, The influence of etching parameters on the sensitivity of plastics. Radiat Effects **5**, 111–116 (1970)
42. Salkauskas M, Stulgiene S, Peculiarities of etching of polyethylene with Cr(VI) solutions in sulfuric acid. Lietuvos TSR Mosklu akademijos darbai (B serija) **2**(87), 3–8 (1975) (in Russian)
43. Stone FGA, Graham WAG, Inorganic Polymers. Academic Press, New York and London, 1992
44. Petersen DD, Improvement in particle track etching in Lexan polycarbonate film. Rev Sci Instrum **41**, 1252–1253 (1970)
45. Apel PYu, Blonskaya IV, Orelovitch OL, Root D, Vutsadakis V, Dmitriev SN, Effect of nanosized surfactant molecules on the etching of ion tracks: new degrees of freedom in design of pore shapes. Presented at the 5th Intl. Symposium on "Swift Heavy Ions in Matter", May 22–25, 2002, Giardini Naxos, Italy
46. Apel PYu, Titova S, Tretyakova S, A study of etching of heavy charged particles in polyarylate. JINR Commun 18-86-788 (1986) JINR, Dubna
47. Daubresse C, Ferain E, Legras R, Energetic heavy ion tracks in PEEK film. Nucl Instrum Methods **B122**, 89–92 (1997)
48. Komaki Y, Tsujimura S, Growth of fine holes in polyethylenenaphthalate film irradiated by fission fragments. J Appl Phys **47**, 1355–1358 (1976)
49. Komaki Y, Matsumoto Y, Ishikawa N, Sakurai T, Heavy ion track microfilter of polyimide film. Polym Commun **30**, 43–44 (1989)
50. Vater P, Production and application of nuclear track microfilters. Nucl Tracks Radiat Meas **15**, 743–749 (1988)
51. Vilensky AI, Oleinikov VA, Markov NG, Mchedlishvili BV, Dontsova EP, Polyimide track membranes for ultra and microfiltration (in Russian). Russian Polymer Sci **A36**, 475–485 (1994)
52. Kravets LI, Dmitriev SN, Apel PYu, Production and properties of polypropylene track membranes. Collect Czech Chem Commun **62**, 752–760 (1997)
53. Apel PYu, Schulz A, Spohr R, Trautmann C, Vutsadakis V, Tracks of very heavy ions in polymers. Nucl Instrum Methods **B131**, 55–63 (1997)
54. Komaki Y, Growth of fine holes by the chemical etching of fission tracks in polyvinylidene fluoride. Nucl Tracks **3**, 33–44 (1979)
55. Daubresse C, Sergent-Engelen T, Ferain E, Schneider YJ, Legras R, Characterization of energetic heavy ion track in PVDF: production of PVDF track-etched membrane and application. Nucl Instrum Methods **105**, 126–129 (1995)
56. Komaki Y, Seguchi T, Scanning electron microscopy of nuclear pore filters in poly(ethylene terephthalate) and ethylene-tetrafluoroethylene copolymer. Polymer **23**, 1143–1146 (1982)

57. Shirkova VV, Tretyakova SP, Physical and chemical basis for the manufacturing of fluoropolymer track membranes. Radiat Meas **28**, 791–798 (1997)
58. Monnin M, Besson H, Sanzelle S, Avan L, Nouveax detecteurs solides de traces nucleaire et nouvelles methods de developpement chimique des detecteurs deja connus. Compt Rendus **264B**, 1751–1752 (1967) (in French)
59. Katz R, Kobetich EJ, Formation of etchable tracks in dielectrics. Phys Rev **170**, 401–405 (1968)
60. Benton EV, On latent track formation in organic nuclear charged particle track detectors. Radiat Eff **2**, 273–280 (1970)
61. Monnin M, Mecanisme de la formation des traces dans les polymeres. Radiat Eff **5**, 69–73 (1970) (in French)
62. Fleischer RL, Price PB, Walker RM, Hubbard EL, Criterion for registration in dielectric track detectors. Phys Rev **156**, 353–355 (1967)
63. Lück HB, A plastic track detector with high sensitivity. Nucl Instrum Methods **114**, 139–140 (1974)
64. Somogyi G, Grabish K, Scherzer R, Enge W, Revision of the concept of registration threshold in plastic track detectors. Nucl Instrum Methods **134**, 129–141 (1976)
65. Trautmann C, Bouffard S, Spohr R, Etching threshold for ion tracks in polyimide. Nucl Instrum Methods **B116**, 429–433 (1996)
66. O'Sullivan D, Price PB, Kinoshita K, Wilson CG, Correlative studies of track-etch behaviour and chemical development of lithographic polymer resists. In: Fowler PH, Clapham VM (eds.), Solid State Nuclear Track Detectors. Pergamon Press, Oxford, 1982, pp. 81–84
67. Ogura K, Hattori T, Hirata M, Asano M, Yoshida M, Tamada M, Omichi H, Nagaoka N, Kubota H, Katakai R, Development of copolymer of CR-39 with high sensitivity to low LET particles. Radiat Meas **25**, 159–162 (1995)
68. Benton EV, Ogura K, Frank AL, Atallah T, Rowe V, Response of different types of CR-39 to energetic ions. Nucl Tracks Radiat Meas **12**, 79–82 (1986)
69. Apel PYu, Didyk AYu, Salina AG, Physico-chemical modification of polyolefins irradiated by swift heavy ions. Nucl Instrum Methods **B107**, 276–280 (1996)
70. Chambaudet A, Romary P, On the variation of some heavy ion track characteristics with the polymeric detector crystallinity. In: Granzer F, Paretzke H, Schopper E (eds,), Solid State Nuclear Track Detectors. Pergamon Press, Oxford, 1978, pp. 307–316
71. Green PF, Ramli AG, Al-Najjar SAR, Abu-Jarad F, Durrani SA, A study of bulk-etch rates and track-etch rates in CR-39. Nucl Instrum Methods **203**, 551–559 (1982)
72. Benton EV, A study of charged particle tracks in cellulose nitrate. USNDRL-TR-68-14, San Francisco, California (1968)
73. Petersen F, Enge W, Energy loss dependent transversal etching rates of heavy ion tracks in plastic. Radiat Meas **25**, 43–46 (1995)
74. DeSorbo W, Ultraviolet effects and aging effects on etching characteristics of fission tracks in polycarbonate film. Nucl Tracks **3**, 13–32 (1979)
75. Schlenk B, Somogyi G, Valek A, A study on the etching properties of electron-irradiated plastic track detectors. Radiat Eff **24**, 247–253 (1975)
76. Endo K, Doke T, Calibration of plastic nuclear track detectors for identification of heavy charged nuclei using fission fragments. Nucl Instrum Methods **111**, 29–37 (1973)

77. Tripier J, Remy G, Debeauvais M, Ralarosy J, Stein R, Determination des vitesses de developpement des traces d'ions lourds dans le Makrofol et application. In: Nicolae M (ed.), Proc 8th Int Conf on Nuclear Photography and Solid State Track Detectors. Institute of Atomic Physics, Bucharest, 1972, pp. 290–297(in French)
78. Chadderton LT, Fink D, Gamaly Y, Moeckel H, Wang L, Omichi H, Hosoi F, Synthesis of buckminsterfullerene in the wake of energetic ions. Nucl Instrum Methods **B91**, 71–77 (1994)
79. Apel PYu, Korchev YuE, Siwy Z, Spohr R, Yoshida M, Diode-like single-ion track membrane prepared by electro-stopping. Nucl Instrum Methods **B184**, 337–346 (2001)
80. Vacík J, Červena J, Hnatowicz Posta S, Fink D, Klett R, Strauß P, Simple technique for characterization of ion-modified polymeric foils. Surf Coat Technol **123**, 97–100 (2000)
81. Komaki Y, Ohtsu H, The effect of coexistent gases during fission fragment irradiation on track etching in polyvinylidene fluoride film. Nucl Tracks **11**, 151–156 (1986)
82. O'Sullivan D, Thompson A, The observation of a sensitivity dependence on temperature during registration in solid state nuclear track detectors. Nucl Tracks **4**, 271–276 (1980)
83. Karamdoust NA, Durrani SA, Effect of registration temperature on the response of CR-39 to alpha particles and fission fragments. Nucl Track Radiat Meas **19**, 179–184 (1991)
84. Apel PYu, Didyk AYu, Fursov BI, Kravets LI, Nesterov VG, Samoilova LI, Zhdanov GS, Registration temperature effect in polymers: Physico-chemical aspects. Radiat Meas **28**, 19–24 (1997)
85. Crawford WT, DeSorbo W, Humphrey JS, Enhancement of track etching rates in charged particle-irradiated plastics by a photo-oxidation effect. Nature **220**, 1313–1314 (1968)
86. Kuznetsov VI, Didyk AYu, Apel PYu, Production and investigation of nuclear track membranes at JINR. Nucl Track Rad Meas **19**, 919–924 (1991)
87. Henke RP, Benton EV, Heckman HH, Sensitivity enhancement of plastic nuclear track detectors. Radiat Effects **3**, 43–49 (1970)
88. Tretyakova SP, Mamonova TI, Influence of gamma-irradiation on detection properties of Lavsan film. Atomic Energy **47**, 261–262 (1979) (in Russian)
89. Komaki Y, Ishikawa N, Sakurai T, Effects of gamma rays on etching of heavy ion tracks in polyimide. Radiat Meas **24**, 193–196 (1995)
90. Mishra R, Electron induced modifications in some polymers. PhD. Thesis, North-Eastern Hill University, Shillong, India, 2001
91. Dwivedi K, Ghosh S, Fink D, Mishra R, Tripathy S, Kulshreshtha A, Kathing DT, Modification in track registration response of PADC detector by energetic protons. Rad Meas **31**, 127–132 (1999)
92. Lück HB, Solvent-induced sensitization of particle tracks in polyester. Nucl Tracks Radiat Meas **19**, 189–195 (1991)
93. Törber G, Enge W, Beaujean R, Siegmon G, The diffusion-etch model. Part I: Proposal of a new two-phase track-developing model. In: Fowler PH, Clapham VM (eds.), Solid State Nuclear Track Detectors. Pergamon Press, Oxford, 1982, pp. 307–310

94. Todorović Ž, Antanasijevič R, A new method for identification of low energy fragments with $8 < Z \leq 18$ in plastic track detector Makrofol. Nucl Instrum Methods **212**, 217–219 (1983)
95. Fujii M, Yokota R, Kobayashi T, Hasegawa H, Sensitization of polymeric track detectors with carbon dioxide. Radiat Meas **25**, 141–144 (1995)
96. Csige I, Post-irradiation sensitization of CR-39 track detector in carbon dioxide atmosphere. Radiat Meas **28**, 171–176 (1997)
97. Sokolova YuD, Machula AA, Milinchuk VK, Zhdanov GS, On the structure of latent tracks in polymers irradiated with heavy ions (in Russian). Colloid J **59**, 395–397 (1997)
98. Somogyi G, Development of etched nuclear tracks. Nucl Instrum Methods **173**, 21–42 (1980)
99. Somogyi G, Szalay SA, Track-diameters kinetics in dielectric track detectors. Nucl Instrum Methods **109**, 211–232 (1973)
100. Enge W, Introduction to plastic nuclear track detectors. Nucl Tracks **4**, 283–308 (1980)
101. Fromm M, Meyer P, Chambaudet A, Ion track etching in isotropic polymers: etched track shape and detection efficiency. Nucl Instrum Methods **B107**, 337–343 (1996)
102. Martin CR, Nishizawa M, Jirage K, Kang M, Lee SB, Controlling transport selectivity in gold nanotubule membranes, Adv Mater **13**, 1351–1362 (2001)
103. Bean CP, Doyle MV, Entine G, Etching of submicron pores in irradiated mica. J Appl Phys **41**, 1454–1459 (1970)
104. Guillot G, Rondelez F, Characteristics of submicron pores obtained by chemical etching of nuclear tracks in polycarbonate films. J Appl Phys **52**, 7155–7164 (1981)
105. Schnoor G, Schütt H, Beaujean R, Enge W, Electrolytical studies of submicroscopic nuclear tracks in plastic detectors. In: Fowler PH, Clapham VM (eds.), Solid State Nuclear Track Detectors. Pergamon Press, Oxford, 1982, pp. 51–54
106. Apel PYu, Conductometric studies of multiply charged ion track structure in various polymers. Nucl Tracks Radiat Meas **19**, 29–34 (1991)
107. Apel PYu, Schulz A, Spohr R, Trautmann C, Vutsadakis V, Track size and track structure in polymer irradiated by heavy ions. Nucl Instrum Methods **B146**, 468–474 (1998)
108. Mazzei R, Bernaola O, Molinari de Ray B, Cabrini R, Replica method for avaluation of submicroscopic nuclear tracks in solid state track detectors. Nucl Tracks **9**, 219–223 (1984)
109. Mazzei R, Bernaola OA, Saint Martin G, Molinari de Ray B, Submicroscopic kinetics of track formation in SSNTD. Nucl Instrum Methods **B9**, 163–168 (1985)
110. Mazzei R, Grasso JC, Bernaola OA, Bourdin JC, Saint Martin G, The submicroscopic track kinetic theory and the variational principle. Nucl Instrum Methods **B34**, 74–80 (1988)
111. Mazzei R, Bernaola OA, Track experimental data related to post-irradiation dynamic processes. Nucl Instrum Methods **B63**, 309–318 (1992)
112. Apel PYu, Dmitriev SN, Root D, Vutsadakis V, A novel approach to particle track etching: Surfactant-enhanced control of pore morphology. Part Nuclei Lett **4** [101]-2000, 69–74 (2000)

113. Apel PYu, Blonskaya IV, Didyk AY, Dmitriev SN, Orelovitch OL, Root D, Samoilova LI, Vutsadakis V, Surfactant-enhanced control of track-etch pore morphology. Nucl Instrum Methods **B179**, 55–62 (2001)
114. Vacík J, Červena J, Hnatowicz V, Fink D, Kobayashi Y, Hirata K, Apel PYu, Strauß P, Study of latent and etched tracks by charged particle transmission technique. Presented at the RADECS Conf. Sept. 1998, Besancon, France
115. Fink D, Alegaonkar PS, Petrov AV, Berdinsky AS, Rao V, Müller M, Dwivedi KK, Chadderton LT, The emergence of new track applications. Proc. of the 21st Intl. Conf. on Nuclear Tracks in Solids, Oct 21–25, 2002, New Delhi, India
116. Tommasino L, Armellini C, Etching technique for damage track detectors. Radiat Eff **20**, 253–256 (1973)
117. Somogyi G, A study of the basic properties of electrochemical track etching. Radiat Eff **34**, 51–56 (1977)
118. Dissado LA, Fothergill JC, Electrical degradation and breakdown in polymers. Peter Peregrinus Ltd., London, 1992

Part II

Applications

5 Applications of Low-Energy Polymer Ion Irradiation

V. Hnatowicz and D. Fink

5.1 Lithography for Electronics

Integrated circuits (ICs) consist of patterned thin films of metals, dielectrics, and semiconductors on a monolithic substrate such as a silicon wafer. The basic functional element of ICs is the transistor. Meanwhile larges-scale integration has become commonplace where many millions of transistors are integrated in a single chip. The manufacture of such integrated circuits has as its primary goal the lowest possible cost per electronic function performed. To achieve this goal, the microelectronics industry has adopted the philosophy of simultaneous fabrication of the circuits side-by-side on a single wafer, and continuing miniaturization of the circuit elements and their interconnections [1].

The production of these microelectronic ICs depends ultimately on the use of lithographic techniques. Circuit patterns in IC wafers are formed first by delineating circuit patterns in an imaging medium called a resist, and then transferring the resist patterns to the substrate wafer by etching and/or deposition processes in a single series of operations.

Lithography consists of three principal processing steps, namely the deposition of the resist onto a semiconductor surface, ion-beam irradiation with visible light, UV light, synchrotron radiation, X-rays, electrons or energetic ions, by which a computer-stored pattern is transferred onto the resist film, and the development by etching in a liquid or gaseous medium, by plasma etching or ion milling. The etching selectively removes either irradiated or unirradiated parts of the resist, leaving a stepped structure with some parts of the substrate free and prepared for subsequent processing steps (dopant diffusion and implantation, replication).

The resists used in IC fabrication are synthetic organic polymers with or without additives that are radiation sensitive [1]. Polymer resists designed and developed for advanced microlithographics are critical to the success of the integrated circuits and consequently to high-speed solid-state electronics. In a typical application, the polymer resist is dissolved in an organic solvent and spin coated onto the surface of the wafer as a thin film of typically 0.1 to 2.0 μm thickness. The resist film is prebaked and exposed to high-energy radiation. The irradiation changes the physical and chemical properties of the resist, especially its solubility in suitable developing media. By exposing the film in a controlled way, e.g., by the light coming through a mask or by a computer-operated electron or ion microbeam, it is possible to transmit an intended circuit pattern to the film and, after development, to the semiconductor substrate. The developed film, with some parts removed, serves as a stencil from which the pattern can be transferred to the semiconductor.

In lithography, different types of resists are commonly used, depending on specific technological needs. Besides some inorganic materials (e.g., oxides, nitrides, glasses), polymers are applied for masking purposes. The resist is called a a negative resist if the solubility of the exposed resist decreases after exposure. Conversely, it is called a positive resist if the solubility of the exposed resist increases after exposure [1]. Negative photoresists (i.e., light-sensitive resists) generally consist of a rubber-like polymer and a cross-linking agent. Upon exposure to irradiation, the cross-linking agent is excited and forms highly reactive species that promote the cross-linking of the basic polymer, thereby decreasing its solubility. Positive photoresists generally consist of a low molecular weight polymer and a so-called dissolution inhibitor. The dissolution rate of the mixture in alkaline aqueous developer is very low. Upon exposure to irradiation, the inhibitor is photochemically converted, which renders the exposed resist soluble in the alkaline developer, thus enabling a positive pattern to be developed [1].

In the case of UV-, electron-beam, ion-beam, and X-ray microlithography the resists are organic polymers without additives. When the polymer resists are exposed to the ionizing radiation cross-linking occurs in the case of negative resists, and chain scission in the case of positive resists. Cross-linking results in a decrease in the resist solubility, chain scission promotes a solubility enhancement.

The main requirements of lithographic resists are high sensitivity, high thermal stability, high resolution, good adhesion to the substrate, and adequate wet and dry etch resistance. Important processing parameters are resist film thickness, resist film prebake and postbake conditions, developer and development conditions, exposure dose, and the substrate onto which the resist film is deposited. For positive resists, the sensitivity is the dose where complete development is obtained, and for negative resists, the sensitivity is defined as the dose required to cross-link the resist such that the exposed area retains 50% of the original thickness after development. The sensitivity of photoresists is largely determined by the sensitivity of the monomeric photosensitizer, rather than by the polymer itself. For electron, ion-beam, UV and X-ray resists, however, the sensitivity also depends on the polymeric component [1]. Sensitivity is commonly expressed as [$\mathrm{J\,cm^{-2}}$] for photo- and X-ray resists, and [$\mathrm{C\,cm^{-2}}$] for electron and ion-beam resists.

High thermal stability is an important requirement of polymer resists when they are used as the etch mask for ion implantation or plasma and ion etching. This implies that they must have a high glass transition temperature T_g, as a resist with low T_g could flow or deform when heated during ion implantation or dry etching. Such deformations eventually set in quite abruptly when elevating the temperature [1]. Most photoresists cannot withstand a processing temperature higher than 150° to 200°C, however, it has been shown that ion irradiation enhances their thermal stability considerably [2]. The care taken with low self-mobility must not be restricted to the base material only, but also be extended to the other components. As it turned out that the spatial resolution of photoresists is given essentially by the finite mobility of the inhibitor, special examinations have been performed to understand and control this effect [3].

The steadily rising need for higher storage capacity and shorter switching times leads to the steady refinements of microelectronic technology. With rapidly increasing packing density of microelectronic circuits, the resolving power of the conventional lithographic technique with visible or UV light has reached its physical limits, which are determined mostly by the radiation wavelength and light diffraction. Deep-UV light makes it possible to produce patterns with typical linewidths down to 0.4 µm. For creation of denser patterns with linewidths down to 0.25 µm electron or X-ray lithography can be used. Linewidths down to 0.1 µm can, in principle, be achieved by ion lithography; at present a 157-nm minimum linewidth has been reported [4].

Unlike the visible or UV light, the electron and ion beams offer several advantages. Besides substantially shorter effective wavelength, they can be easily generated, accelerated, deflected and electromagnetically focused into narrow beams, the position of which on the specimen can be precisely controlled. The beam energy and flux can also be easily adapted to specific needs.

In comparison with electrolithography, ion-beam lithography (IBL) offers both advantages and disadvantages resulting from different mechanisms of interaction with matter. In contrast to photons and electrons the ions exhibit much stronger and better spatially localized interaction with matter. An interesting feature of IBL at high fluences concerns polymer ablation that involves the decomposition of the polymer into volatile components. This signifies self-development to some extent. However, in the case of low-energy ions, the typical electrostatic lenses used have higher aberrations than magnetic lenses. Secondly, for highly charged ions, the space-charge effect is greater than for electrons. Thirdly, ion sources are much less efficient than electron sources and their emittance is poorer. Fourthly, the short range of low-energy ions may cause nonuniformity problems in exposure. On the other hand, increasing the ion energy produces uniform mask exposure at the expense of damage of the underlying semiconductor [5].

However, as compared to all other techniques, ion lithography has the best aspect ratio ζ (i.e., the ratio between depth of the micrometric structure and its lateral width). For visible light, $\zeta \sim 1$, for electrons, $\zeta \sim 10$, for X-rays, $\zeta \sim 10^3$, but for energetic ions, $\zeta \sim 10^2$ to 10^5, depending on the ion type and polymer. Furthermore, in contrast to electrons, ions have only negligible backscattering probability from the underlying substrate into the resist, which prevents an exposure of shielded areas. The interaction of heavy ions with matter gives rise only to low-energy secondary electrons with typical ranges of 5–10 nm, which do not deteriorate the patterns lateral contrast significantly. Last, but not least, the energy density deposited by an ion is very high (typically 0.1–10 keV/nm), so that the choice of the resist is much less limited by its radiation sensitivity and the spectrum of eligible resists widens. Therefore it seems that ion-beam lithography will become important for the generation of microstructures in the nm range, and will probably help in realizing quantum-effect devices [6].

When considering the present main strategies in electronics that essentially still follow the present trend to larger-scale integration of smaller and faster elements by means of masks of increasing refinement, one should bear in mind that this concept might be rendered useless already long before one runs into the technological and fundamental limits such as quantum and statistical fluctuations, as wiring and power dissipation – and also the cost of production – will set earlier limits. In microprocessors that were offered at the end of the last milliennium, already 2 km of wires and 10 W power dissipation were found in each cm^2.

Therefore it has been suggested [7] to put the future emphasis rather on nonbinary (analog or the like) and nonserial (massively parallel) computing elements, in a similar way as the human brain works that in total consumes only 4 W. For both, ion-beam-assisted plastic electronics seems to be ideal as an enabling base technology, combining cheap 3D parallel analog computing sections with sensors and actuators as well as with plastic fibers or optical waveguides towards intelligent machines of the future. However, it is still too premature to draw conclusions concerning the long-term future electronic technology with certainty. At any rate, silicon devices, including their polymeric masks, are expected to dominate microelectronics for another 20–30 years [5].

5.2 Micromachining by Focused Ion Beams

Integration of microelectronics and micromechanical components into one microelectromechanical system (MEMS) represents one of the modern trends in miniaturization. Several techniques for the production of three-dimensional (3D) components of MEMS systems have been developed that make use, e.g., of synchrotron X-radiation for transmitting a desirable pattern onto a resist (e.g., high-density PMMA). However, the use of synchrotron radiation is disadvantageous because of the high cost of sychrotron time and the need for special masks. Free of these disadvantages is the production of 3D structures using focused ion beams, the position of which is computer controlled [8,9]. In [10] 2-MeV proton beams with a spot size of 1 μm^2 and typical ion-beam currents from 1–100 pA were used for the production of cogs, gears and other structures with typical lateral dimensions of a few mm in PMMA and SU-8 resists. The height of the structures produced is equal to the range of protons in the resist, i.e., tens of μm. The exposed resists were further processed by standard chemical etching. To produce well-defined smooth structures, a high-resolution beam-scanning system is essential.

In this connection, polysilanes might also gain some importance, as they can easily be transformed into hard silicon carbide or silicon carbonitride [11, 12] tools by either the "classical" pyrolysis or by ion-beam irradiation. The same holds for organometallic polymers [13]. Also, polysiloxanes can be tailored by ion irradiation for technical applications [14, 15].

Apart from these already existing applications, new possibilities show up. In a recent study [16] it was shown that isolated Pd atoms, being produced in situ in PMMA by the thermal decay of a suitable organometallic compound embedded in the polymer matrix, migrate upon thermal annealing towards damage centers that had been created previously by low-energy Ar^+ irradiation, to precipitate there and to form metallic clusters. In this way, buried conducting layers can be created. Using a focused ion beam for this purpose, and varying both the ion beam energy and position on the sample surface according to a predetermined program, one can thus "write" three-dimensional damage patterns into the organometal-containing matrix, at which the metal atoms are trapped, after being released from their organocompounds. Instead of using polymers with dissolved organometals for this purpose, one can also work with metal nanocluster-containing polymer films that are produced, e.g., by codeposition of metal and polymer [17]. In fact, the principle possibility to write 3D micropatterns into polymeric solids has recently been demonstrated with a self-focusing laser beam impinging onto a UV-sensitive photopolymerizable resin, exploiting here the two-photon process [18]. The patterns produced in this way (spirals, rays, etc.) are, however, not conducting and show up only under optical observation.

5.3 Applications of Ion-Beam-Modified Surfaces

5.3.1 Tribological Applications

The growing application of polymers in various machinery components and in prostheses comprising mechanical couplings of polymer with polymer, metal or ceramics increases the demand for wear-resistant polymers. Different approaches have been attempted to improve the wear properties of the various polymers, namely structural design during polymer synthesis, formation of composites reinforced with fibrous or particulate materials and plasma, γ-ray- or UV-light-stimulated polymerization. However, the alternative approach of surface modification by ion irradiation has been explored relatively little for the wear improvement of polymers. So it appears that practical applications of ion beams in this technology branch are at present still largely missing.

5.3.2 Enhancement of Metal–Polymer Adhesion

A thin metallic coating of polymers is of great importance for many current technologies. In the inverse situation, many applications depend on the protective coating of metals by polymers. Excellent and permanent adhesion of the coating is essential in such applications. The adhesion should be strong, i.e., resistant to peeling, scratching and delamination. The adhesion performance is determined by the structure of the interface region. It depends especially on the character of chemical bonding, interface morphology and some other factors [19]. Ion beams may be used for adhesion enhancement in a variety of different ways, namely:

1. Ion-beam mixing, when energetic ions are passed through an interface of a preformed polymer–metal structure. The ions passing through the interface break atomic bondings, and disorder and mobilize the atoms near the interface. In the resulting random distribution there can be an elevated probability that local clusters of atoms will form complex chemical bonding structures linking film and substrate.
2. Reactive-ion implantation in the interface region, in order to enable or enhance the formation of chemically bonded complexes comprising both film and substrate atoms together with those implanted.
3. Substrate presputtering uses a low-energy beam of inert ions to clean the substrate surface and change its morphology prior to the film deposition. Commonly, 0.1–1-keV Ar ions are used, penetrating to depths of a few nm and treating therefore only a few monolayers of the substrate.
4. Ion-beam-assisted deposition (IBAD) employs simultaneous implantation of low-energy ions and deposition of coating materials. The ion bombardment serves not only to pretreat a substrate, but also to affect the interface formation and subsequent deposition of the film.

5.3.3 Aerospace Technology: Applications that Require Enhanced Chemical Resistance

Polymeric materials are used in spacecraft technology as thermal-control coatings, insulations, structural composites, encapsulants etc. Therefore the processes of polymer degradation in outer space are of great concern [20]. On the one hand, the polymer is subject to intense radiation essentially of electrons and protons of several MeV energies, with additional stress of UV radiation, as the annual surface dose of a satellite is as high as 25 MGy [21]. Furthermore, the polymer must withstand temperature fluctuations and high vacuum.

Another application concerns the reduced susceptibility to oxidation and chemical attack of ion-irradiated polymers. Atomic oxygen has been of great concern for polymers in low earth orbit satellites. Also ozone and other chemically active species may cause serious adverse effects on polymers. Therefore, in [22] the etching of pristine and ion-irradiated polyimide (Kapton) in an oxygen plasma is described. It was found that the specimens irradiated simultaneously with 200-keV B and 300-keV N ions to the fluences of 4×10^{15} cm^{-2} exhibit six times less erosion than those that remained unirradiated. Plasma-immersion ion implantation of Al ions was used for protection of PI (Kapton) and PET (Mylar) from oxidative environments in [22]. Metal plasma-immersion processing is an attractive technique for surface processing of polymers. In this technique the polymer surface is bombarded with metal ions with a continuous spectrum of energies (up to tens of keV) and ion charges. Due to the lower ion energy the carbonization and undesirable specimen darkening, common at standard ion implantation, are reduced significantly. In treatment with highly reactive oxygen plasma, the polymers modified by the plasma-immersion technique exhibit erosion rates two orders of magnitude less than for pristine polymers. So the plasma-immersion-implantation technique drastically improves both oxidation and erosion resistance of common polymers, leaving at the same time both the surface morphology and bulk thermomechanical properties of the polymers practically unchanged.

Specifically PI, important as a construction material for solar-array blankets [24], shows significant improvement against oxidation after ion irradiation, with erosion rates reduced by a factor of 6 [25].

5.3.4 Nuclear-Waste Management: Applications that Require Reduced Permeability

The application of radioactivity in the 20th century has caused large amounts of radioactive waste with varying concentrations in long-lived radionuclides. In order to optimize the packaging and storage of such products – essentially the gaseous daughter product radon – it has been proposed to embed the waste in successive barriers of sealed polymeric membranes. Laboratory tests have shown that PVC greatly reduces the Rn emission from a radioactive source. In order to check the durability of such barriers, PVC samples have been subject to different accelerated aging processes by UV and γ irradiation. The analysis indicates that the permeation is decreased from 2.41×10^{-16} cm^2 s^{-1} to values lower by factors between 3 and 10, due to cross-linking effects. This result indicates that the sealing property of PVC improves during the aging and prolonged exposure to the α-emitting radon [26].

5.4 Applications of Changes of Optical Properties of Irradiated Polymers

Despite numerous individual studies on the influence of irradiation on the changes in the polymers' optical properties, a general theory is still missing, and therefore the possible optical applications of ion-irradiated polymers are still largely based on empirical knowledge. Optical signal processing in telecommunications requires different components such as waveguiding, modulating and switching devices. These components can be fabricated from suitable inorganic materials, e.g., $LiNbO_3$. Several years ago polymers also gained interest as alternative materials for these purposes. Since polymers exhibit a wide variety of molecular structures, they are supposed to be able to meet the requirements of various optical applications. Their properties can be modified relatively easily by different techniques including irradiation with energetic ions. They exhibit low optical losses, and their optical properties, mechanical and thermal stability can be modified easily at low substrate cost.

The optical applications of polymers modified by ion irradiation focus mainly on passive devices, mostly waveguides. An optical waveguide is a region in a transparent material that has a higher refractive index than its surrounding. A small index diference of less than 0.1% is sufficient to confine the light into a waveguide (an optical well) [27]. In polymers such changes of the refractive index may be created by ion irradiation at low or moderate fluences. The main advantage of this technique is the possibility to control easily the value and spatial profile of the refractive index. In optoelectronic devices the typical waveguide length is often less than a centimeter, so that the requirements on optical losses are not as strong as for common optical fibers. This implies that optically important irradiation-induced defects that adversely affect the losses can be kept at an acceptable level. The ion irradiation can be used in several ways depending on the effect that it has on the substrate. The question whether the refractive index will increase or decrease upon irradiation depends on at least two counteracting effects – namely the changes in material density and chemical changes – and therefore it is difficult to predict for a particular case. If the index increases then the irradiation may be used directly to form the waveguide. If the index falls then the ion beam can be used to create waveguide boundaries.

In a study by Brunner [28], optical waveguides were produced in PMMA by irradiation of the substrate through a mask with 230-keV H^+ ions at a fluence of $5 \times 10^{13}\,cm^{-2}$. The mask was prepared from a silicon wafer by electron lithography. Strip waveguides, branches and Mach–Zender interferometers were fabricated in this way. A new technique for the measurement of the refractive index as a function of depth was developed. The optical measurements performed at 633 nm and higher wavelengths reveal good transparency of the prepared structures. It was concluded that the increased refractive index is mainly due to the electronic energy loss of the H^+ ions. Also, ion irradiation of silicone rubber has been considered for this purpose [29].

5.5 Applications of Changes of Electrical Conductivity of Irradiated Polymers

Polymeric conductivity can be obtained either by suitable chemical strategies or by irradiation with energetic ions. Ion bombardment leads to creation of amorphous conducting graphitic-like structures. Furthermore, adequate implants enhance the conductivity by acting as dopants. The sensitivity of the conductive properties of, e.g., irradiated PI and PES polymer films to moisture uptake is significantly increased so that one can use such films as biomorphic humidity sensors [30]. Also, polymeric light-emitting diodes have already been formed by ion irradiation, etc. [21].

The two principal disadvantages of most conducting polymers are their limited stability and low electronic mobility, as compared with inorganic semiconductors. Only in very recent years has a significant improvement of these characteristics been achieved that now finally allow commercial applications of conducting polymers in electronics to become more realistic. Polymers with enhanced conductivity are likely to be commercially significant in applications where processing of large-area components at low material cost is important, but they are not likely to compete in areas where large mobilities are required. Here eventually ion-irradiated polymers may fill the gap.

An obviously possible application of polymers with a conductive surface layer created by the ion irradiation is the fabrication of planar resistors. The advantage of the ion-irradiation technique is the possibility to control the resistivity in a very broad range from 10^{-2}–$10^{16}\,\Omega\,\mathrm{cm}$ by changing the ion fluence and polymer type. Since the change of the sheet resistivity is closely related to deep structural and compositional changes of the polymer, the resulting resistor structures are expected to exhibit high stability. Other applications are mentioned, e.g., in [30–33]. In the latter two studies it was shown that by implantation of, e.g., 100-keV light ions a buried conductive layer can be created in the polymer, i.e., that a sandwiched structure of the type: {porous low-conducting surface layer / buried conductive layer / insulating bulk} is formed. This structure makes it possible to control the transconductance of the buried layer by applying an external electric field. This technology seems to open a way for fabrication of field effect transistor-like electronic switches. It also points to sensor applications by allowing specific materials to fill the porous surface layer and thus to modify their electrical characteristics. Ion-beam-induced electrical conduction of PI films can find potential applications for encapsulation of microelectronics.

Because of their exclusive electrical properties, conjugated polymers are promising materials for various applications. A major problem in commercial applications is still their production in processible and stable form, but substantial progress in this area has been achieved in recent times. Polypyrrole (PPR), polyaniline (PANI), and polythiophenes are the most commonly used conjugated polymers because of their relatively high stability. Current commercial applications of such conducting polymers use intermediate conductivity levels and include battery electrodes, conductive coatings for electrostatic speakers or for subsequent polymer metallization in galvanic baths, transparent conductive coatings, electrolyte replacement in electrolytic capacitors, antistatic protection coatings, etc.

Generally, the conductivity enhancement that can be achieved by the ion implantation in conjugated polymers is less than or equal to that achieved by common chemical and electrochemical doping techniques. However, ion implantation is advantageous for the potential fabrication of microelectronic devices in thin films of conductive polymers where spatially selective or localized conductivity enhancement is required.

In some instances, the ion implantation can be combined with chemical doping. In [35,36] the polyacetylene (PA) conductivity was changed by chemical doping with $FeCl_3$ or by implantation of 30-keV K^+ ions. A large increase in p-type (chemical doping) and n-type (implantation) was observed. By combining both p-type and n-type PA the authors constructed a diode that exhibited the common current–voltage characteristics but degraded spontaneously due to diffusion of dopants.

Ion-beam modification (50–180-keV B^+, 10^{13} to $10^{16}\,cm^{-2}$ [29]) of polymers may also be useful for the production of biomorphic humidity sensors, as it increases the sensitivity of, e.g., PI or PES to moisture uptake.

Another possibility to exploit the sensing properties of irradiated organic matter has been detected recently by Naddaf [37]. Whereas lead-phthalocyanine (Pb–PC) monomers show a strong nonlinear sensitivity to moisture but none to alcohol, polymeric Pb–PC surfaces act as stable sensitive alcohol sensors but lose their response to H_2O and other gases such as H_2, Cl_2, CO, CO_2, etc. The sensing properties are expected to stem from the interaction of electropositive compounds with the positive metal ion in the center of the phthalocyanine ring before irradiation, and from the interaction of electronegative compounds or acceptor gases with the peripheral ion-activated sites of the relatively large Pb–PC monomeric units after the irradiation. Such polymerization has been performed by Naddaf by exposure of Pb–PC to an ECR plasma. If using more energetic ions instead to initiate polymerization, the thickness of the gas-sensing layer should increase, and hence also the sensitivity of that sensor.

High-fluence irradiation of previously baked novolac positive-tone photoresists (e.g., by $10^{16}\,cm^{-2}$ Ar and P ions) gives rise to electrical conductivity that can be exploited for field-emission devices. The emission current (which sets in at currents of typically $40\,\mu A$) is fairly stable as compared with the emission characteristics of synthethized diamond [38].

It had been suggested [39] to implant metal ions at a high dose – typically $3 \times 10^{16}\,cm^{-2}$ – into polymers such as PS or PE to obtain a matrix rich in conducting nanoparticles for optical applications, similarly as has been realized successfully with other insulating targets. However, this suggestion appears to be problematic as one should expect complete target carbonization at these doses and energies exceeding some 100 eV, though the authors of [39] claimed that their targets did not yet show any significant degradation. Meanwhile, Faupel et al. [17] have found a more realistic way to produce radiation-damage-free metal-cluster/polymer composites by coevaporation of metal and polymer, so that the aforementioned high-dose ion-implantation concept is rendered useless for that application. However, this concept makes sense for other applications where high hardness, high conductivity, and high field-emission capacity are required, and where carbonization effects are of minor importance [39, 40].

Apart from the attempts to modify polymeric conductivity properties, ion irradiation of polymers also finds an application for cross-linking wire and cable insulation in order to improve the abrasion resistance and softening point. It is also used in the production of shrink film, tubing and packaging materials. Furthermore, radiation-induced degradation is used to form powdered Teflon that is applied in technology as a lubricant. Degradation of Teflon produces perfluoro intermediates required in production of fluoro surfactants, fluorinated dielectrics and fluorinated finishing agents for the textile industry.

5.6 Ion-Beam-Modified Polymers for Medicine and Biotechnology

The problem of bioadhesion is not only restricted to medical problems of implants. Other examples that show striking similarities are the microbial fouling processes that occur in oral cavities, cooling-water structures, and processing equipment for various diary products. The deposited biofilms contain living cells and are formed when micro-organisms and their metabolic products start to adhere and to reproduce on the surface of materials. They differ from conventional thin films by their aibility for self-reproduction. On the one hand, they are intentionally generated on the surfaces, e.g., of chemical and biological reactor vessels in biotechnology, waste-water treatment, the food industry and medical applications. On the other hand, there are cases when their formation is unwanted, e.g., on medical instruments, on vessels in the food industry, or on vessels for biological or medical research. In many of the cases cited above the substrates of the (wanted or unwanted) biofilms are polymers.

In both cases, irradiation with ions, electrons, or γ-rays may be applied usefully. On the one hand, high-fluence ion irradiation transfers polymer surfaces into a chemically highly activated carbonaceous material that enhances the biocompatibility. On the other hand, irradiation treatment of polymer surfaces initiates the grafting of components that may be bioactive or biocompatible, or biorejectant. And finally, polymer-surface irradiation enables its sterilization by killing unwanted living species.

If a substrate material, e.g., a polymer, contacts blood, the first response in the initial seconds to minutes is the adsorption of proteins. This is followed by platelet, white blood cell, and red blood cell interactions [41]. The living species will be attached to the surface by adsorption if there is an attractive force that pulls the micro-organisms to the material's surface. The micro-organisms move or diffuse only slowly or not at all over the surface.

This holds for blood–polymer interaction as well as for polymer–bacteria interaction, as given for polymers embedded in waste water. The number of deposited micro-organisms depends on the chemistry and surface structure of the substrate, and on the concentration of the organisms in the liquid. In the case of abundant food in the liquid, the adsorbed organisms will grow and multiply, thus forming chains, clusters or networks or even continuous biofilms on the substrate. The specific complex growth patterns of biofilms can already be elucidated by a relatively simple statistical model with respect to the specific growth behavior of biological systems, in which the sticking probability, the chain-formation probability and the side growth (branching) probabilities are the only decisive parameters [42].

Clotting of blood in contact with inserted polymers in the human organism is one of the greatest obstacles for protheses of all kinds. To overcome this problem, anticoagulant drugs such as heparin are injected. Because side effects are reported among patients who continuously need such drugs, grafting of heparin on medical polymer surfaces is supposed to be a feasible solution to this problem. The anticoagulant activity of heparin is endowed by its ability to form strong complexes with some blood-clotting factors, and this binding property is largely due to the amount and the distribution of functional groups such as the N-sulfate group. Therefore in [43] attempts were undertaken to graft such N-sulfate groups onto polysulfone foils. Indeed, the XPS spectra suggested the formation of polar groups on aromatic rings in the polymer chain. These polar groups might be N-sulfate, O-sulfate as well as amino groups. The modified surfaces display a tenth of the platelet adhesion of untreated polysulfone, indicating a much increased hydrophilicity of the surface.

Synthetic polymers are of growing importance in medicine and biology. They can be used for construction of tissue and organ transplants, fabrication of cell-growth supports and for experimental studies of interaction between cells and an extracellular matrix. For these purposes, usually common nondegradable polymers (e.g., polystyrene, polyethylene, polypropylene, polyurethane) are used. The excellent mechanical properties of these polymers, especially their elasticity and low specific weight, make them attractive for soft-tissue surgery. They can be used for repair or replacement of blood vessels, valves, injured skin or parenchymatous organs. After reinforcement with metallic, ceramic or carbon fibers the polymers can be used for the construction of artificial bones and joints. Biodegradable polymers are used as carriers for local delivery of cytostatics, antibiotics, hormones, etc. Hydrophilic polymers are used for the production of artificial eye lenses. Vesicles from diblock copolymers (made from PEO, PEE) are constructed for potential gene therapy. Last, but not least, synthetic polymers provide a good support widely used in cell and tissue culture. Also silicon-based polymers have been used for various medical [13, 44, 45] applications.

Modern artificial transplants are fabricated as a three-dimensional biodegradable scaffold. The cells migrate through the transplant, proliferate, resorb the polymeric components and replace them by newly synthesized extracellular matrix. In an ideal case, the whole artificial substitute is replaced by original well-functioning tissue. Materials supporting cell adhesion and growth are preferred for fabrication of artificial blood-vessel prostheses. The undesirable adhesion of platelets and formation of thrombus can be prevented by lining the internal surface of the prostheses by a cofluent layer of endothelial cells. However, the material supporting adhesion and growth of endothelium may be attractive also for other types of cells, particularly vascular smooth muscle cells (VSMC). These cells can colonize the prosthesis and thanks to their large potential to migrate and proliferate they can obliterate the prosthesis lumen. To prevent the obliteration by VSMC several methods are used, e.g., local or systemic application of various drugs. An alternative approach is to reduce the VSMC growth by increasing the number and spreading of these cells initially adhered to the prostheses material. The idea is based on the fact that if the number of initially adhered cells and their spreading are sufficiently high, their subsequent growth is slowed down by multiple and large cell–matrix and cell–cell contacts. The adhesion ability of cells depends on the physicochemical properties of the material, which can be changed, e.g., by ion implantation. Compared to the pharmacological treatment, the mere reduction of VSMC growth is a more physiological approach, since the VSMC cells could also have a positive impact on the prosthesis functioning and its integration with the surrounding tissue.

Standard concepts of wetting, spreading, surface energy, polarity, electric conductivity, roughness, crystallinity, porosity, and the presence of chemically active groups [46] as applied in nonliving systems allow significant correlations with bioadhesive phenomena as well. The initial adhesion of cells, their subsequent growth, differentiation and survival are strongly affected by these physical and chemical properties of the support surface. To obtain the appropriate properties, many biological or chemical procedures have been developed, e.g., coating the surface with proteins of a natural extracellular matrix, treatment with formaldehyde or acetic acid. These methods, however, are associated with the high risk of pathogen transfer, release of cytotoxic agents or immune reaction. In this respect, the physical methods of surface modification, such as ion implantation, seem to be more appropriate.

In general, one can state that surfaces rich in polar moieties, especially carboxyl groups, seem capable of bonding most strongly to living biomass. For artificial hearts surface enrichment of methyl groups is regarded as being most biocompatible. Baier et al. [47] denoted it as "universality of the principle of bioadhesion" that polymers with low surface tensions of the order of 20 to 30 $\mathrm{dyne\,cm^{-2}}$ prevent successful interface conversion by the spontaneously depositing glycoproteinaceous macromolecules. This induction of a poor "primer coat" inhibits the long-term adhesion of biological cells, even those cells specialized in surface colonization.

The irradiation of polymers with low- (including gas plasmas) and medium-energy ions leads to irreversible structural and chemical changes in the polymer surface layer, in particular to splitting of macromolecular chains, formation of low-mass degradation products and products with excessive, unsaturated bonds. Some of the volatile degradation products, usually hydrogen rich, may escape from the polymer surface layer leaving behind them excessive free volume. At higher ion fluences the escape of hydrogen-rich species leads to gradual carbonization of the irradiated material. The degradation products with increased chemical reactivity are the sites where some components from the ambient atmosphere (e.g., oxygen) may be captured. Thus, excessive oxidized structures may be created spontaneously on the polymer surface. Due to their abundant free volume, the irradiated polymers exhibit an increased permeability to liquids or gases, which may penetrate the irradiated material and eventually be captured on chemically active sites. This phenomenon can be used advantageously for further tailoring of the surface properties by chemical doping. These changes on the microscopic level – discussed in more detail in the earlier chapters of this book – lead to significant changes in the polymers' macroscopic properties, such as surface polarity, wettability, electrical conductivity, porosity and surface morphology that in turn affect the polymers' biocompatibility.

The first attempts to affect the biocompatibility of common polymers by ion irradiation were performed about ten years ago. The biocompatibility of various synthetic polymers irradiated by different ions with energies from a few eV up to hundreds of keV has been investigated in both in vitro and in vivo experiments with different biological objects. (To our knowledge, irradiation at elevated energies in the MeV to GeV range was hardly performed in this connection.) The effects of ion irradiation on the biocompatibility of silicon-based polymers have been studied in [48–52]. In [48] it was shown that the irradiation of medical silicone rubber with Na^+ ions changed the amount of plasma proteins that bonded to the specimens. In a later study [50] the silicone rubber was irradiated with a broad spectrum of 150-keV ions (H^+–Kr^+) in order to improve its blood compatibility. The chemical and physical structure of the irradiated specimens was investigated in relation to antithrombogenicity using different techniques. The platelet accumulation and thrombosis were studied in vivo in rats and dogs. It was found that the platelet accumulation was strongly reduced on specimens implanted with H^+ ans O^+ ions, so that the ion implantation is a useful technique to improve silicone-rubber antithrombocity. In [51] the irradiation of polyhydroxy-methyl-siloxane (PHMS) with 5-keV Ar^+ ions was shown to induce a remarkable enhancement of fibroblast cell adhesion and proliferation. The ion-beam-induced cytocompatibility was related to chemical modifications of the polymer surface characterized by means of XPS and contact-angle measurement. The adhesion and proliferation of fibroblast cells was found to be an increasing function of the ion fluence and the measured contact angle that characterizes the specimen's surface polarity. The irradiation changes the initial chemical composition of PHMS into a complex SiC_xO_y phase, which is rich in $[SiO_4]$ units. At the same time the initially hydrophobic surface becomes more hydrophilic.

The effect of the ion irradiation on the tissue compatibility of polystyrene (PS) and polyurethane (PU) was studied in [50] and [51]. The polymers were irradiated with 50–150-keV N^+, O^+, Na^+, Ne^+, Ar^+ and Kr^+ ions, and the adhesion of endothelial and HeLa cells was studied in vitro. Various techniques (XPS, FTIR, Raman spectroscopy) were used for the characterization of the chemical and physical properties of the irradiated polymers. The ion-irradiated PS and PU exhibit remarkably higher or complete adhesion and spreading of endothelial cells in comparison with unirradiated specimens [50]. The ion-irradiated PS exhibited a dramatic improvement of adhesion of HeLa cells that seems to be caused by new carbon structures and radicals induced by the ion irradiation [52]. In [46] and [53] polystyrene (PS), polyethylene (PE) and polypropylene (PP) were irradiated with 150-keV N^+, F^+ and Ar^+ and the adhesion of vascular smooth muscle cells on the modified polymers was studied in vitro. Two well-pronounced adhesion maxima were observed at ion fluences of $5 \times 10^{12}\,\mathrm{cm}^{-2}$ and $5 \times 10^{14}\,\mathrm{cm}^{-2}$ that are explained by an increase of the polar component of the free surface energy and by progressive carbonization, respectively. The second maximum was characterized also by an increased spreading of the cells and the strength of their adhesion.

The ion-assisted reaction with 0.5–1.2-keV Ar^+ ions in an oxygen atmosphere was used to affect the cytocompatibility of polystyrene (PS) in [54]. Newly formed hydrophilic groups were identified on the modified PS using the XPS technique, and an enhanced growth of rat pheochromocytoma cells on the modified surface was observed.

The effect of 4-keV to 1-MeV irradiation with He^+, Ga^+, and Xe^+ of PMMA at fluences between 10^{12} and $10^{16}\,\mathrm{cm}^{-2}$ in relation to mammalian cell adhesion was examined in [55]. It was found that cells adhere well on the areas ion-bombarded at fluences of 4×10^{14} to $5 \times 10^{15}\,\mathrm{cm}^{-2}$, but not at all on pristine PMMA. This effect is believed to be produced by a preferential protein adsorption on the bombarded areas and surfactant adsorption on the untreated ones. Also, removal of surface oxygen functionalities by the ion beam might play a role, as they are known to contribute to the hydrophilic character of the surface. Furthermore, there are speculations about the influence of differences in surface morphology and of long-living radicals on the cell behavior. This research aims at applications such as nerve reconstruction.

The adhesion of cells to artificial growth supports, including synthetic polymers, is mediated by the adsorption of extracellular matrix molecules (vitronectin, fibronectin, collagen. laminin, etc.) to these materials and the subsequent binding or cell surface adhesion of molecules of the integrin superfamily to specific amino-acid sequences of the adsorbed proteins. Direct binding of cells to synthetic polymer molecules is possible but, in fact, very rare. The cytocompatibility of the ion-irradiated polymers can further be increased by coating of irradiated polymers, which are expected to have higher adsorption capacity, with adhesive proteins or by grafting with amino-acid sequences representing ligands for cell integrin receptors. The adsorption of specific proteins (albumin, collagen and fibronectin) was studied on different synthetic polymers (PET, PU, PS, etc.) irradiated with ion beams as a function of the irradiation-induced compositional, structural and morphological changes in [56]. Ion irradiation was found to increase the surface free energy and to trigger the specific adsorption behavior of the proteins. The protein adsorption rate depends, however, in a complex way on the ion fluence. A similar study was accomplished on collagen-coated polystyrene irradiated with 150-keV He^+ ions to the fluences up to $10^{16}\,cm^{-2}$ [57]. The adsorption rate of the thrombogenic proteins, fibronectin and fibrinogen, on ion-beam-modified collagen was found to have a complex dependence on the ion fluence with a minimum at the fluence of $10^{14}\,cm^{-2}$. Enhanced adsorption was observed at fluences above $10^{15}\,cm^{-2}$. A similar technique was applied in [58] to small-diameter artificial vascular grafts made of PTFE with a collagen-coated inner surface. The ion irradiation ensures enhanced antithrombogenicity and excellent graft patency. In [59] collagen-coated PS irradiated with 50-keV He^+ and Ar^+ ions at fluences from 10^{13} to $10^{16}\,cm^{-2}$ exhibits enhanced attachment strength of bovine aorta endothelial cells at medium ion fluences. For high fluences the attachment strength suddenly decreases. The enhanced attachment strength is explained by immobilization of collagen on PS due to ion-beam mixing.

The modification of bioproperties of synthetic polymers and other organic materials by ion bombardement remains one of most challenging problems in view of the promising results obtained up to now and the possible application of modified materials and structures [56, 60]. Not only the need for biocompatible polymers for tissue-engineering but also potential applications as biosensors or biomicroelectronic devices stimulate an enormous interest in realization of hybrid systems based on controlled adhesion of cells or proteins on polymers. Up to now, however, the related experiments have been accomplished on a few but very different systems, and from the existing, rather scattered experimental data it is not easy to derive general conclusions on the relationships between physicochemical changes induced by the ion irradiation and observed change in biocompatibility.

It has been demonstrated that the cytocompatibility induced by the ion irradiation has a rather well-defined ion-fluence dependence, with an onset of massive cell adhesion in the interval from 5×10^{14} to $10^{15}\,cm^{-2}$. However, in some instances a much more complex dependence has been reported. At such fluences, the polymer structure is altered dramatically and a considerable carbonization occurs. In several studies the biocompatibility enhancement has been assigned to the formation of an amorphous carbon phase both for silicon-based and carbon-based materials. This opinion, at least for carbonaceous materials, is in accord with recent experimental results showing strongly enhanced cell adhesion on pyrolitic carbon [61] and the well-known fact that carbon-enriched or carbon-covered artificial materials are known to support adhesion of various cell types in vitro as well as in vivo. It is also supposed that the biocompatibility enhancement may be affected by the presence of radicals and new chemical structures (e.g., new oxidized units) on the surfaces of ion-irradiated polymers. These structures are also responsible for the observed transition from the hydrophobic (unirradiated polymer) to a hydrophilic (irradiated polymer) behavior. The biocompatibility enhancement is obviously connected with this transition [52].

Apart from the improvements in biocompatibility, ion irradiation of polymeric prosthesis material – e.g., ultrahigh molecular weight polyethylene (UHMWPE) – is also used to improve the wear resistantace, increase the wettability, and reduce the friction of joint cups with titanium balls [62].

A recently suggested application points at using MeV ion beams at elevated fluence for improved sterilization of biomedical polymers. Whereas the typical 25 kGy applied to polymers in conventional sterilization via UV or γ-ray irradiation leads to already substantial material degradation [63, 64], an equivalent dose with energetic ions does not yet lead to apparent polymer modifications. Furthermore, repeated sterilization processes with ions will prevent degrading chain-scissioning effects [25].

References

1. Lai JH, Polymers for Electronic Applications. CRC Press, Boca Raton, Florida, 1989
2. Guimarães RB, Amaral L, Behar M, Zawislak FC, Fink D, Range measurements and thermal stability study of AZ111 photoresist. J Appl Phys **63**, 2502–2506 (1988)
3. Fink D, Klett R, Behar M, Sánchez G, Kaschny JR, Hertlein WG, Changes in the photo-resist inhibitor distribution after ion irradiation and thermal treatment. Nucl Instrum Methods **B132**, 660–670 (1997)
4. Metha GK, Materials engineering with heavy ions. 21th Intl. Conf. on Nuclear Tracks in Solids, New Delhi, Oct. 21–25, 2002; presentation YS24/IT4
5. Ngô NG, Rosilio C, Lithography for semiconducting technology. Nucl Instrum Methods **B131**, 22–29 (1997)
6. Schlegel L, Schnabel W, Polymers in X-ray, electron beam and ion beam lithography, Radiat Curing Polym Sci Technol **1**, 119–192 (1993), and many references therein
7. Xu JM, Plastic electronics and future trends in microelectronics. Synth Met **115**, 1–3 (2000)
8. Springham SV, Osipowicz T, Sanchez JL, Gan LH, Watt F, Micromachining using deep ion beam lithography. Nucl Instrum Methods **B130**, 155–159 (1997)
9. Sanchez JL, van Kan JA, Osipowicz T, Springham SV, Watt F, A high resolution beam scanning system for deep ion beam lithography. Nucl Instrum Methods **B136/138**, 385–389 (1998)
10. van Kan JA, Sanchez JL, Xu B, Osipowicz T, Watt F, Micromachining using focussed high energy ion beams: Deep ion beam lithography. Nucl Instrum Methods **B148**, 1085–1089 (1999)
11. Vaahs T, Brück M, Böcker WDG, Polymer-derived silicon nitride and silicon carbonitride fibers. Adv Mater **4**, 224–226 (1992)
12. Ceramics from polymers; Polysilazane ET 70. Material's Description, Hoechst High Chem Series, Hoechst AG, Frankfurt/Main, Germany, 1990

13. Peuckert M, Vaahs T, Brück M, Ceramics from organometallic polymers. Adv Mater **2**, 398–404 (1990)
14. Hagiwara M, Polymers and track membranes. In: Yu.Ts. Oganessian, R. Kalpakchieva (eds.): Proceedings of the VIth Intl. School-Seminar on "Heavy Ion Physics", Dubna, Russia, Sept. 22–27, 1997, World Scientific Publ., Singapore, 1998, pp. 754–760
15. Julio CA, Higa OZ, Modificaçãoes de borracha de silicone pela tecnica de enxertia utilizando o metodo da pre-irradiacao em fonte de 60-Co, Anais do 4. Congresso Geral de Energia Nuclear, **V.1**, 331–334 (1992); available from: Biblioteca da Comissão Nacional de Energia Nuclear, Rio de Janeiro, Brazil (in Portuguese)
16. Fink D, Müller M, Nakao Y, Hirata K, Kobayashi Y, Behar M, Kaschny JR, Vacík J, Hnatowicz V, Ion-induced redistribution of palladium in polymethyl methacrylate. Nucl Instrum Methods **B166/167**, 610–614 (2000)
17. Faupel F, Biswas A, personal communication; to be published (2002)
18. Kawata S, Okamoto K, Shoji S, Photon-induced micro/nano fabrication, manipulation, and imaging with unconventional photo-active systems. Mol Cryst Liqu Cryst **314**, 173–178 (1998)
19. Baglin JEE, Ion beam enhancement of metal-insulator adhesion. Nucl Instrum Methods **B65**, 119–128 (1992)
20. Milinchuk VK, Klinsport ER, Tupikov VI, Foundations of radiation resistance of organic materials. Energoatomizdat Moscow, 1994
21. Das A, Bera S, Dhara S, Patnaik A, Physical and chemical implications of 100 keV H^+ implantation of laser ablated PPS thin films. Nucl Instrum Methods **B134**, 377–384 (1998)
22. Lee EH, Rao GR, Lewis MB, Mansur LK, Ion beam application for improved polymer surface properties. Nucl Instrum Methods **B74**, 326–330 (1993)
23. Iskanderova YA, Kleiman JI, Gudimenko Y, Tkachenko A, Tennyson RC, Brown IG, Monteiro OR, Metal ion implantation and dynamic ion mixing for the protection of high-performance polymers from severe oxidation environment. Nucl Instrum Methods **B148**, 1090–1096 (1999)
24. Srinivasan V, Banki BA, Materials degradation in low earth orbit. In: Warrendale, Pa (ed.): Joint Corrosion and Environmental Effects Committee, The Minerals, Metals, and Materials Society, ISBN:0-87339-152-7 (1990)
25. Lee EH, Rao GR, Mansur LK, Super-hard-surfaced polymers by high-energy ion-beam irradiation. Trends Polym Sci **4**, 229–237 (1996)
26. Tomasella E, Labed V, Klein D, Robé MC, Cetier Ph, Chambaudet A, Effect of UV and gamma irradiation on Rn222 permeation through PVC. Application to the packaging of radium sources for the purpose of storage. Nucl Instrum Methods **B105**, 285–290 (1995)
27. Townsend PD, Ion implanted waveguide lasers in insulators. Nucl Instrum Methods **B89**, 270–276 (1994)
28. Brunner S, Rück DM, Frank WFX, Linke F, Schösser A, Behringer U, Waveguide structures in polymers by ion implantation. Nucl Instrum Methods **B89**, 373–378 (1994)
29. Kaneko L, Ueno Y, Tanahashi S, An optical waveguide facet formation method for siloxanepolymer waveguide on ceramic substrate. Proc of the 7th Intl. "Plastic Optical Fibres" Conference, Berlin, Oct. 5/8, 1998, AKM AG Basel (1998)

30. Günther M, Gerlach G, Suchaneck G, Sahre K, Wolf KJ, Deineka A, Jastrabik L, Ion-beam induced chemical ans structural modification in polymers. Surf Coat Technol **158-159**, 108–113 (2002)
31. Odzhaev VB, Kozlov IP, Popok VN, Sviridov DV, Ion Implantation of Polymers, Belorussian State University, Minsk (1998)
32. Popok VN, Odzhaev VB, Kozlov IP, Azarko II, Karpovich IA, Sviridov DV, Ion beam effects in polymer films: structure evolution of the implanted layer. Nucl Instrum Methods **B129**, 60–64 (1997)
33. Popok VN, Odzhaev VB, Azarko II, Kozlov IP, Sviridov DV, Hnatowicz V, Vacík J, Červená J, Multistage ion implantation of polyimide-6 films. Nucl Instrum Methods **B166/167**, 660–663 (2000)
34. Zhukova SI, Ponaryadov VV, Popok VN, Prikhod'ko ZhL, Paramagnetic defects in multistage ion-implanted polyamide films. In: Anishchik VM (ed.): Proc 3rd Intl. Conf. on Interaction of radiation with solids. Minsk, Oct. 1999. Vol.1, 1999, pp. 142–144
35. Lin SH, Sheng KL, Rong TW, Bao JR, Wang WM, Wan HH, Zhou ZH, Zhu XF, Yang FJ, Electrical properties of chemical-doped and ion-implanted polyacetylene films. Nucl Instrum Methods **B59/60**, 1257–1262 (1991)
36. Wang WM, Wan HH, Rong TW, Bao JR, Lin SH, Diode characterstics and degradation mechanism of ion implanted polyacetylene films. Nucl Instrum Methods **B61**, 466–471 (1991)
37. Naddaf M, Chakane S, Jain S, Hullavarad SC, Bhoraskar SV, Mandah AB, Modification od sensing properties of metallophthalocyanine by ECR plasma, Nucl Instrum Methods, submitted in 2001
38. Asano T, Sasaguri D, Higa K, Ion beam modified photoresist – a new class of field emitter materials for large area devices. IEICE Trans Electron, vol. E81-C, 1998, pp. 1715–1720
39. Umeda N, Vasilets VN, Bandouko VV, Kishimoto N, In-situ photon emission spectroscopy and ex-situ surface analyses of polymers irradiated with 60 keV Cu^- ions. Nucl Instrum Methods **B191**, 708–713 (2002)
40. Hiwa C, Nakagawa T, Hirai M, Baba A, Development of fuctional composite using material using metallic fibers. J Soc Mater Sci Jpn **39**, 261–265 (1990), and: Baba A, Higa K, Asamo T, Jpn Appl Phys **38**, 261 (1999)
41. Khang G, Park JB, Lee HB, Prevention of platelet adhesion on the polysulfone porous catheter by saline solution perfusion. I. In vitro investigation. Bio-Med Mater Eng **6**, 47–66 (1996)
42. Richter A, Smith R, Ries R, Growth of biological films – microscopical investigations and computer simulations. Mater Sci Eng **C8-C9**, 451–462 (1999), and: Richter A, Smith R, Ries R, Wildenauer F, Features of biofilm growth on solid surfaces analyzed by Monte-Carlo simulations. phys stat sol **A176**, 953–967 (1999)
43. Xu G, Hibino Y, Awazu K, Tanihara M, Imanishi Y, Specific modification of polysulfone with cluster bombardment with assistance of Ar ion irradiation. Rev Sci Instrum **71**, 789–792 (2000)
44. Satriano C, Conte E, Marletta G, Ion beam-induced chemical modifications and enhanced cell adhesion on polysiloxane surfaces. Nucl Instrum Methods **B148**, 1079–1084 (1999)
45. Suzuki Y, Kusakabe M, Iwaki M, Suzuki M, Surface modification of silicone rubber by ion implantation. Nucl Instrum Methods **B32**, 120–124 (1988)

46. Bačáková L, Svorčík V, Rybka V, Míček I, Hnatowicz V, Lisá V, Kocourek F, Adhesion and proliferation of cultured human aortic smooth muscle cells on polystyrene implanted with N^+, F^+ and Ar^+ ions: Correlation with polymer surface polarity and carbonization. Biomaterials **17**, 1121–1126 (1996)
47. Baier RE, Meyer AE, Aspects of bioadhesion. In: Lee L-H (ed.): Fundamentals of Adhesion. Plenum Press New York and London, 1999, pp. 407–425
48. Suzuki Y, Kusakabe M, Iwaki M, Kusakabe K, Akiba H, Sato S, Effects of ion implantation on protein adsorption onto silicone rubber. Mater Res Soc Symp **110**, 669–679 (1989)
49. Suzuki Y, Kusakabe M, Kusakabe K, Akiba H, Iwaki M, In vivo evaluation of anti-thrombogenicity for ion implanted silicone rubber using indium-111-tropolene platelets. Nucl Instrum Methods **B59/60**, 698–704 (1991)
50. Suzuki Y, Kusakabe M, Iwaki M, suface analysis of antithrombogenic ion-implanted silicone rubber. Nucl Instrum Methods **B59/60**, 1300–1303 (1991)
51. Suzuki Y, Kusakabe M, Iwaki M, Akiba H, Kusakabe K, In vivo evaluation of anti-thrombogenicity and surface analysis of ion-implanted silicone rubber. Radiat Phys Chem **39**, 553–560 (1992)
52. Satriano C, Marletta G, Conte E, Cell adhesion on low-energy ion beam-irradiated polysiloxane surfaces. Nucl Instrum Methods **B148**, 1079–1084 (1999)
53. Švorčík V, Rybka V, Hnatowicz V, Bačáková L, Lisá V, Kocourek F, J Mater Chem **59**, 27–30 (1995)
54. Kim K-H, Cho J-S, Choi D-J, Koh S-K, Hydrophilic group formation and cell culturing on polystyrene petri-dish modified by ion-assisted reaction. Nucl Instrum Methods **B175–177**, 542–547 (2001)
55. Lhoest JB, Dewez JLK, Bertrand P, PMMA surface modification under keV and MeV ion bombardment in relation to mammalian cell adhesion. Nucl Instrum Methods **B105**, 322–327 (1995)
56. Kurotobi K, Kaibara M, Suzuki Y, Iwaki M, Nakajima H, Kaneko S, Plasma protein adsorption onto cell attachment controlled ion implanted collagen. Nucl Instrum Methods **B175–177**, 791–796 (2001)
57. Marletta G, Satriano C (2000), Irradiation-controlled adsorption and organisation of biomolecules on surfaces: from the nanometric to the mesocsopic level. Invited talk I–10. Presented at the 12th Int. Conference on Ion Beam Modification of Materials, Canela-Rio Grande do Sul, Brazil, Sept. 3–8, 2000, unpublished
58. Suzuki Y, Iwata H, Nakao A, Iwaki M, Kaibara M, Sasabe H, Kaneko S, Nakajima H, Kusakabe M, Ion implantation into collagen for the substrate of small diameter artificial grafts. Nucl Instrum Methods **B127/128**, 1019–1022 (1997)
59. Yokoyama Y, Tsukamoto T, Suzuki Y, Nakao A, Osada M, Kaibara M, Iwaki M. Presentation at the 12th Int. Conference on Ion Beam Modification of Materials, Canela-Rio Grande do Sul, Brazil, Sept. 3–8, 2000, unpublished
60. Kurotobi K, Kaibara M, Suzuki Y, Iwaki M, Nakajima H, Plasma protein adsorption rate onto all attachment controlled ion implanted collagen. Nucl Instrum Methods **B175-177**, 791–796 (2001)
61. Cenni E, Granchi D, Arciola CR, Ciapetti G, Savarino L, Stea S, Cavedagna D, Di Leo A, Pizzoferato A, Adhesive protein expression on endothelian cells after contact in vitro with polyethylene terephthalate coated with pyrolytic carbon. Biomaterials **16**, 1223–1227 (1995)

62. Rieu J, Pichat A, Rabbe LM, Rambert A, Chabrol C, Robelet M, Ion implantation effects on friction and wear of joint prosthesis materials. Biomaterials **12**, 139–143 (1991)
63. Xu XL, A new approach to microporous materials – application of ion beam technology to polyimide membranes. Mater Res Soc Symp Proc **354**, 351–356 (1995)
64. Lee EH, Rao GR, Mansur LK. In: Surfaces in Biomaterials Symposium. Surfaces in Biomaterials Foundation, P.O.Box 2611, Minneapolis, MN 55426-0111, USA,
and: Lee KL, Baker RW, Lonsdale HK, Membranes for power generation by pressure-retarded osmosis, J Membr Sci **8**, 141–171 (1999)

6 Ion-Track Manipulations

D. Fink

There appears to be great potential for applications of ion tracks in polymers far beyond what has been realized already up to now. To expand the use of tracks one has, however, to modify the tracks, which can be performed by steps such as doping, etching, annealing, adsorption or chemisorption, galvanic or electrodeless material deposition, γ, electron, or low-energy ion irradiation, lithographic or evaporation processes, and others. As in the previous chapters, we also have to distinguish here between manipulations of as-implanted (latent) and etched tracks.

6.1 Manipulation of Latent Tracks

At least four basic strategies show up for manipulations of latent tracks, three for exploiting latent tracks themselves as objects for practical applications, and one for properly preparing latent tracks for etching. To the first group belong the material's chemical conversion (e.g., from polysilanes to silicon carbide) and phase transitions (e.g., the conversion of carbonaceous materials such as diamond and fullerite to sp2 enriched conducting carbon in the case of non-polymeric materials), the trapping of intentionally introduced impurities at latent tracks, and the use of ion-induced excess free volume [1]. Two of them require additional manipulation processes. The other group includes track sensitization and fading of latent tracks as tools to produce well-defined etched-track patterns.

6.1.1 Material Chemical Conversion and Phase Transitions

It was reported above that swift heavy-ion-irradiated polysilanes are transformed to a silicon-carbide-like matter, roughly characterized by SiC_xH_y, with $x \approx 1$, and y being as yet unknown [2]. As this residue often turns out to be quite flexible (see Fig. 4.29) it appears to be useless for applications. However, it has been found that upon annealing for a prolonged time at elevated temperatures, the radiochemical product apparently crystallizes and gains mechanical stability, so that it peaks out pronouncedly from the molten unirradiated environment, see Fig. 8.16 in Vol. I, Sect. 8.4.3). Such nanostructures might be useful for micromechanical and microelectronic applications.

6.1.2 Trapping of Migrating Impurities at Latent Tracks

In Chap. 5 of this volume it was mentioned that one can let isolated mobile Pd atoms in a polymer (being produced in situ in PMMA by the decay of a suitable organometallic compound embedded in the polymer matrix) precipitate at damage centers that had been created previously by low-energy Ar^+ irradiation [3]. In a similar way, it is possible to use swift heavy-ion tracks, which are zones of linearly extended polymer damage and therefore act as nucleation centers for the precipitation of diffusing metal or semiconductor atoms, or of (semi-)conducting clusters embedded earlier in a polymer matrix. In this way, it should be possible to produce (semi-)conducting nanowires within polymers for electronic applications. Such experiments are presently underway [4].

6.1.3 Use of Ion-Induced Excess Free Volume

Swift heavy ions produce excess free volume as a result of chain scissioning and release of volatile reaction products. This free volume can be filled with materials of interest, e.g., with suitable electrolytes. In this way, conducting liquid nanowires can be formed that might have some future application potential in medical electronics.

Apart from this concept, one might think of galvanic or chemical deposition of, e.g., metals along latent ion tracks, in a similar way to what has been done already in etched tracks. The main precondition is to select a polymer with sufficient free volume along the tracks. Such an approach has not yet been tried. Finally, grafting of suitable materials along latent tracks is a way to produce nanowires with new properties, for whatever applications.

6.1.4 Manipulation of Latent Tracks for Production of Etched-Track Patterns

It has been articulated for the first time by Dehaye and coworkers[1] that, as a preparatory step towards patterned polymeric etch-track devices, swift heavy-ion-irradiated but still unetched polymer foils could be first globally faded by a thermal treatment, then sensitized in the desired regions by, e.g., UV light through a suitable patterned mask, and then etched. In this way, an etched track pattern shows up only in the regions of interest, and nowhere else.

[1] Dehaye F, personal communication (2002)

6.2 Manipulation of Etched Tracks

Etched tracks can be filled, in principle, with any material to obtain any new nanostructure. The tracks can either be filled completely or partly. In the first case, one obtains cylindrical rods (wires, fibers) of the embedded matter. Partial filling can either signify that a rod is not embedded along the whole track length, or that a tubule is embedded, so that the central track zone is still free of matter. The filling of the tracks can either be accomplished from the gaseous or from the liquid phase. In the first case a solid residue can be deposited from a gas decomposing within the tracks (chemical vapor deposition), or one can evaporate matter onto the inner track walls, or one can use plasma deposition processes.

In the second case, the penetrating liquid can be subsequently (fully or partly) solidified, or solid matter can precipitate from that liquid. This precipitation inside the etch tracks can be initiated chemically or galvanically. When one wants to create tubules of the penetrant, the inner etch-track walls have to be wetting to enable good adsorption. This adsorption can be improved considerably by additional physical or chemical nucleation centers to initiate bonding reactions (chemisorption). The wetting itself can be improved by suitable agents, e.g., by the use of octadecyl tricholorosilane.

Before working with very narrow etched tracks, one should verify whether these objects really still exhibit open space after their preparation. Capillary contraction is a well-known effect of "healing" of narrow pores during drying of microporous objects impregnated with a well-wetting liquid in the usual conditions. The drying process reduces the etch-track radius r by about 5 nm, for whatever r. This means that pores below 5 to 7 nm radius virtually vanish during drying, and a second etching of a dried microporous sample virtually begins from zero track radius. This track compression is not a surface but a bulk effect [5]. Hence, in order to avoid this effect, one should not dry the tracks before their manipulation starts, but store them, e.g., in distilled water.

6.2.1 Formation of Track Templates from the Vapor Phase

Tubule Formation by Evaporation

Usually an evaporation source emits atoms or clusters isotropically within a given solid angle. Particles originating from the evaporation source and being emitted from a point along the track axis with trajectories within a solid angle Ω_{trans} will be transmitted through the tracks, particles emitted outside the solid angle Ω_{dep} will be deposited on the microporous foil surface, and particles within Ω_{dep} and Ω_{trans} will be deposited on the inner track walls, thus forming tubules (Fig. 6.1a). As the mathematical formulation of this problem is exactly complementary to that of ion transmission spectrometry

(ITS), including the emission from nonaxial points of origin (Sect. 3.3), we refer to the equations derived there.

The tubule layer thickness is determined by the cosine of the angle of incidence ζ between the atomic/molecular beam and the surface normal of the inner track wall. As for narrow cylindrical tracks the maximum angle $\zeta_{\max}$ of incidence that is reached for Ω_{dep}) is usually very small (e.g., $\sim 3 \times 10^{-4\,\circ}$ in the case of a track with a 1 μm diameter in a 10-μm thick foil at a distance of 10 cm from the evaporation source), only negligible tubule thicknesses can be obtained in this case. In the example given, a tubule wall thickness of only ∼ 1 nm would be obtained, while at the same time already ∼ 3 μm would have been deposited on the foil surface. Therefore, this evaporation geometry has not gained any importance for cylindrical tracks.

The situation improves when evaporating at tilted angles, Fig. 6.1b. For example, evaporation at 45° enables one to obtain tubule thicknesses equal to the layer thickness on the foil surface, however, only down to depths corresponding to the track diameter, and only in one direction. This approach is useful for producing electrical contacts of, e.g., sensor material embedded within the etched tracks. In order to obtain circular contacts, either the evaporation source or the sample have to be rotated against each other during evaporation.

The case of evaporation in the axial direction onto conical and hyperboloidal tracks, Fig. 6.1c, is similar to that of the tilted angle geometry of Fig. 6.1b. For example, in the case of tracks with opening angles of 5° and for perpendicular evaporation geometry, tubules of 100 nm wall thickness can be grown within the tracks when simultaneously layers of ∼ 1.2 μm have been deposited on the foil surfaces.

The eventual disadvantage of evaporating matter into etched tracks due to angular obstacles is compensated by the great advantage that nearly any material (such as B, Al, Si, Cr, GaAs, HTCs etc.) can be deposited in this way – in contrast to the restrictions of galvanic or chemical deposition. Also, evaporation can be considered as being a more contamination-free process than liquid deposition techniques.

Please keep in mind, however, that metal deposition from the vapor phase onto polymers is usually accompanied by some metal diffusion into the plastics with subsequent metal precipitation [6], i.e., one might expect the buildup of a "halo" of metallic nanoparticles within the polymeric environment of the formed tubule. Also be aware that deposition of matter by evaporation lead to deposition of heat on the sample. In the case of excessive evaporation speed this may lead to intolerable target heating especially for sensitive polymers such as PET or PC, eventually resulting in track deformations. Therefore PI foils are recommended here as track carrier materials.

If the track backside is covered by, e.g., a foil or a Si wafer, the transmitted particles will be deposited thereupon, thus forming nanodots, with their diameters equal to the track dimensions. This may be useful for quantum electronic applications.

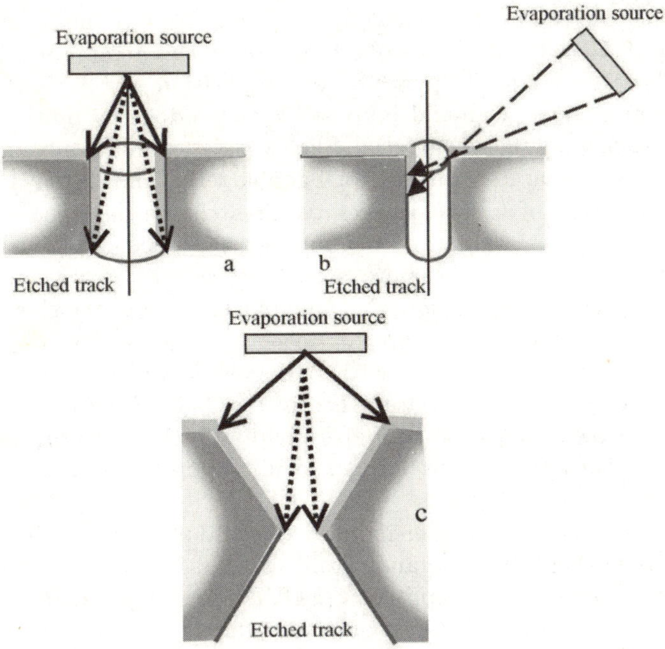

Fig. 6.1. Illustration of tubule production by evaporation. (**a**) Evaporation point source aligned with cylindrical tracks; (**b**) tilted angle geometry. In (*a*) the *dashed arrows* correspond to Ω_{trans} and the solid arrows correspond to Ω_{dep}. The *arrows* in (*b*) illustrate up to what depth layers can be deposited within etched tracks. (**c**) In the case of conical tracks with sufficiently large opening angle – leading to large angles ζ of impact of the atomic/molecular beams – thick layers can be evaporated up to great depths

Tubule Formation by Chemical Deposition from the Gas Phase

Chemical-deposition reactions on inner track walls can be initiated from the gas phase. Two concepts arise here. Either the reaction products are deposited onto the walls of conical or hyperboloidal tracks similarly as in the case of evaporation, or the microporous membrane is placed in a high-temperature furnace and an organic gas such as ethene or propene is passed through the membrane. Thermal decomposition of the gas occurs in the latter case throughout the pores, resulting in carbon tubules. This process is usually limited to those porous membranes that withstand the high temperatures required (around 700 to 900°C), which signifies that no organic polymer but only inorganic (i.e., Al_2O_3 [7], mica, SiO_2, or similar) membranes can be used for this purpose. However Fe- or Ni-catalyzed chemical vapor deposition (CVD) enables one to reduce the reactor temperature to $\geq 545°C$, using ethylene or pyrene as the gas to be decomposed to carbon within the pores, and annealing the carbon-containing membrane subsequently at 500° for 36 h to enable the conversion of the carbon nanostructures into highly ordered and conducting (0.1 to 3 S cm^{-1}) graphite [8]. As at this temperature PI membranes suffer only moderate pyrolytic decomposition and etched tracks still preserve their shapes, it should be possible to produce graphitic nanotubes also within such a polymeric material. Such tubules could then be intercalated with Li to enable charge transport for nanobatteries and displays.

The application of CVD to pores requires that a gas such as ethylene or propylene passes through the membrane that is heated to a high temperature in a furnace. Furthermore, low deposition rates are important, as otherwise the surfaces of the pores become blocked before the chemical vapor can traverse the length of the pore.

If a tiny catalyst particle (e.g., Ni) is deposited within a nanopore, then buckytube growth can be initiated therein by CVD. Suppose such catalyst particles are grown on the bottom of individual etched tracks, then regular patterns of buckytubes can be formed in microporous foils, e.g., for use as field-emitting displays.

Similarly as CVD, also laser-induced deposition of metal films (such as Ni, Pd, Pt and others) from organometallic precursors can be applied [9], and also plasma chemical vapor deposition (PCVD), for the production of tubules of, e.g., indium tin oxide (ITO) or silicon oxynitride should, in principle, be feasible.

6.2.2 Formation of Massive Track Templates (Rods, Wires, Fibers) from the Liquid Phase

Capillaric Filling by Wetting Penetrants

A basic precondition of the capillaric penetration of liquids into etched tracks is the wetting of the inner track walls by the penetrant (or its precursor) solution to enable good adsorption. This adsorption can be improved by additional chemical-bonding reactions. In order to obtain the best adsorption, the acidic-basic correlation of the three components – material, solvent, polymer – of the given system must be optimized.

The acidity of a solid's surface corresponds to the ability of the surface to convert an adsorbed base into its conjugate acid. Accordingly, the basicity of a solid surface corresponds to its ability to convert an acid into its conjugate base. As the deposition of matter onto the polymeric surfaces occurs in the given cases through a solvent, the latter may compete for access to basic and acidic sites on the polymeric surface with the material to be deposited. If, e.g., PMMA is cast from THF solvent (which is more basic than PMMA) onto SiO_2, the solvent molecules are more readily absorbed onto the surface SiOH groups than PMMA, resulting in poor PMMA adhesion. On the other hand, a more acidic solvent tends to bond strongly to the molecules of the basic polymer so that their bonding to the less-basic solid will be poor.

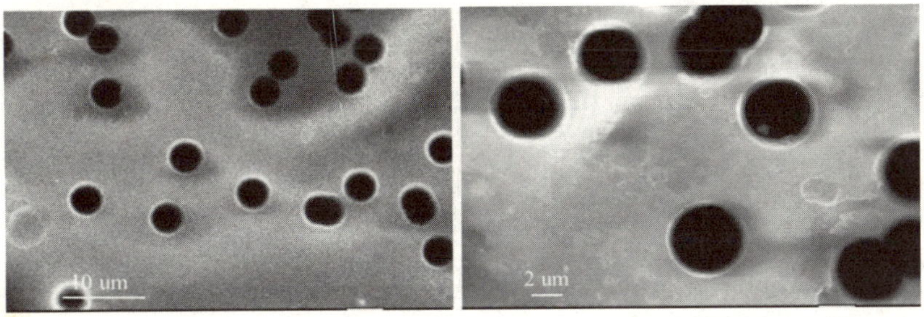

a b

Fig. 6.2. Etched ion tracks in PET with embedded tubules of (**a**) PMMA, (**b**) PAN [13]

Therefore, usually the best results are achieved with neutral solvents. If water is taken, it interacts acidic in contact with basic surfaces (e.g., MgO), basic in contact with acidic surfaces (e.g., SiO_2), or mixed, basic and acidic in contact with neutral surfaces such as Fe_2O_3 and Al_2O_3. Therefore, from the corresponding bond formation energies it follows that the best durability for adhesive polymeric bonds under water attack is given for the following polymeric functional groups: carboxylic acids or amines for neutral deposited matter, amines for acidic matter, and carboxylic acids and phenols with basic matter [10]. What has been derived in [10] for solids such as SiO_2, Al_2O_3, and others holds, of course, for polymers, too.

If an ion track is wetting for a given liquid penetrant, the latter will penetrate by capillaric forces. In this way, e.g., a supersaturated KCl solution was allowed to enter etched tracks, in which subsequently massive or hollow KCl crystals precipitated [11]. Other examples of this process are tubules of fullerene (Fig. 7.3); [12], of PMMA (Fig. 6.1a), of PAN (Fig. 6.1b), or of conducting polymers such as PEOT. In these cases, the microporous polymer foils are inserted into the C_{60}-, PMMA-, or PEOT-saturated solutions, and then dried. For the C_{60} tubules shown in Sect. 7.2.6, ITS revealed that the tubules had diameters of 0.2 to 4 µm and wall thicknesses of up to 200 nm, whereas the C_{60} layer deposited at this time on the surface had a thickness of about 250 nm.

The first attempts have also been made to produce structures of photoresist inside etched tracks. Irradiation of such structures within radiation-insensitive polymeric matrices with visible or UV light, electrons, or low-energy ions should make part of the tubules soluble (or insoluble, respectively, depending on wther the resist is positive or negative) in suitable developer solutions, so that one could extend known lithographic techniques to 3D track manipulations. Until now, only massive wires of photoresist could be obtained within the tracks, as the attempt to produce photoresist tubules failed due to too strong capillaric forces. Only if air is blown through the tracks with a pressure Δp_{blow} exceeding the capillaric pressure $\Delta p_{\mathrm{capill}} = 2\sigma \cos(\Theta)/r_{\mathrm{pore}}$, can one expect the opening of the photoresist liquid column adsorbed to the track. For $\Delta p_{\mathrm{blow}} \approx 1\,\mathrm{atm}$, this is given for pores with $\sim 50\,\mu\mathrm{m}$ diameter or more.

Pressure Injection of Nonwetting Penetrants

Similarly, in the case of nonwetting penetrants one has to apply high pressure to force the liquid into the tracks. This so-called pressure injection is a technique that has been applied successfully for some time to insert nonwetting liquids into porous materials. Specifically, this method has been applied for low-melting metals (such as Sn, SnGe, Bi, and (doped and undoped) Bi_2Te_2 [14]) into porous ceramic Al_2O_3 foils. The latter products served for application in thermoelectric devices. In spite of the success of this technique it has been occasionally reported that only a minor fraction of the pores (typically $\sim 10\%$) was actually filled. Treatment of the substrate with sulfuric acid improved its wetting, and hence the metal's permeation probability. This technique has not yet been applied to polymeric microporous foils, but should be, in principle, easily possible at least for microporous PI foils that withstand $\sim 500°\mathrm{C}$.

Tubule Formation by Chemical Deposition Reactions

Gases and liquids can be incorporated in etched ion tracks to initiate therein chemical reactions for practical applications. In this way, e.g., templates can be produced. The correponding reactions can be chemical deposition (electrodeless deposition, ELD) processes, electrochemical reactions, polymerization reactions, or just any other chemical reaction. Chemical deposition comprises deposition from the vapor phase (chemical vapor deposition, CVD, see above) and deposition from the liquid phase. The latter is subdivided into electrodeless deposition (ELD) and chemical bath deposition (CBD), which is also denoted as chemical solution deposition (CSD). In order to form continuous layers of precipitates on the track walls via reduction of the penetrant, it is additionally necessary to provide a sufficient areal density of nucleation centers.

Performing chemical reactions in the confined volume of a micro- or a nanopore such that an etched track signifies that the pores are transformed into temporary micro- or nanoreactors. Several such reactions have already been performed on different occasions, such as:

1. $LiNO_3 + Mn(NO_3)_3 + T \rightarrow LiMn_2O_4$ [15], [2]

2a. $AgNO_3 + NaBr \rightarrow AgBr\downarrow + NaNO_3$, [3,4]

[2] Asmus T, Fink D (2000), unpublished work

[3] Esser M (1999), personal communication

[4] Fink D, Vacík J, Červená J, Hnatowicz V, Klett R (1996), unpublished

2b. $AgBr + h\nu \rightarrow Ag + Br\uparrow$ [3,4]

3. $LiJ + KF \rightarrow LiF\downarrow + KJ$ [4]

4. Organometal Ni + T $\rightarrow$ Ni [8]

5. $Fe(CO_5) - PVF_2 + T \rightarrow \gamma\text{-}Fe_2O_3$ [16]

($h\nu$ = light, T = heat treatment at 400 to 500°C). The reactions took place either in etched tracks in PC or PET, or inside natural self-aligned pores of Al_2O_3 or of PPy. In the first case, the objective was to form 200-nm thick nanotubules of crystalline $LiMn_2O_4$ spinel as cathodes in rechargable Li batteries. The second case served as an attempt to form a pressure-sensitive sensor by depositing dispersed metal particles in a track, and the third case served as a test for precipitation and leaching reactions under confinement. (It turned out that the emerging LiF depth distributions and desorption kinetics are highly complex and partly not yet understood.[4])

Of special interest are also chemical reactions such as those of types (4) and (5) where organometallic compounds decay by heat aplication to release metal or metal oxide films [17, 18]. For example, the aforementioned decomposition of the Fe-carbonyl compound $Fe(CO_5)$-PVF_2 was used to produce nanorods of γ-Fe_2O_3 of about 100 nm diameter in PMMA and PC [16]. Also, e.g., SnO alcohol sensors [18] could, in principle, be produced within tracks by decomposition of organometallic precursors.

Two points have to be observed when applying organometals to polymeric ion tracks. One is the question of inserting the material into the tracks. To accomplish this, the organometal should either be liquid or gaseous, or it should be soluble in some liquid that does not attack the polymer. The other point of importance is the decomposition temperature of the applied organometal that must not exceed the range of stability of the host polymer used. This means, in the case of PC and PET the decomposition temperature should not exceed ∼100 to 150°C – a criterion that is fulfilled for only a few organometals, and in the case of PI hosts the decomposition temperature should not exceed ∼ 500°C to prevent the polymer's pyrolysis. This means that the production of, e.g., tin oxide tubules from organometallic precursors in tracks should be feasible, but the production of, e.g., HTC wires or tubules [17] within polymeric tracks from organometallic precursors (that requires much higher decomposition temperatures) appears at present to be impossible.

Be aware that the emerging nanometric objects may be highly reactive and eventually even tend to interact with the host polymer. For example, the above-mentioned reaction (4) has been used as a catalyst to reduce the temperature of CVD deposition of carbon from organic vapor [8], and γ-Fe_2O_3 nanostructures have been found to initiate chain scissions in a PC host matrix.

Electrochemical Deposition of Matter along Etched Tracks

Under the influence of a potential applied to an electrode immersed in an electrolyte, ions in the vicinity of that electrode surface are rearranged resulting in the formation of an electrical double layer, the so-called Helmholtz layer, which is further joined by a diffusion layer. Cations migrate through these layers towards the cathode that is covered by the microporous membrane. In this way, the galvanic deposition of conducting or semiconducting matter (metals, conducting polymers, semiconductors) inside insulating ion tracks or other porous matter leads to the formation of conducting rods (wires, fibrils). The formation of massive templates (e.g., of conducting polymers) as thin as 3 nm in diameter has been reported [19]. In this way, nanofibrils of PPy, PMT (poly(3-methylthiophene), PAni, Cu, Ag, Au, Ni, Pt, [7], Fe [20], Co [21], and FeCo [22] have been produced – in the latter case with oxygen-free nitrogen as protective gas.

The galvanic process is complicated as it depends on many, often poorly controllable parameters such as the temperature, pH, concentration, agitation rate, purity of solvent and solute, nature of additives, magnitude and distribution of the electrical current, gas formation on the electrodes, etc. The solutions must be tailored so that they have good microthrowing power, i.e., that also fine details of the templates can be replicated well. The electrodeposition of noble metals such as Au or Ag on base metals like Cu needs much care as the adhesion of the plated metal is found to be poor. In general, the metallic nanowires are polycrystalline, with the grains generally extending along the full width of the wire and the grain boundaries usually aligned perpendicularly to the growth direction at regular spacing, typically in the range of 5 to 50 nm or more [21]. If, however, all parameters are optimized, even single-crystalline rods can be produced within etched tracks [11].

It was reported that only 10% of all metallic wires that are produced along etched tracks by galvanic metal deposition can be contacted properly, though the tracks are 95% filled with these nanowires. The reason as found by SEM is that quite a number of tracks are not filled with these metal wires throughout up to the polymer foil surface. Therefore Hjort [23] developed a technique of partial ashing of polymeric surfaces. In this way, all metallic nanowires were exposed and could be contacted. An alternative possibility is to apply chemical deposition to deposit a continuous metal film onto both track walls and polymer foil surface. In the case of necessity, the film can be subsequently thickened by prolonged chemical or galvanic deposition.

The structure of metallic deposits is usually columnar or fibrous along the growth direction. If their reduction potentials are sufficiently far apart from each other, one can electrodeposit separately several atomic species from one bath within an etched track. In this way, Cu/Co multilayer wires were produced [24].

Galvanic techniques also allow one to produce porous metal membranes as the direct replica of polymeric microporous membranes [25]. In this case

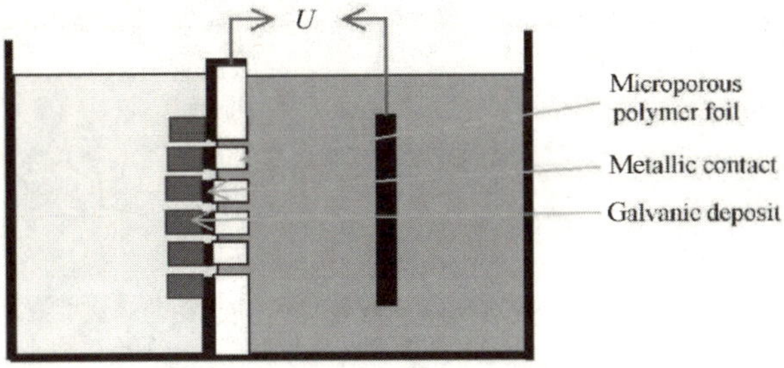

Fig. 6.3. Principle sketch of production of metallic microporous foils, according to [25]

the electroplating electrolyte must be contacted through the pores of the template with a solvent or solution, which increases the concentration of chemical polarization in the vicinity of the pore orifices. Thus, deposition of metal over the pores is prevented, and the lateral growth of the deposit is significantly reduced. The shape and size of the pores depend on the nature and concentration of the counter-electrolyte and on its hydrostatic pressure onto the plastic membrane since this pressure determines the flow rate through the template pores, Fig. 6.3. The pores in such a membrane have the same appearance as those in the microporous carrier polymers foils.

Tubule Formation by Polymerization Reactions

The deposition of conductive polymers from a liquid monomer solution onto etch-track walls can be realized by either oxidative electropolymerization or electrodeless oxidative polymerization. In the case of a suitable surface energy, the newly formed polymer deposits on the etch-track walls and thus forms tubules. Up to now, tubules of PEDT,[5] PEOT,[5] PPy [26], PAN [26],PPO (polyphenyl oxide) [27, 28], and PMT (poly(3-methylthiophene)) [26] have been formed. In the case of electropolymerization, the deposition process can be controlled by cyclic voltammetry (e.g., in the case of PPO deposition between -0.2 and $+0.8$ V [27, 28]).

It has frequently been found that the etched tracks are filled irregularly. Whereas some tracks remain empty, some are filled completely, and in others tubules develop [38]. Furthermore, polypyrrole nanotubules have been found to be quite irregular in their thickness [7].

The key to the conducting polymer tubule formation is the adsorption of the nascent polyheterocyclic chain to the etch-track wall by electrostatic attraction. Heterocyclic polymers are synthetisized via oxidation of the monomer and subsequent coupling of the radial cations, thus a cationic polymer is obtained that bonds well to the anionic polymer track walls. This

[5] Asmus T, Fink D (1999/2000), unpublished work

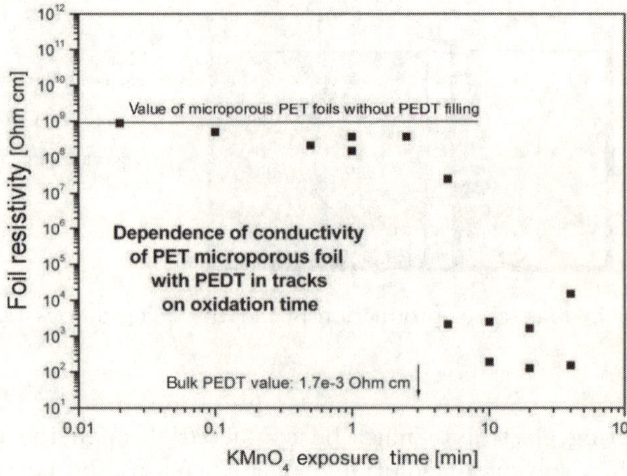

Fig. 6.4. Dependence of the resistivity of microporous PET foils filled with PEDT nanotubules on the degree of oxidation of the PET host polymer [5]

was found to be important for the production of PPy tubules, and also for the deposition of PEDT on PET track walls the basic precondition is the existence of an oxidant layer (e.g., MnO_2) that triggers the polymerization process. An improved wetting is a positive side effect. This is done by, e.g., boiling the sample in hot (90°C) concentrated (5 mole/l) $KMnO_4$ solution before the deposition of the EDT monomer. Figure 6.4 illustrates how crucially the EDT polymerization – hence the PEDT tubule formation, and consequently the overall sample conductivity – depend on the MnO_2 deposition time.[5] MnO_2-triggered oxidizing reactions have also been useful for initializing pyrrole polymerization to promote the deposition of PPy polymers via the formation of a $LiMn_2O_4$ interface layer. The tubule thickness is controlled by the polymerization time.

In the case of conducting etch-track walls, PPy can be electrochemically deposited from an aqueous methanol solution of $NaIO_4$ as supporting electrolyte. The IO_4^- is incorporated as dopant for PPy [28]. For PPy, the continued polymerization process inside the tracks ultimately leads to a "closing-up" of the tubules to form solid fibrils [26]. This is not observed for PAni [26] and PEDT.

The polymer nucleation and growth on the etch track walls are determined by two counteracting properties of the conductive polymers: (1) an electrostatic component because the polymers are cationic and there are anionic sites on the polymer walls [29], and (2) a solvophobic component to the interaction between polymer and wall, as the polycation forms of these polymers are completely insoluble, though the monomers are soluble. This illustrates the importance of fixing the material on the track walls by activation centers (also denoted as "molecular anchors" by Martin [26]).

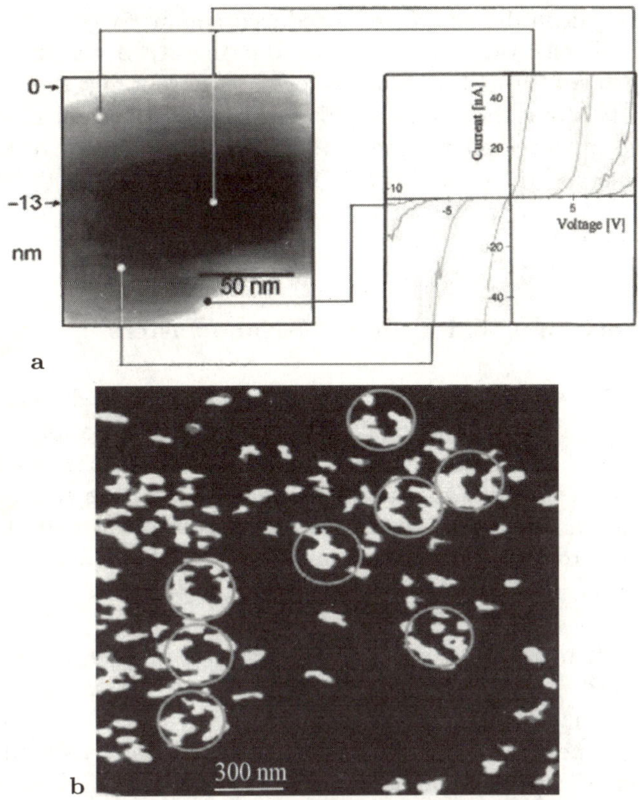

Fig. 6.5. (**a**) *left:* AFM image of a PEDT tubule embedded in a track in PET. The tubule center shows up as a black hole, (**a**) *right:* Current–voltage curves as measured at different positions around the tubule. Some quantum oscillations show up in the I–U diagram. (**b**) Distribution of breakthrough events along PEDT tracks, as measured by AFM with a conducting tip and high field strength applied. Smaller breakthrough events are filtered away electronically so that only the major zones of conductivity are left. The tracks (marked by the *circles*) can be identified by depressions in the corresponding AFM images (not shown here)

Also electrically insulating tubules (conductivity less than $10^{-9}\,\mathrm{S\,cm^{-1}}$ [28]) can be synthetized within etched tracks. For example, polyacrylonitrile (PAN) tubules can be prepared by immersing a porous membrane foil into a solution containing acrylonitrile and a polymerization initiator [7,30], see Fig. 6.1b. These PAN tubules, if deposited, e.g., within Al_2O_3 porous membranes, can be transformed by pyrolysis at 700°C to graphitic carbon tubules [7,30] with a conductivity of $10^{-6}\,\mathrm{S\,cm^{-1}}$ [28].

Figure 6.5a (right) shows some current–voltage curves as measured at different positions around a PEDT-filled tubule, which is displayed in Fig. 6.5a, left) by AFM. In Fig. 6.5b, the track shows up as circular conducting zones, within which a deep depression with reduced conductivity is found. By applying a strong electric field at the conducting AFM tip that is so high that electric breakthrough discharges occur, one obtains the distributions of conductivity along the tubules. This measurement reconfirms that in the present stage of development the obtained homogeneity of polymeric conducting tubules is still poor. In fact, various experimentalists have already complained that their tracks are only inhomogeneously filled, or occasionally not filled at all by the conducting polymers.

Formation of Wires and Tubules of Nanoclusters

Small clusters of solid matter dissolved in liquids, so-called colloids, possess a number of peculiarities that make them attractive for applications. With decreasing size of, e.g., metallic clusters, their electronic bandgap is widened, until term schemes with well-separated levels emerge, the energies of which increase. This leads to a change of these clusters from the metallic via semiconducting towards an oligomeric nonmetallic state. Such nanoparticles show a blue shift of the light absorption with decreasing particle size as the bandgap is widened.

These small metal clusters (e.g., of Ag) are, on the one hand, highly sensitive towards oxygen, and, on the other hand, act as efficient electron acceptors from nucleophilic substances. The latter causes the Fermi level to shift to a more negative electrochemical potential. Radicals readily transfer their electrons to metallic clusters in solution so that the latter can be used for charge storage. For example, a 10^{-3} molar solution of 6-nm Ag colloids has an electrical capacity of about 10 Farad per liter. The electrons stored on such metal particles can be used for reduction of other metal ions, or as catalysts for unusual free-radical reactions. They can reduce ions of a second metal in contact with them [31]. The detailed mechanism of charge storage in metallic or semiconductor nanocrystals is still unknown. Also, electron emission into the ambient upon absorption of light has been observed [31].

Colloids can be produced in various ways. Apart from the classical chemical approach of sol (= suspensions of colloidal particles) production there is the possibility to produce small clusters of, e.g., metals by sublimation into Ar or He atmospheres, and subsequent trapping of the formed nanoclusters in cooled organic solvents. To adjust the size distribution of these clusters, and to maintain them in solution, the coagulation must be counteracted by dispersion processes. The coagulation is governed by a combination of van der Waals attraction and Coulombic repulsion forces among charged particles. The Schultze–Hardy rule states that an increment of ion concentration and/or the addition of high-valence ions into a well-dispersed solution induces rapid coagulation. The coagulation process can be accelerated strongly by irradiation of light. The dispersibility is a function of metal species and the type of solvent. Gold is one of the metals with the most positive electron

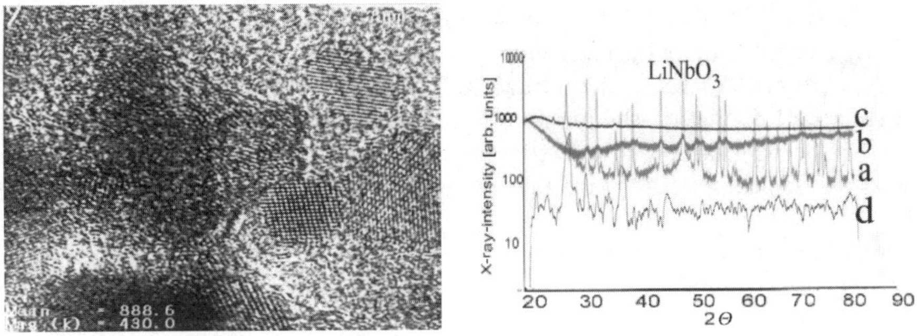

a b

Fig. 6.6. (**a**) TEM image, and (**b**) XRD spectra of of ~ 20-nm large $LiNbO_3$ colloid particles. The crystalline nature of these colloids is clearly seen. The XRD spectra show signals from *a* micrometer-sized commercial $LiNbO_3$ powder; *b* dry nanoparticles after ball milling, and **c** $LiNbO_3$ colloids (= nanoparticles clad with amylacetate). For the sake of clarity, *d* shows the difference of spectrum c and pure amylacetate. Thus even the tiny nanocrystal signals become clearly visible, indicating the crystalline nature of these colloidal particles

affinity and thus disperses easily, with alcohols being the best solvents for this system. Such solutions can remain stable over years [32].

Chemically produced colloids are usually amorphous. If, however, they are formed by grinding of crystalline precursor material and subsequent dissolution in a liquid by attaching appropriate organic ligands, they maintain crystallinity, however, with some distortion, Fig. 6.6a,b.

Colloidal clusters can be modified in their properties by chemical surface modification. For example, chemisorption of OH^- and excess Cd^{2+} can block surface states of CdS colloids so that they change their fluorescence properties. Binding of thiol onto CdS leads to structures with an inorganic semiconductor core and an organic outer shell. Thiol-capped Au colloids, electrophoretically deposited onto planar surfaces form oriented two-dimensional structures (e.g., of hexagonal symmetry in the case of Au colloids) [31].

All the above-mentioned properties make colloids also attractive for incorporation within etched tracks, as numerous possible applications show up in this way. The first experiments of this kind have already been performed; e.g., TiO_2, ZnO, WO_3 [7], SiO_2 and,[6] $LiNbO_3$ sols have been filled into polymeric and ceramic porous membranes, and then transformed to gel (aggregated sol particles) on the pore walls. As an example, Fig. 3.18 shows different stages of penetration of a $LiNbO_3$ colloidal solution into ion tracks. Longer immersion times of the tracks in the gel may lead to their complete filling, i.e., to fibril formation.

[6] Chemseddine A, Fink, D (1998), unpublished work

Annealing at about 450°C for 1 h of gel-filled membranes transforms the gels to sintered, or eventually even to single cystalline (e.g., in the case of TiO_2) ceramics, or to semiconducting tubules [7]. For example, for SiO_2, the sol-gel reaction of, e.g., the precursor sol $RSi_n(OR)_{4-n}$ (with R being, e.g., an alkyl) proceeds via hydrolysis towards the formation of Si–OH groups, and the latter are then being transformed to Si–O bonds at elevated temperature. It is interesting to note that a self-assembly takes place during this process, which allows the parallel chains to arrange alternatingly according to [33]:

–Si O–
–O Si–
–Si O–
–O Si– .

A precondition for the absorption of the sol particles on the negative pore wall is that they are positively charged. The gelation rate is higher in the pores than in bulk solution, due to the enhancement in the chemical concentration of the sol particles due to absorption on the pore walls.

Summarizing, the first studies on the incorporation of colloids in etched tracks have revealed that there exists a great potential for applications due to their promising properties.

6.2.3 Tubule Formation by Chemical Bath and Electrodeless Deposition Processes

Besides the chemical bath deposition (CBD) process, the electrodeless deposition (ELD) technique is widely used for metal deposition in the sense that it is a chemical process that does not involve electron exchange with a conducting substrate as in the electrodeposition technique, and therefore ELD does not require conducting substrates. ELD is an oxydo-reduction process, the presence of the reducing element in the film being a side effect. In CBD there is no change in the oxidation state of the elements in solution and in the film. They are deposited together as in precipitation reactions. The basic requirement in CBD is that the material of interest precipitates from solution.

The nucleation mechanisms in solution belong to two categories [34]:

- . Here a new phase is formed in a solution that originally did not contain any foreign phase.
- *Heterogeneous nucleation.* Here a foreign interface is present on which the growth proceeds selectively.

For a *homogeneous* nucleation, the change in Gibb's free energy ΔG accompanying a phase transition between n molecules in a solution and in a cluster is composed of a term describing the chemical potential in the solution and in the solid, and of a term that describes the variation of the surface energy.

Table 6.1. Summary of chalcopyrites that have already been produced by ELD [34]

Binary chalcogenide compounds					Ternary chalcogenide compounds		
As_2S_3	Ag_2S	Bi_2S_3	CdS	CoS	$Cd_xPb_{1-x}S$	$Cd_xZn_{1-x}S$	$CuBiS_2$
Cu_xS	In_2S_3	MnS	NiS	PbS	Cu_3BiS_3	$CuInS_2$	$Hg_xPb_{1-x}S$
PdS_2	Sb_2S_3	SnS_x	TlS	ZnS	$Cd_xZn_{1-x}Se$	$CuInSe_2$	CdS_xSe_{1-x}
Bi_2Se_3	CdSe	CoSe	Cu_xSe	NiSe	ZnS_xSe_{1-x}	InS_xOH_{1-x}	SnS_xOH_{1-x}
PbSe	Sb_2Se_3	$SnSe_x$	TlSe	ZnSe	ZnS_xOH_{1-x}		
CdTe	PbTe						

ΔG is also the activation energy of that homogeneous nucleation process. At equilibrium the chemical potential in the solid equals that in solution, from which a critical cluster size can be determined [34]. The nucleation rate depends sensitively on both the supersaturation ratio S^* of the solution and the surface tension σ at the solution/solid interface. Below a critical value of S^* (around 12) the nucleation rate drops to very low values. Also an increasing σ leads to markedly decreasing S^*. The surface tension can be tailored through modification of the pH value, the ion nature and by additives, as it is related to the adsorption of species at the surface.

When a foreign surface is additionally present, the surface tension at the interface between substrate and the precipitating nucleus also has to be considered. This additional term reduces the overall surface energy needed for the nucleation process. This means, in the presence of a foreign substance that matches with the substrate material, the nucleation energy barrier tends to be suppressed. As a consequence, *heterogeneous* nucleation will, in principle, take place once S^* exceeds unity, hence at values of S^* far below those required for homogeneous precipitation. In the case of heterogeneous growth from the vapor phase the growth mode of the deposit has to be related to the interfacial tension between the two materials. For very low tension there is no nucleation barrier and the growth takes place via a 2D layer-by-layer mode [34].

The more stable the precipitate, the more negative is ΔG and the smaller the solubility product. As many of the materials of interest in chemical deposition (metals, chalcopyrites, etc.) are highly insoluble, very low concentrations of their ions will already lead to high supersaturation and hence to homogeneous precipitation. Free ions generally form complex species with ligands present in solution – e.g., in aqueous solution OH^- for metal cations, or H^+ for chalcogenide ions. By selected complexing agents, additional ligands can be introduced. As a consequence, additional chemical equilibria between ligands and the ions to be precipitated have to be considered, which leads to their increased solubility after complexation. Typical complexing agents are NH_3, ethylene diamine (EN), diethylene triamine (DEN), triethylene triamine (TRIEN), nitrilotriacetate (NTA), ethylene diamine tetraacetate (EDTA) and thiourea (TU) [34].

ELD allows for more uniform metal deposition than electrochemical plating techniques [35]. Metals such as Cu, Ag, Au, Pt, and Ni can be chemically deposited. As this procedure is commonplace in today's technology, even commercial solutions for ELD are available [36]. Recently, the first ELD of metallic alloys such as Co–W–P, Ni–W–P, and Ag–W were reported [37]. Table 6.1 summarizes all chalcopyrites that have been deposited by ELD.

Tubule Formation by Electrodeless Deposition Processes

If polymers with etched ion tracks are used as substrate material, ELD processes of, e.g., metals or chalcogenide compounds leads to their deposition on the etch-track walls, which signifies the production of tubules of this matter.

For a chemical deposition process, the ratio of the deposition speed of matter onto the inner track walls to the deposition speed onto the outer surface of the foil in which the tracks are embedded will depend on the ratio of the speed v_{penetr} of penetration of the precursor liquid into the tracks to the speed v_{surf} of arrival of the metal ions at the foil surface. For $v_{\mathrm{penetr}} \gg v_{\mathrm{surf}}$, the tubule thickness will be comparable to the layer thickness on the foil surface; for $v_{\mathrm{penetr}} \ll v_{\mathrm{track}}$ the tubule growth will be negligible, and the deposition of matter will block the nanopore entrance. Therefore, high penetration speeds are important.

Depending on the amount of material deposited on the etch-track walls, one can either form a continuous film or discontinuous islands. In the first case, which occurs after prolonged exposure to the depositing chemical reaction, massive self-supported tubules emerge that can be isolated by dissolving the polymeric host matrix in a suitable solvent. In the latter case, which emerges after short-time exposure, the dispersed deposited islands of matter on the ion track walls may allow, e.g., to tailor devices with specific electronic transport properties such as given by two-dimensional thermionic electron emission, electron tunneling or percolation. This might be exploited, e.g., for use as pressure or hydrogen sensors.

The production of nanotubules is still at its beginning. Though Cu, Ag, Au, Pt, and Ni nanotubules [7] and tubules of conducting polymers have already been prepared, tubules of other, e.g., semiconducting compounds (such as given, e.g., Table 6.1) have not yet been formed. Here a great application potential is still present.

Table 6.2 gives an overview of some of the recipes for deposition of metals and conducting polymers for tubule production. For more recipes, see, e.g., [39] for Cu and Ni, and [40] for Fe. Please note that there are some large differences in the parameters given by different experimentators.

Sensitization and Activation

The basic requirement for any CBD or ELD technique is a good wetting of the polymer surface, followed by sensitization and activation, i.e., the formation

Table 6.2. Some recipes for electrodeless deposition of metals within ion tracks

Depos. Matter	Recipe	pH	Temp. [°C]	Exposure [min]	Ref.
Sensitization + activation step	0.2% $SnCl_2$ in 0.02 M HCl	< 7	RT ?		[41] [26]
	0.025 mole/l $SnCl_2$ in 0.07 mole/l trifluoroacetic acid.	?	RT	45	[42]
	$SnCl_2$ + 1,3 dihydroxy benzene + HCl + $[Pd\ (NH_4)_4]Cl_2$	5.5	43 ± 7	10	[36]
Cu	Cu_2SO_4 + formaldehyde + NaOH	> 7	24 ± 4	2 µm/h	[36]
Ag	Aqueous ammonical $AgNO_3$ solution + reducting agent (glucose, formaldehyde, or sodium hypophosphite)	> 7	≈ 20	5–15...	[36]
Au	K-Au^{I} cyanide	$6 \pm .2$	85 ± 5	10–20	[36]
	Commercial Au plating solution + 0.127 mole/l Na_2SO_3 + 0.625 mole/l formaldehyde + 0.025 mole/l $NaHCO_3$	12, and stepwise decrease to 10	5		[42]
Ni	Ni_2SO_4 + Na acetate + various agents[7]	$5 \pm .1$	88 ± 1	15–20 µm/h	[36]

of suitable nucleation centers on the etch-track walls. For example, for Au deposition it is recommended to use acetonitrile for wetting improvement, (2-cyanoethyl)-triethoxysilane for sensitization, and 0.03 mole/l $AgNO_3$ in NH_4OH solution for activation. Other common activators are Sn or Pd metal atoms, attached chemically to the polymer surface, or dangling bonds introduced via ion or laser irradiation. For Pd, let us consider, e.g., the sensitizer complex $[Pd\ (NH_3)_4]^{2+}$ (with, e.g., Cl^- as counterion) that, if reduced by a suitable agent, can bind the Pd atom to the polymer via complexation with surface amine, carbonyl, and/or hydroxyl, thus forming groups such as –COOPd. The sensitized membrane is then activated by exposure to Ag^+ [7], Sn^{2+}, or Pd^+ [36], resulting in the formation of discrete metallic particles on the polymer surface. The sample is then immersed in a Cu, Ag, or Ni plating bath containing the metallic ions as a reducing agent that results in metal plating on all foil surfaces, including the etch-track walls, according to:

$$\mathrm{Sn}^{\mathrm{II}}_{\mathrm{surface}} + 2\mathrm{Cu}^{\mathrm{I}}_{\mathrm{solution}} \rightarrow \mathrm{Sn}^{\mathrm{IV}}_{\mathrm{surface}} + 2\mathrm{Cu}^{0}_{\mathrm{solution}} \tag{6.1}$$

for the case of Sn activation.

7 Composition kept in confidence by the producer

As it has been shown that organocyanides strongly chemisorb to Au, the latter can also be used as molecular anchors to absorb to –OH groups on the polymeric pore wall according to [43]:

$$\begin{array}{llll} \mathrm{OH} & & -\mathrm{O} & \\ & & \quad\backslash & \\ -\mathrm{OH} + & (\mathrm{CH_3CH_2O})_3\mathrm{SiCH_2CH_2CN} \rightarrow & -\mathrm{O} - \mathrm{SiCH_2CH_2CN}\,. & (6.2) \\ & & \quad / & \\ -\mathrm{OH} & & -\mathrm{O} & \end{array}$$

In the case of electrodeless Au plating, it is recommended to deposit the latter metal onto preformed Cu, Ag, or Ni nuclei, according to a surface redox reaction similar to (6.1):

$$\mathrm{Cu}^0_{\mathrm{surface}} + \mathrm{Au}^{\mathrm{I}}_{\mathrm{solution}} \rightarrow \mathrm{Cu}^{\mathrm{I}}_{\mathrm{solution}} + \mathrm{Au}^0_{\mathrm{surface}}\,. \qquad (6.3)$$

The surface-bound Au nanoparticles are said to be good autocatalysts for the reduction of $\mathrm{Au^I}$ to $\mathrm{Au^0}$ using formaldehyde as the reducing agent. As a result, Au deposition begins at the pore walls, so that Au tubules emerge [35].

Eventually, preimmersion of the sample in methanol for 5 min before all preparation and deposition steps is helpful. High pH values in the plating baths yield high plating rates, but plating proceeds preferentially on the faces of the membranes. By contrast, lower pH plating baths result in nanotubules with more uniform inside diameters [42].

The crucial point in chemical deposition is the preceding "activation" or "derivatization" stage, which means the formation of nucleation centers of high areal density on the inner etched-track surfaces. In Fig. 6.7 curve a shows that the exposure of the inner etch-track surfaces to an activator solution, which dramatically increases the density of nucleation centers ("derivatization"), gives rise to an enhanced metal layer growth after some characteristic incubation time during which the metallic nuclei form and grow.

Instead of producing the nucleation centers chemically, one can also prepare them by laser or ion irradiation [44]. For example, an energetic laser beam impinging onto a PI surface embedded in an oxidizing wet medium can activate a ring structure so that oxygen is attached, the latter enabling the metal addition [45] according to the scheme:

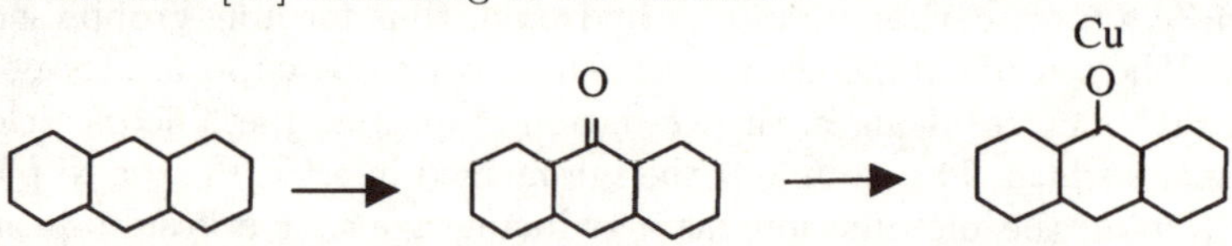

Ion irradiation is known to modify the polymeric surfaces chemically and physically [46]. Therefore, systematic ion-irradiation experiments have been performed with very low energy ions at high fluence from plasmas [47] as well as with ions of slightly higher energies (30 keV) [13]. The experiments indicate that, as a result of nucleation-center formation, an enhanced metal–polymer adhesion takes place. Surface nucleation-center formation is already

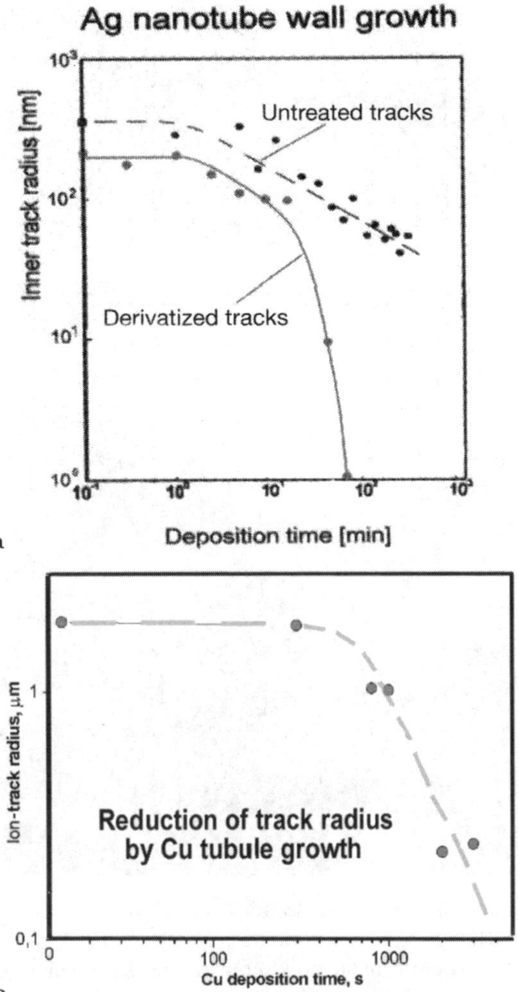

a b

Fig. 6.7. Decrease of the radius of activated (derivatized) etched tracks with deposition time for ELD of (**a**) Ag and (**b**) Cu. Additionally it is shown in (*a*) how the nanotube wall growth is affected if the initial activation step is missing

observed at ion fluences as low as $\sim 10^{12}\,\mathrm{cm}^{-2}$. The formation of nucleation-centers appears to be the result of electronic rather than of nuclear-energy transfer, consistent with the laser irradiation results where electronic excitation processes are triggered by the energetic photons. As the subsequent chemical deposition is most efficient immediately after the irradiation and less efficient after a longer delay time, one may conclude that some of the electronic defects anneal out already at ambient temperature.

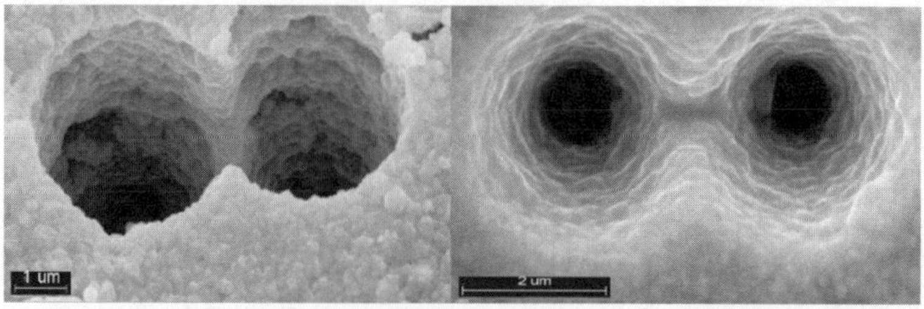

Fig. 6.8. Comparison of Ag layers deposited onto etched 1-GeV U tracks in PET after chemical activation (*left*) and ion-induced activation (*right*). In the first case, acidic $SnCl_2$ solution was used, in the second case the sample was ion irradiated with 300-keV Ne^+ ions at a fluence of $10^{15}\,cm^{-2}$, after the ion-track etching and before the Ag deposition [13]

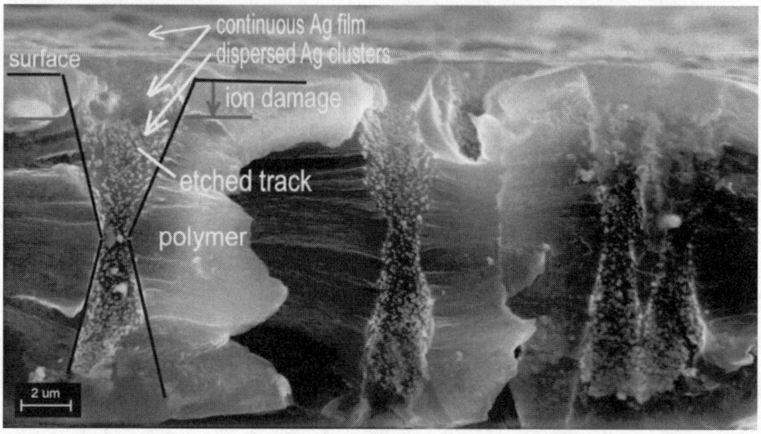

Fig. 6.9. Cross-sectional view of double-cone shaped etched tracks in PET that were activated with 300-keV Ne^+ ions at $1 \times 10^{15}\,cm^{-2}$ before the chemical deposition of Ag. The thereupon formed continuous Ag layer (*top*) ranges only from the sample surface up to a depth of $\sim 700\,nm$ inside the etched tracks. Below that depth, some Ag is seen to precipitate at intrinsic defects on the track wall as isolated nanocrystals. The polymer is seen to fracture quite irregularly between the perforations

The higher nucleation-center density obtained by ion irradiation as compared with chemical activiation leads to smoother deposited layers, as illustrated in Fig. 6.8. This is important for obtaining multilayer structures of good quality.

As ion-induced activation ranges only up to the projectiles' range, this approach enables one to tailor the metal deposition along a track. In this way it is possible to deposit cylindrical metal contacts on both sides of a track, which are electrically insulated against each other. Figure 6.9 shows the result of a first test experiment undertaken for this purpose.

Recent experiments with swift heavy-ion-irradiated (300-MeV Xe^{17+}; $10^{11}\,cm^{-2}$) PI [13] have shown that the efficiency of Pd nucleation-center formation is also improved by the ion irradiation, probably due to easier Pd–polymer bonding at radiation defects. If, however, the (pristine or irradiated) PI surfaces are pretreated with HNO_3, the picture reverses. Now metal is chemically deposited on the unirradiated surfaces rather than on the irradiated regions. This is understood as the result of some passivation of radiation-induced radicals against subsequent chemical deposition. It hence appears that one can tailor the tubule formation by specific treatment with energetic ions before the activation and/or chemical deposition processes.

In this connection it is interesting to note that nucleation center formation by ion irradiation can also take place on substrates other than polymeric, e.g., on Si wafers, where recently VO_2 nanoparticles have been deposited.

Metallic nanotubule formation does not necessarily need electrodeless plating techniques. If molecular anchors are attached to the pore walls metal preferentially deposits there also by electrodeposition. For example, Au tubules could be prepared electrochemically after applying cyanosilane to the pore wall [7].

Tubule Growth

Figure 6.10 shows for the recipes given in Table 6.2 that the tubule growth is faster for Ni deposition than for Cu deposition. One can further see that in the given example the growth slows down and approaches saturation, which is either due to the exhaustion of the ELD solution in this case, or due to complete track filling. Finally, the comparison of the metallic layer growth in

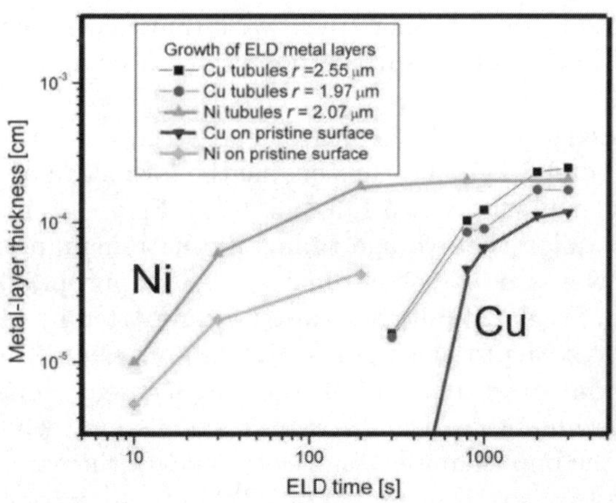

Fig. 6.10. Growth of the layer thickness of chemically deposited metallic nanotubules in etched tracks in PI, in comparison with the layer growth of the same metals onto pristine Pi foils under otherwise the same conditions

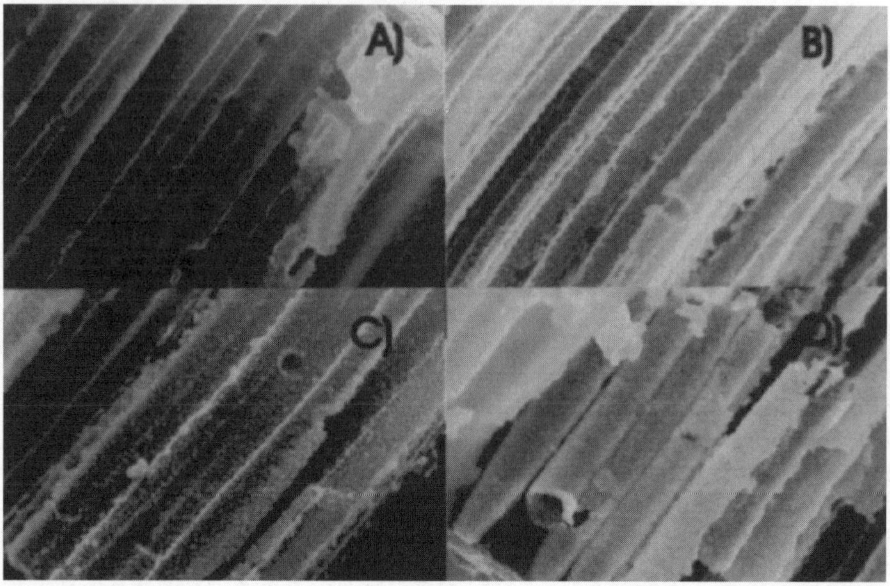

Fig. 6.11. Illustration of the different stages of the growth of Ag tubules in tracks. In the given example, the polymer was preactivated with acidic $SnCl_2$ solution. SEM side views. Exposure times to the Ag solution: (**a**) 1 min; (**b**) 2 min; (**c**) 5 min; (**d**) 10 min. Average size of the Ag precipitates: (**a**) < 20 nm; (**b**) 20 nm; (**c**) 50 nm; (**d**) > 50 nm. Only in the last stage have the tubules obtained sufficient mechanical stability to withstand mechanical fracture of the polymer host foil [13]

the tracks and on the foil surface gives the astonishing result that the metal layers grow faster in the tracks than on pristine foils – possibly due to the tiny curvature radius of the inner track walls that makes them energetically favored due to their reduced surface potential, as compared with a planar surface.

Upon continued growth, the heterogeneous precipitates rapidly touch each other (Figs. 6.10 and 6.11), and thus enable the formation of thin continuous conducting tubular films, Figs. 6.11a and 6.15). Figures 6.10 illustrates that it is possible to tailor the average radius of the remaining inner pore free volume down towards nanometric dimensions by appropriate timing of the metal deposition. Similar results have also been reported by Martin et al. [35], who even claim to have produced pores with sub-nm sizes. One should keep in mind, however, that these inner tubule radii are average values as determined by appropriate techniques (e.g., ITS or gas transmission), which do not reveal anything about the pore shape (cylindrical, conical), the variance of the inner pore radius, and the pore-wall roughness. Taking into account that chemical deposition leads to nanocrystal sizes of up to 50 nm or more when applying chemical activation (Figs. 6.9, 6.11a, 6.12a, the local fluctuation of the inner tubule radii is considerable. To obtain pores of meaningful small inner radii,

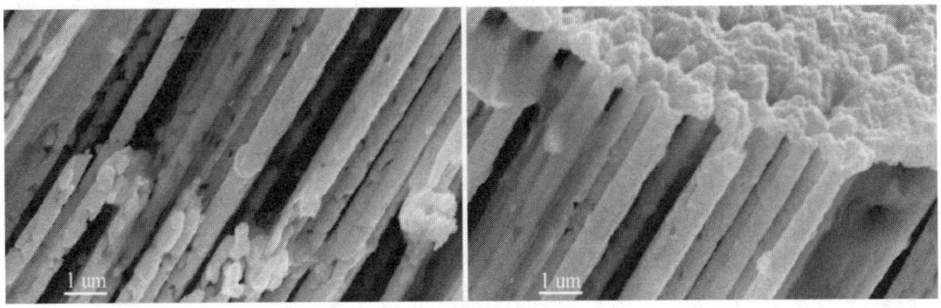

Fig. 6.12. SEM images of Ag tubules formed (**a**) without, and (**b**) with $SnCl_2$-preactivation. Whereas in the first case, large isolated crystals develop along the tracks, tubules of reasonable consistency emerge in the second case [13]. Polymer matrix dissolved after tubule formation, for the sake of better visibility

it is therefore absolutely necessary to ensure extremely smooth inner walls. According to our present experience, this can be accomplished only by ELD after ion-beam-induced activation or by evaporation.

The sizes of the particles that form upon chemical deposition of, e.g., Ag onto a polymeric surface are found to vary over a great range, from about 5 nm to about 10 µm in diameter. Some of the smaller clusters show a shaded outer rim in SEM images, which might point to some limited oxidation. A low nucleation-center density, hence a larger center-to-center distance, implies that the precipitates can grow to larger sizes before they touch each other, to form continuous films. In the extreme case, when no activation is performed at all (i.e., when only a few dispersed intrinsic defects act as nucleation centers), one obtains after long chemical deposition times extremely large precipitates that, nevertheless, touch each other only poorly, Fig. 6.12a.

In some cases it is even desirable to form layers of dispersed metallic nanoparticles. On the one hand these particles exhibit interesting electronic properties when going to very small particle radii (showing the transition towards non-metallic, and acting as electron storage), they exhibit strong catalytic activity, and eventually even become photoconductive. Due to the large particle-to-particle distance, layers of these particles also exhibit quite a high pressure sensitivity, with a reproducible region within the limits of polymeric elasticity, and a nonreproducible response at higher pressures when plastic polymer flow sets in, Fig. 6.13.

The many additional preparative steps such as "conditioning" (= etching of the plastic surface to improve its wettability), "generation" and "stabilization", recommended by commercial companies (see, e.g., [36] for ELD) have proven to be partly fatal for tubule formation in tracks in, e.g., PI, PET and PSiO, as in these cases the alkaline agents dissolve the etch-track walls (including the previously deposited nucleation centers), rather than improve the metal layer deposition [13].

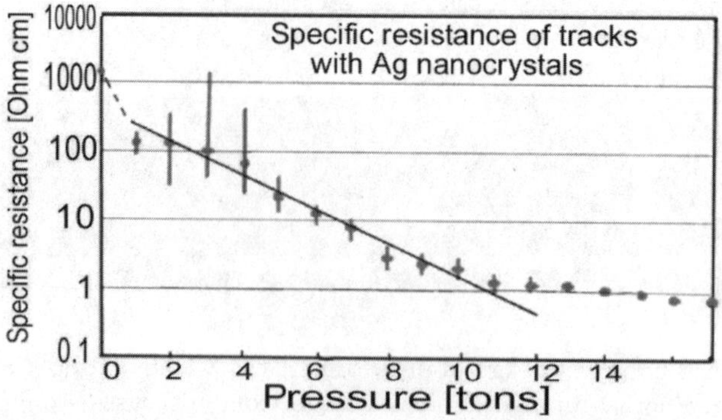

Fig. 6.13. Pressure sensitivity of etched tracks filled with isolated Ag crystallites. A reproducible region within the limits of polymeric elasticity (*dashed part*; not shown in detail here) is followed by a nonreproducible plastic response at higher pressures. The adjustment of the polymer to the new high-pressure ambient proceeds in many individual relaxation steps that show up by strong current fluctuations that persist for several days each. The maximum amplitude of these fluctuations is illustrated by the given bars

Occasionally, the metallic tubules or wires are only loosely embedded in the polymeric pores, so that solution can creep into this junction that eventually degrades the applicability of these nanostructures. For this case Hulteen and Martin [7] showed that heating stretch-oriented polymeric foils above their glass temperature (e.g., in the case of PC: above $\sim 150°C$) makes the membranes relax and shrink so that the junctions between the pores and the metallic nanostructures can be sealed.

Ag tubules have been found to be thermally stable in shape and orientation in PET foils up to $\sim 200°C$ and in PI up to at least 700°C. At $\sim 250°C$ the tubules appear to lose their perfect alignment in PET, probably due to the gradual shrinking of the host polymer that widens the pores. Ag tubules in PI foils, however, are still well aligned even when the host matrix has already been carbonized by pyrolysis at 700 to 800°C [13].

6.3 Formation of Complex Nanostructures

By sequential galvanic deposition of different materials, one obtains rods with different matter at different axial positions. For example, metallic and/or semiconducting multilayers have been deposited to produce nanowires exhibiting perpendicular giant magnetoresistance (GMR) [48], Resonant tunneling diodes [49, 50], or microdiode arrays [51]. In principle, galvanic, ELD, and CBD processes could be arbitrarily combined with each other to obtain structures of high complexity; however, such experiments have not yet been performed systematically.

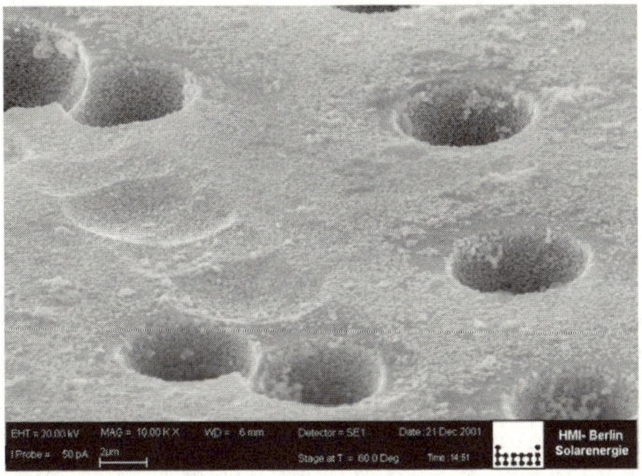

Fig. 6.14. SEM image of a Ag/TiO_2 double layer structure deposited onto the surface and the etch-track walls of a 20-μm thick microporous PI foil.[8] Magnification 5000×. It is seen that during the deposition process the TiO_2 particles preferentially bond to those surface areas with lowest potential, i.e., with highest curvature radius (i.e., the track interiors and entrance regions)

The strategy for producing concentric multilayer nanotubules is described, e.g., in [27, 28]. These composites have very high interfacial surface areas between concentric layers of materials. In such a way, TiO_2/PPy, TiO_2/Au and TiS_2/Au nanocomposites have been produced [7], which are excellent photocatalysts. The disadvantages of TiO_2 as a photocatalyst and TiS_2 as a Li^+-intercalation material due to its low electrical conductivity is overcome by a central conducting polymeric electrode.

It has been found that the deposition of TiO_2 sols onto the inner walls of Au nanotubules requires the previous immersion of the Au tubules in a 2% solution of 3-mercapto propyl trimethoxysilane in ethanol at pH = 5, adjusted with acetic acid [28], to make the Au surface wetting for the sol. If, by contrast, TiO_2 or SiO_2 colloids are deposited on the inner walls of Ag tubules, no such treatment has been found to be necessary. Dense layers with a thickness equal to the TiO_2 particle diameter were deposited in this case, Fig. 6.14. In both cases the underlying metal tubule provides individual current collecting electrodes for each catalyst TiO_2 particle.

The performance of several subsequent different deposition steps leads to tubules with concentric multilayer structures. Once conducting tubules have been formed, the subsequent deposition of other conducting matter onto the inner walls of these tubules can either be accomplished by ELD or galvanically. It is reported [35] that electroless plating techniques allow for more uniform metal deposition than galvanic techniques, and should hence be favored. Furthermore, ion-beam activation has turned out to yield layers of greater

[8] Rojas-Chapana J, Fink D, Tributsch H, Küpers J, Petrov A, Wilhelm M, unpublished work; patent pending (2003)

smoothness than chemical activation, and should therefore be preferred when forming complex structures. Hitherto, {Cu/Ni}, {Ni/Cu}, {SiO_2/TiO_2}, {Ag/TiO_2} [13], {TiO_2/PPy}, {TiO_2/Au} [28], Au/PPO/PPy [27], and {carbon/PAN/Au} [7] structures have already been prepared; more refined structures are under development.

Also, {ZnO/Au} microstructures were prepared [28]. Here the thiol chemisorbs onto the Au tubules, providing three surface –OH groups per chemisorbed –SH. These surface –OCH_3 groups help to initiate the growth of the subsequently deposited sol-gel ZnO_2 film [27, 28].

Similarly, TiS_2/Au tubules were produced. TiS_2 is a Li^+-intercalation material for Li-based rechargable batteries (reaction: $TiS_2 + xe^- + x$Li+ $\rightarrow$ Li_xTiS_2). TiS_2/PPy composite nanostructures reversibly intercalate and deintercalate Li^+. Concentric carbon/PAN/Au structures may act as nanocondensors [7]. Another concentric microstructure reported in the literature was Au/PPO/PPy (PPO – polyhenylene oxide) [27, 28]. In this case, $\sim$ 50-nm thick defect-free layers of the insulating polymer poly(2,6-dimethylphenol), PPO, were electropolymerized onto the inner surfaces of the Au tubules. Prior to the deposition of the inner PPO tubules the Au-tubule-containing membrane was immersed in an aqueous solution of the redox-active species $Fe(CN)_6^{3-}$ [27].

A simple bilayer structure with intimate phase contact was obtained by Nishizawa et al. [15] by pouring a pyrrole/$HClO_4$ solution into preformed $LiMn_2O_4$ nanotubules. The manganese dioxide component of the spinel tubules worked as an oxidizing agent for the in situ polymerization of pyrrole to promote the deposition of PPy onto the $LiMn_2O_4$ nanotubules.

Once nanotubules or nanofibrils of any kind have been produced inside etched tracks, they can be isolated by dissolving the polymeric host matrix in a suitable solvent, e.g., methylene chloride for dissolving PET [26], hexaflouro isopropanol for poylester [27], or dimethylether or dichloromethane for PC [52].

The electrodeless deposition techniques described here include deposition on both the inner etch-track walls and the membrane surfaces. The latter deposition can be removed by either subsequent polishing with a laboratory tissue dampened with water or a suitable organic solvent, or by exposure to O_2 plasma [27]. Another, more elegant technique would be to evaporate some thin hydrophobic film onto the sample surface before the material deposition takes place – which then would occur only inside the tracks but not on the surface. (The realization of this approach has not yet been reported.)

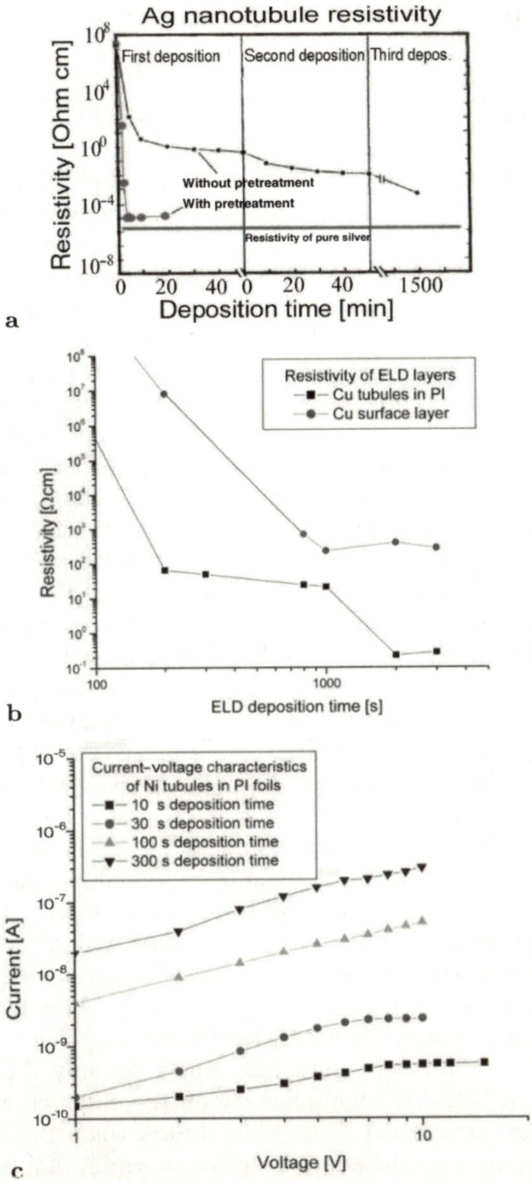

Fig. 6.15. (**a**) Resistivity of Ag tubules formed in microporous PET foils without, and with preactivation. Whereas with application of $SnCl_2$ activator, the resistivity rapidly drops to the characteristic values of thin Ag films, the deposition without preactivation never gives films of good conductivity, even after prolonged exposure times and several renewals of the Ag complex salt solution [13]. (**b**) Comparison of resistivities of Cu tubules and planar Cu deposits on pristine sample surfaces as a function of deposition time; (**c**) current–voltage characteristics of Ni tubules in PI, as a function of the deposition time

6.4 Electrical and Magnetic Properties of Nanowires and Nanotubules

Tubule Resistivity

The tubule resistivity is better than the resistivity of bulk matter deposited on polymeric surfaces (Fig. 6.15b), possibly due to the denser layer deposition. The current–voltage characteristics of tubules reveals nonlinear relationships, essentially after short deposition times corresponding to discontinuous films (Fig. 6.15c). Such measurements, together with measurements on the temperature dependence of the tubule conductivity, will reveal in closer detail the conductivity mechanism in dependence of the deposition conditions so that available theoretical predictions (see below) can be controlled.

The resistivity drops dramatically with activation time τ_{act} (which determines the nucleation center density) up to $\tau_{act} \sim 10\,s$, and thereafter remains constant for constant metal-deposition time, due to the saturation in nucleation-center density. Figure 6.15a shows clearly that activation before ELD rapidly yields tubules of good conductivity, whereas in nonactivated tracks discontinuous tubules are formed (as shown in Fig. 6.12a), with poor conductivity, even after several renewals of the ELD solution used.

In between dispersed clusters the charge propagation is dominated by the conductivity of the inner track-wall surfaces. The latter appears to be an ionic conductivity due to absorbed moisture and ions retained from the etching process. Consequently, the conductivity of dispersed metal films also exhibits a pronounced humidity dependence. Insertion of materials sensitive to other ambients should enable one to develop various types of ion-track sensors.

Relating the measured tubule resistances [Ω] to the tubule cross-section (as derived, e.g., from ITS) and to the tubule length, one obtains the resistivity [Ω cm] of the metal deposited along the tubule. The latter drops dramatically with increasing deposition time (i.e., with layer thickness) from the value of the etched polymeric host material towards a value that approaches the value of the bulk metal more closely, thereby passing various intermediate stages. These stages are determined by the growth mechanism.

For metal deposition onto polymers, this is usually the *Volmer–Weber* growth type where small clusters nucleate on the polymer surface and grow into islands of the condensed phase; this occurs when the adatom–adatom (cohesive) energy exceeds the adatom–substrate–atom (adhesive) energy. On an untreated polymer, the nucleation density is greatly influenced by substrate surface defects since the adatom binding energy at a defect is greater than on the open surface. The original clusters are quite small, e.g., 4 or 5 atomic diameters in the case of Pt or Ni films on a-C [53]. These clusters are single crystals, usually oriented along a main crystal axis direction (e.g., in the case of Ag on PET: the $\langle 100 \rangle$ direction) towards the polymer surface.

With continued deposition these islands grow in size, come into contact with each other to form particle chains – this defines the percolation threshold – until finally a continuous film with thickness t_{cont} emerges. The other important growth type, the *Frank–van de Merwe* growth, which leads to a smooth deposition of monolayer by monolayer, is not found here as in metal–polymer systems the adhesive interaction energy never exceeds the cohesive one.

The resistance of a discontinuous film of islands is much larger than the bulk value and has a negative temperature coefficient of resistance (TCR). Its exponential temperature dependence indicates a thermally activated process. Thermionic emission and electron tunneling are the principle modes of conduction in this stage, with the tunneling activation energy E_{a} being given by either:

$$E_{\mathrm{a}} = (e^2/K)[a^{-1} - (a+b)^{-1}] \tag{6.4a}$$

or by:

$$E_{\mathrm{a}} = \hbar^2 2\pi^2 (2m\,a\,d)\,, \tag{6.4b}$$

(a = particle size, b = particle separation, e = electron charge, K = dielectric constant of the intervening medium, $\hbar$ = Planck's constant divided by 2π, m = electron mass, and d = atomic spacing), depending on the chosen model [54].

To calculate the resistivity of a two-dimensional porous network, one can describe the latter by a resistance R_{s} and a parallel capacitance C_{s} in series with a resistance R_{i}. Then the complex impedance Z (i.e., the inverse of the admittance, that is defined as the ratio of current to the applied voltage at the given frequency) of that circuit is given by [55]:

$$\begin{aligned} Z = {} & [R_{\mathrm{s}} + R_{\mathrm{i}}(1+\omega^2 R_{\mathrm{s}}^2 C_{\mathrm{s}})]/(1+\omega^2 R_{\mathrm{s}}^2 C_{\mathrm{s}}) \\ & - \mathrm{i}\omega R_{\mathrm{s}}^2 C_{\mathrm{s}}/(1+\omega^2 R_{\mathrm{s}}^2 C_{\mathrm{s}})\,. \end{aligned} \tag{6.5}$$

For the example of a vapor-deposited Pt film of $t_{\mathrm{l}} = 0.5\,\mathrm{nm}$ nominal thickness on a glass substrate, the DC conductivity is about 3 orders of magnitude larger than that of the glass, and the admittance increases above 100 kHz by a factor of 200 with frequency, until it reaches saturation at about 5 MHz. At $t_{\mathrm{l}} = 0.6\,\mathrm{nm}$, when the transition from the discontinuous film to a continuous film takes place in that example, the sample does not show any frequency dependence, and for $t_{\mathrm{l}} = 2$ to $3\,\mathrm{nm}$, the admittance even decreases by 15% at 5 MHz. The DC resistivity σ of this thin layer is higher than the bulk value σ_0 by:

$$\sigma/\sigma_0 = 3d/(4\lambda_0)\ln(\lambda_0/d)\,, \tag{6.6}$$

as long as the mean free path λ_0 of the electrons in the conduction band exceeds the layer thickness d [55].

At the onset of a continuous film the lateral island dimensions define the grain size D_{g} of the film that, in turn, determines the grain-boundary scattering contribution to the in-plane film resistivity ρ_{f}. For chemically deposited films, D_{g} is typically of the order of several 10 nm, hence larger than what is usually achieved by film evaporation.

Resistivity of Nanowires

Quite a lot of work has been invested to understand the resistivity of nanowires. The latter differs greatly from that of thin two-dimensional films (as given approximately also in nanotubules, see above). The length determining the contribution to the resistance of a nanowire by electron–electron interaction effects is the so-called *thermal length* L_T, given by $L_T = (\hbar D/k_B T)^{1/2}$, with D being the electron diffusion constant in the material and k_B being the Boltzmann constant. For long wires ($\geq 90\,\mu m$) the resistance scales with $T^{-1/2}$, for shorter wires the resistance seems to approach a constant as $T \to 0$. As an example, Au-Pd wires of about 50 nm width and 250 nm to 90 µm length yielded: $D = 12\,cm^2\,s^{-1}$ and $L_T \approx 90$ nm at 1 K [56]. Other important lengths are the Fermi wavelength λ_F and the electrons' mean free pathlength l_e. For $\lambda_F \ll l_e \ll L$ the electron transport through the nanowire is a diffusive one, for $\lambda_F \ll L \ll l_e$ the electrons follow a ballistic path through the wire, being reflected at its boundaries.

Au contacted galvanically produced nanowires of, e.g., Cu are not stable in their characteristics with time; their resistivity gradually increases. Whereas they exhibit ohmic characteristics after short times, their characteristics resemble that of two counteracting diodes after prolonged times. This is explained by nanowire corrosion, described by, e.g., the transition from Cu to CuO so that Au/CuO and CuO/Au Schottky diodes emerge at the contacts. The increase of the temperature of Cu nanowires decreases their resistivity [57].

Wires of conducting polymers have also been produced for a long time. Frequently they show a pronounced fibrillar fine structure [58]. The conductivity of fibrils or tubules of these conducting polymers increases with decreasing etch-track size, due to the proceeding alignment of the polymeric chains along the etch-track axis on the track walls. This effect becomes important for track diameters below some 200 to 800 nm diameter (depending on the synthesis temperature, which counteracts the alignment), and reaches values that exceed the bulk conductivity values by a factor of around 20 [38, 59]. (Similar effects have also been observed for two-dimensional microfluidic water channels on surfaces of silicon wafers where filamentous biomolecules and self-assemblies were aligned along the channels, the degree of alignment depending on both the channel width and the persistence length of the biomaterials [60].)

Nanowires of magnetic materials such as Fe, Ni or Co also exhibit, of course, also magnetic properties. For a ferromagnetic thin film, the layer geometry dictates that the demagnetizing factors are approximately 0 and 4π for fields parallel and perpendicular to the film plane, respectively. Consequently, the shape anisotropy tends to force the magnetization M to be in the film plane. It takes an external field of at least $4\pi M$ to align M perpendicular to the film plane. Similarly, for a ferromagnetic rod, the demagnetizing factors for fields applied along and perpendicular to the rod axis are approximately 0 and 2π, respectively. In the absence of an external magnetic field this shape anisotropy compels the magnetization to be along the axial direction. Hence in arrays of nanowires being parallel to the surface normal of the membrane, the nanowire direction leads to magnetization perpendicular to the membrane surface. In other words, the technologically highly desirable but elusive perpendicular magnetization is a natural consequence of the geometry of the ferromagnetic nanowire arrays.

Furthermore these ferromagnetic nanostructures exhibit greatly enhanced (by typically two orders of magnitude) magnetic coercivity because of their single-domain nature, the enhancement increasing with decreasing wire diameter. However, if the wire diameter decreases below 50 nm, the coercivity in some cases decreases again due to imperfections in the nanostructures.

6.5 Nonstatistical Ion-Track Distributions

The greatest obstacle for the future application of polymeric ion tracks for miniaturized devices is the statistical distribution of the ion tracks on the target areas, which leaves some sample areas without any, and others with overlapping tracks. Of course this can be overcome by writing sequentially ion-by-ion onto predetermined positions [61], but the excessive time and cost of this operation render this "single ion track approach" rather obsolete for commercial mass production. Therefore one will rather have to adopt the "classical" approach to irradiate ions through masks with a regular array of holes, at a fluence corresponding to multiple track overlapping. In this way, even small pixels on the sample surface will contain at least some ion tracks. Conventional masks of good quality and narrow size are, however, extremely costly and hence prohibitive for the envisaged polymeric electronics that should establish a cheaper alternative to the present silicon-based technology.

Therefore, the first test experiments have been undertaken to see whether it might be possible to use masks of self-organizing materials such as cheap commercially available Al_2O_3 ceramic filter foils for this purpose. Such foils have thicknesses of typically 100 µm and pore sizes of 20 to 200 nm. In fact, the first ITS tests showed that ion transmission through such structures is indeed possible [1], Fig. 6.16. The high transmission yield of those particles that do not touch the pore walls (and consequently suffer neither any energy loss nor any deviation in their trajectories), see the high-energy peaks, reveals the excellent parallelism of the pores, even with diameters as small as 20 nm, corresponding to an aspect ratio of up to 5000. Particles that hit the alumina membrane itself are fully absorbed due to the membrane's large thickness ($\sim$ 100 µm). There is, however, still a third fraction of particles that are transmitted through the membrane with lower energies, due to their scattering from the pore walls. These can be eliminated easily from the transmitted

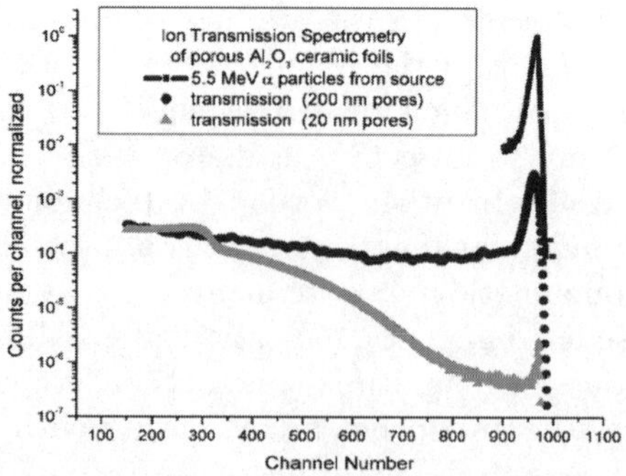

Fig. 6.16. ITS of 5.5-MeV α particles, transmitted through commercial Al_2O_3 foils of $\sim$ 100 μm thickness and 20 and 200 nm pore size, respectively. The energy spectra show that ion transmission through the nanopores is, in principle, possible, so that their application as masks for future swift heavy-ion-track electronics appears feasible

ion beam due to their divergence, by keeping some distance in between the porous ceramic foil and the target. Based on this positive experience, more tests with swift heavy-ion transmission are underway.

6.6 Approaches Competing with Ion Tracks for Applications

There exist many materials that compete with track-based polymeric microporous membranes for applications using, e.g., the template technique to produce nanowires and nanotubules. For example, it has been found recently that one can produce 3-nm wide hollow pores in NiO after irradiation with 350-MeV Au ions, without the necessity of track etching.[9] These pores could be filled with matter as described in this chapter.

Apart from using alumina as a self-assembling inorganic matter, one can also, in principle, take zeolites (e.g., $AlPO_4$-type crystals [62], CVD-deposited fullerene films, or self-assembling organic matter such as certain lipid molecules in alcohol and added water [63], or proteins derived from a bacterium [64], polypeptides [65], Langmuir–Blodgett films, [66] or block-copolymers [67]. Self-assembly has also been achieved with colloids of heavy atoms (e.g., Ag, Au) in light matrices (e.g., SiO_2), being produced by ion exchange or high-fluence ion implantation, with subsequent annealing.

Many of these materials have the advantage that well-ordered arrays of pores are readily obtained in large scale, which is by far superior to the

[9] Klaumünzer, S (2003), personal communication

statistical etch-track distributions for many applications. Partly these arrays are available even as thin films, and some of them are thermally highly stable. The applications of these self-ordering materials are rapidly expanding; at contemporary conferences (e.g., the MRS Spring or Fall Meetings) several sessions are largely devoted to these materials.

Al_2O_3 ceramic foils of about 100 µm thickness and pore diameters between about some ten and hundred nm are already commercially available. Complementarily, pore diameters in the 0.3 to 1 nm range can be obtained by using zeolites. The pores in these foils exhibit an excellent parallelism even at high length/pore diameter ratios (e.g., 100 µm/200 nm = 500 and 100 µm/20 nm = 5000 in the example of Fig. 6.16), and they can be used as templates for nanotubule or nanowire production. By careful engineering of the topology of the Al surface onto which the porous Al_2O_3 structures are grown, defect-free regular pore arrays with up to $10 \times 10\,cm^2$ total area have been obtained [68]. A combination of nanoimprint methods with self-organization effects allows one to get monodomain pore arrays with lattice constants smaller than that of the imprint stamp [69]. In particular, the combination with pyrolysis enables one to produce e.g. carbon or SiC nanotubes from precursor tubules of organic or silicon-based polymers. The pyrolysis temperature being of the order of 450 to 600°C, only high-temperature stable polymer foils such as PI are competitive. Some of the above-mentioned products have, however, also disadvantages such as high brittleness or low resistance to, e.g., acids or alkalines (e.g., of the porous alumina membranes).

Recently, a glass with periodically ordered nanochannels with diameters as small as 3.3 and 8.7 nm and pore densities up to $3 \times 10^{10}\,cm^{-2}$ has been described [70]. These mesoporous glasses are prepared by using a structure-directing agent together with a precursor for the wall material. The surfactant is alkyl trimethyl-ammonium (C_nTAB with different chain lengths n from 10 to 18). Materials of the SBA-15 type, with pore widths of up to 10 nm, can be prepared using amphihilic triblock copolymers (Pluronic ®) [71] to direct the polymerization of silica precursors to periodically ordered mesoporous structures.

Finally, one can also drill the required holes by laser beams into thin solid films. However, the pore size is usually limited to around 50 to 100 µm diameter, the holes are of irregular size, and neighboring holes, e.g., in polymeric targets tend to merge with each other. Another, more precise approach is to use a combination of laser irradiation and photolithographic etching to produce microporous membranes of, e.g., photoresist/Si/glass sandwiches [72].

Attempts are being made to also use buckytubes for nanoelectronics. One-electron transistors have been produced in semiconducting individual nanotubes, but the electronic properties are strongly dependent on the specific material's properties of the tubule used, and it has been difficult to progress carrier-free tubules towards larger arrays. Therefore attempts are in progress to mount these tubules on regularly spaced islands of Si [73] or in

Al_2O_3 pores [74]. Two such arrays, opposing each other, should in principle be useful to produce memories as well as logical circuits.

If (hydrophilic) MgF_2 is evaporated through a longitudinally patterned mask onto a hydrophobic surface of a given substrate, and if subsequently water vapor is condensed onto such cooled substrates, microscopic water channels emerge. By tailoring the degree of water-vapor saturation, localized water blobs emerge that eventually lead to a connection of neighboring water channels (morphological wetting transition). It is speculated that these connections could be shifted intentionally by suitable applied electrical fields. In such a way, materials might be transported, and the planar microchannels might be used as microreactors to examine reactions of minute amounts of matter; [75]. Taking polymerizable liquid matter instead of water, one could envisage creating 2-dimensional networks that could form part of future micromechanical machines [17].

Another approach to producing planar arrays of one-dimensional parallel channels is to evaporate metastable $\langle 110 \rangle$ oriented NaCl films onto $\langle 110 \rangle$ NaCl substrates, and then to relax the films homoepitaxially at 250°C, whereby $\sim$ 20-nm wide $\langle 110 \rangle$ and $\langle 010 \rangle$ oriented ridges and valleys emerge. Carbon replicas from this template containing the ridge/valley structure can be used for various applications, e.g., to host one-dimensional strings of Au nanoparticles for quantum-electronic devices [76]. For more literature on competing nanoprocessing and nanoelectronics, we refer the reader to [77].

The present trend to self-ordered materials should, however, not obscure the great advantages that ion-track-based microporous polymer foils still offer:

- High flexibility and, in the case of rubbers, also high elasticity.
- Micro- or nanostructured areas as large as a few $100\,m^2$ can be produced.
- Good resistivity against many aggressive chemicals (except for very high pH ambients and some organic solvents).
- By suitable choice of polymer, ion specie and energy, and etching conditions, various etch-track shapes can be tailored easily.
- By suitable choice of the ion fluence, the pore density can be tailored.
- By suitable choice of the etching time, the pore diameter can be tailored. The variance in the pore diameter is extremely low.
- Well-known lithographic and evaporation techniques can be applied to polymeric microporous foils.
- Many nuclear-track polymers can be readily deposited (e.g., spin-coated) onto, or attached to Si wafers, so that in principle hybrid structures can be produced, combining conventional semiconductor electronics with track-based polymeric electronics.
- Nucleation centers necessary for subsequent chemical deposition processes can be formed on polymeric surfaces not only by chemical methods, but also by laser methods on ion-beam irradiation. The latter are very sharp tools – essentially if using microbeams – that enable one to tailor nucleation-center

distributions in three dimensions in both macroscopic and microscopic scales (i.e., tailoring the overall track distribution, or individual track segments), which is expected to give rise to a multitude of complex ion-track-based nanostructures.

- The main disadvantage of ion tracks – their statistical distribution in the irradiated foils – can, in principle, be overcome by irradiation at a higher fluence through suitable masks. These masks can be either conventionally produced lithographic grids or self-ordered ceramic foils.

References

1. Fink D, Petrov A, Stolterfoht N, Wilhelm M, Hoffmann V, Richter A, Behar M, Farenzena L, Hirata K, Kobayashi Y, Chadderton LT, Schulz A, Fahrner WR, Creation of nanoscale objects by swift heavy ion track manipulations. Proc 2nd Intl Symp on Material Chemistry in Nuclear Environment, March 13–15, 2002, Tsukuba, Japan, 2002
2. Herden V (2001) Das Verhalten von lichtinduzierten Ladungsträgern in Polysilanen unter besonderer Berücksichtigung von Dotiereungen und strahlenchemischer Vernetzung. PhD. Thesis, Technical University Berlin, No. D83 (in German)
3. Fink D, Müller M, Nakao Y, Hirata K, Kobayashi Y, Behar M, Kaschny JR, Vacík J, Hnatowicz V, Ion-induced redistribution of palladium in polymethyl methacrylate. Polym Instr Meth **B166/167**, 610–614 (2000)
4. Biswas A, Awasthi DK, Kanzow J, Ding SJ, Fink D, Gupta R, Zaporojtchenko V, Faupel F, Nanostructural modifications in Au cluster arrays distributed in Teflon AF layers upon MeV heavy ion impact. To be published in 2003
5. Apel PYu, Kuznetsov VI, Ovchinnikov VV (1985) Capillary contraction of pores in polymeric nuclear membranes. Kolloidnyi Zhurnal **49**, 537–538 (in Russian);
   and Apel PYu, Ovchinnikov V, Capillary Contraction of small pores and latent track parameter measurements in polymers. Rad Eff Defects Solilds **126**, 217–220 (1993)
6. Faupel F, Metal diffusion during metallization of high-temperature polymers. In: Mittal KL (ed.): Metallized Plastics. Marcel Dekker Inc., New York 1998, pp. 79–83
7. Hulteen JC, Martin CR, Template synthesis of nanoparticles in nanoporous membranes. In: Fendler JH (ed.): Nanoparticles and Nanostructured Films. Wiley-VCH, Weinheim, 1998, pp. 235–262
8. Che G, Lakshmi BB, Martin CR, Fisher ER, Ruoff RS, Chemical vapor deposition based systhesis of carbon nanotubes and nanofibers using a template method. Chem Mater **10**, 260–267 (1998)
9. Boag NM, Dowben PA, Design of organometallics for vapor phase metallization of plastics. In: Mittal KL (ed.): Metallized Plastics, Marcel Dekker Inc., New York, 1998, pp. 1–7
10. Gutowski W, Thermodynamics of adhesion. In: Lee L-H (ed.): Fundamentals of Adhesion. Plenum Press, New York and London (1991)

11. Dobrev D, Vetter J, Neumann R, Growth of potassium iodide single-crystals using ion track membranes as templates. Nucl Instrum Methods **B146**, 513–517 (1998)
12. Berdinsky AS, Fink D, Petrov AV, Müller M, Chadderton LT, Chubaci JF, Tabacnics MH, Formation and conductive properties of fullerite in etched ion tracks in a polymer film. Proc MRS Fall Meeting, Boston 27.11.–1.12. 2001, Contribution Y4.7, 2001
13. Petrov A, Production of micro- and nanoelectrotechnic devices based on ion tracks in insulators. PhD. Thesis, Fernuniversität Hagen, 2004
14. Ragan R, Sih V, Atwater HA, Non-lithographic fabrication of dense Sn nanowire arrays. Proc MRS Boston, Fall Meeting. Contribution V4.3 (2001); and Borca-Tasciuc DA, Chen G, Thermal characterisation of nanowire array in α-Al_2O_3 matrix. Proc MRS Boston 27.11.–1.12.2001, Contribution V2.7, 2001
15. Nishizawa M, Mukai K, Kuwabata S, Martin CR, Yoneyama H, Template synthesis of polypyrrole-coated spinel $LiMn_2O_4$ nanotubules and their properties as cathode active materials for lithium batteries. J Electrochem Soc **144**, 1923–1926 (1997)
16. Tannenbaum R, Self-assembly of metal nanoclusters in block copolymers. Proc MRS Boston 27.11.–1.12.2001, Contribution V2.1, 2001
17. Mantese JV, Micheli AL, Hamdi AH, Vest RW, Metalorganic deposition (MOD): a nonvacuum, spin-on, liquid-based, thin film method. MRS Bull, Oct 1989, pp. 48–53, and references therein;
and: Hamdi AH, Mantese JV, Micheli AL, Laugel RCD, Dungan DF, Zhang ZH, Padmanabhan KR, Formation of thin film high T_c superconductors by metallorganic deposition. Appl Phys Lett **51**, 2152–2154 (1987)
18. Micheli AL, Chang SC, Hicks DB, Tin oxide gas sensoring microsensors from metalor-ganic deposited CMOD thin films. Ceram Eng Sci Proc **8**, 1095–1105 (1987); and: Xu JJ, Shaikh AS, Vest RW, Indium tin oxide films from metallorganic precursors. Thin Solid Films **161**, 273–280 (1988)
19. Wu CG, Bein T, Conducting polyaniline filaments in a mesoporous channel host. Science **264**, 1757–1758 (1994)
20. Al Mawlawi D, Coombs N, Moskovits M, Magnetic properties of Fe deposited into anodic aluminium oxide pores a a function of particle size. J Appl Phys **70**, 4421–4425 (1991)
21. Whitney TM, Jiang JS, Searson PC, Chien CL, Fabrication and magnetic properties of arrays of metallic nanowires. Science **261**, 1316–1319 (1993)
22. Shao I, Searson PC, Cammarata RC, Vereecken PM, Chien CL, Electrochemical deposition of FeCo and FeCoV thin films and nanowire arrays. Proc MRS Boston 27.11.–1.12.2001, Contribution M5.4
23. Hjort K, The European network on ion track technology. Presented at the 5th Intl. Symposium on "Swift Heavy Ions in Matter", May 22–25, 2002, Giordano Naxos, Italy
24. Enculescu I, Spohr R, Electrodeposition of Cu/Co multilayered nanowires. Proc of the Workshop on European Network on Ion Track Technology, Caen, France, 24–26 Feb. 2002
25. Dobrev D, Neumann R, Angert N, Vetter J, Preparation of metal membranes by direct electroplating of ion-track filters. Proc of the Workshop on European Network on Ion Track Technology, Caen, France, 24–26 Feb. 2002

26. Martin CR, Nanomaterials: a membrane-based synthetic approach. Science **266**, 1961–1966 (1994)
27. Cepak VM, Hulteen JC, Che G, Jirage KB, Lakshmi BB, Fisher ER, Martin CR, Chemical strategies for template syntheses of composite in micro- and nanostructures. Chem Mater **9**, 1065–1067 (1997)
28. Cepak VM. Chem Mater **8**, 2109 (1996)
29. Martin CR, Template synthesis of polymeric and metal microtubules. Adv Mater **3**, 457–459 (1991)
30. Parthasarathy RV, Phani KLN, Martin CR, Template synthesis of graphitic nanotubules. Adv Mater **7**, 896–897 (1995)
31. Henglein A, Electronics of colloidal nanometer particles. Ber Bunsenges Phys Chem **99**, 903–913 (1995)
32. Satoh N, Hasegawa H, Tsujii K, Kimura K, Photoinduced coagulation of Au nanocolloides. J Phys Chem **98**, 2143–2147 (1994)
33. Kuroda K, Shimojima A, Formation of ordered silica-organic hybrids by self-assembly of hydrolyzed organoalkoxysilanes with long organic chains. Proc MRS Boston 27.11.–1.12.2001, Contribution V2.8
34. Lincot D, Froment M, Cachet H, Chemical deposition of chalcogenide thin films from solution. Adv. Electrochem Sci Eng **6**, 167–235 (1999), and references therein
35. Martin CR, Nishizawa M, Jirage K, Kang M, Lee SB, Controlling transport selectivity in gold nanotubule membranes. Adv Mater **13**, 1351–1362 (2001)
36. AMI DODOCO GmbH & Co., Geschäftsfeld Oberflächentechnik, Im Altgefäll 12, 75181 Pforzheim, Technical Information sheets, 2001
37. Inberg A, Shacham-Diamand Y, Rabinovich E, Golan G, Croitoru N, Electroless-deposited Ag-W films for microelectronics applications. Thin Solid Films **389**, 213–218 (2001)
38. Cai Z, Martin CR, Electronically conductive polymer fibers with mesoscopic diameters show enhanced electronic conductivities. J Am Chem Soc **111**, 4138-4139 (1989)
39. Vaughn GD, Selective metallization of optically varible devices. In: Mittal KL (ed.): Metallized Plastics, Marcel Dekker Inc., New York, 1998, pp. 71–77
40. Haga Y, An H, Sato Y, Yosomiya R, Yosomiya T, Magnetic properties of metal layer generated by reduction of polymer containing metal ion. In: Mittal KL (ed.): Metallized plastics. Marcel Dekker Inc., New York, 1998, pp. 85–93
41. Brumlik CJ, Menon VP, Martin CR, Template synthesis of microtubule ensembles utilizing chemical, electrochemical, and vacuum deposition techniques, J Mater Res **9**, 1174–1182 (1994)
42. Jirage KB, Hulteen JC, Martin CR, Effect of thiol chemisorption on the transport properties of gold nanotubule menbranes. Anal Chem **71**, 4913–4918 (1999)
43. Brumlik CJ, Martin CR, Template synthesis of metal microtubes. J Am Chem Soc **113**, 3174–3175 (1991)
44. Hwang SD, Kher S, Spencer JT, Datta S, Dowben PA, Laser-induced selective copper deposition on polyimides and semiconductors from solution. In: Mittal KL (ed.): Metallized Plastics. Marcel Dekker Inc., New York, 1998, pp. 9–13
45. Ogawa Y, Tokunaga H, Murahra M, Photochemical nucleation of copper on polyimide surface with 10 ns laser irradiation. Proc MRS Boston 27.11.-1.12.2001, Contribution S5.8

46. Gheorgiu M, Popa G, Pascu M, Vasile C, Chemical and physical surface modifications of polymers by ion beam treatments. In: Mittal KL (ed.): Metallized Plastics. Marcel Dekker Inc., New York, 1998, pp. 269–279
47. Ibidunni AO, Brunner RJ, Metal/polymer adhesion: Effect of ion bombardment on polymer interfacial reactivity. In: Mittal KL (ed.): Metallized Plastics. Marcel Dekker Inc., New York, 1998, pp. 281–289
48. Liu K, Nagodawithana N, Searson PC, Chien CL, Perpendicular giant magnetoresistance of multilayered Co/Cu nanowires. Phys Rev **B51**, 7381–7384 (1995)
49. Chakarvati SK, Vetter J, Template synthesis – a membrane-based technology for generation of nano/micro materials: a review. Rad Meas **29**, 149–159 (1989)
50. Biswas A, Awasthi DK, Singh BK, Lotha S, Singh JP, Fink D, Yadav BK, Bhattacharya B, Bose SK, Resonant electron tunneling in single quantum well heterostructure junction of electrodeposited metal semiconductor nanostructures using nuclear track filters. Nucl Instrum Methods **B151**, 84–88 (1999)
51. Klein JD, Herrick RD II., Palmer D, Sailor MJ, Brumlik CJ, Martin CR, Electrochemical fabrication of cadmium chalcogenide microdiode arrays. Chem Mater **5**, 902–904 (1993)
52. Brumlik CJ, Menon VP, Martin CR, Template synthesis of microtubule ensembles utilizing chemical, electrochemical, and vacuum deposition techniques. J Mater Res **9**, 1174–1182 (1994)
53. Maaroof A, Evans BL, Onset of electrical conduction in Pt and Ni films. J Appl Phys **76**, 1047–1054 (1994)
54. Herman DS, Rhodin TN, Electrical conduction between metallic microparticles. J Appl Phys **87**, 594 (1996)
55. Klett R, Klose HA, Morawetz K, Müller-Jahreis U, Noack R-A, Thiele P, Optical and electrical in-situ characterization of electron-beam and ion-beam deposited PtSi-nanolayers. Personal communication; to be published (2003)
56. Masden JT, Giordano N, Finite-size effects in the electrical conduction of thin wires. Phys Rev **B36**, 4197–4202 (1987)
57. Neumann R, Ion induced modifications in solids: basic aspects and applications in nanoscience. 21st Int. Conf. on Nuclear Tracks in Solids, New Delhi, 21-25 Oct. 2002
58. Burford RP, Tongtam T, Conducting polymer with controlled fibrillar morphology. J Mater Sci **26**, 3264–3270 (1991), and references therein
59. van Dyke LS, Martin CR, Fibrillar electronically conductive polymers show enhanced rates of charge transport. Synth Met **36**, 275–281 (1990)
60. Pfohl T, Kim JH, Yasa M, Miller HP, Wong GCL, Bringezu F, Wen Z, Wilson L, Kim MW, Li Y, Safinya CR, Controlled modification of microstructures silicon surfaces for confinement of biological macromolecules and liquid crystals, Langmuir **17**, 5343–5351 (2001);
and: Li Y, Pfohl T, Kim JH, Yasa M, Wen Z, Kim MW, Safina CR, Selective surface modification in silicon microfluidic channels for micromanipulation of biological macromolecules. Biomedical Microdevices **3**, 239–244 (2001)
61. Heiss M, Fischer B, Cholewa M, The ion-microprobe at GSI and its applications. Proc European network on ion track technology, Caen, 24–26.2.2002;
and: Cholewa M, Fischer BE, Heiss M, (2002) Preparatory experiments for a second-generation radiation-biological single hit facility. Presented at the 5th Intl. Symposium on "Swift Heavy Ions in Matter", May 22–25, 2002, Giordano Naxos, Italy

62. Wang N, Tang ZK, Mono-sized and single-walled 4-Angstom Carbon nanotube arrays. Proc MRS Boston 27.11.–1.12.2001, Contribution V12.8;
and: Beck JS, Vartuli JC, Roth WJ, Leonowicz ME, Kresge CT, Schmitt KD, Chu CTW, Olson DH, Sheppard EW, McCullen SB, Higgins JB, Schlenker JL, A new family of mesoporous molecular sieves prepared with liquid crystal templates. J Am Chem Soc **114**, 10834–10854 (1992)
63. Pool R, Physicists tackle theory, tubes, and temperature. Science **247**, 1410–1412 (1990)
64. Douglas K, Devaud G, Clark NA, Transfer of biologically derived nanometer-scale patterns to smooth substrates. Science **257**, 642–644 (1992)
65. Clark TD, Ghadiri MR, Supramolecular design by covalent capture. Design of a peptide cylinder via hydrogen-bond-promoted intermolecular olefin metathesis. J Am Chem Soc **117**, 12364–12369 (1995)
66. Ozin GA, Nanochemistry: synthesis in diminishing dimensions. Adv Mater **4**, 612–649 (1992);
and: Schöllhorn R, Intercalation systems as nanostructures functional materials. Chem Mater **8**, 1747–1757 (1996)
67. Tokarev A, Sidorenko A, Minko S, Stamm M, Organized array of metallic nanoclusters via self-organization of block-copolymers. Intl. Workshop on Nanostructures for Electronics and Optics, ID:41, August 18–21, 2002, Dresden
68. Nielsch K, Wehrspohn RB, Barthel J, Kirschner J, Gösele U, Fischer SF, Kronmüller H, Schweinböck TH, Weiss D, High-density hexagonal nickel nanowire arrays with 65 and 100 nm-period. Proc MRS Boston 27.11.–1.12.2001, Contribution Y9.3;
and: Nielsch K, Hertel R, Wehrspohn RB, Barthel J, Kirschner J, Gösele U, Fischer SF, Kronmüller H, Schweinböck TH, Weiss D, Switching behavior of single nanowires inside dense nickel nanowire arrays. Proc IEEE International Magnetics Conference, Piscataway, NJ, USA, IEEE 2002, p.GB2 of vii and 513
69. Gösele U, Nielsch K, Reiche M, Müller F, Wehrsporn R, Choi J, Two-dimensional ordered nanopore arrays, Proc Intl. Workshop on nanostructures for electronics and optics, Dresden, Aug. 18–21, 2002, contribution # 40
70. Tonucci RJ, Justus BL, Campillo AJ, Ford CE, Nanochannel array glass. Science **258**, 783–785 (1992)
71. Grün M, Unger KK, Matsumoto A, Tsutsumi K, Ordered microporous MCM-41 adsorbents: novel routes in synthesis, product characterization and specification. Royal Society of Chemistry. Characterization of Porous Solids **IV**, pp. 81–89 (1997)
72. Zhao D, Feng J, Huo Q, Melosh N, Fredrickson GK, Chmelka BF, Stucky GD, Triblock copolymer syntheses of mesoporous silica with periodic 50 to 300 Angstrom pores. Science **279**, 548–552 (1998)
73. Rückes Th, Kim K, Joselevich E, Tseng GY, Cheung C-L, Lieber CM, Carbon nanotube-based nonvolatile random access memory for molecular computing. Science **289**, 94–97 (2000)
74. Kyotani T, Tsai LF, Tomita A, Preparation of ultrafine carbon tubes in nanochannels of an anodic aluminium oxide film. Chem Mater **8**, 2109–2113 (1996)
75. Herminghaus S, Mikroskopische Wasserkanäle für die Nanotechnologie. Presse-Information PRI C1/99(5), Wirsing B, Frese W, Hintsches E, Trepte A (eds.), Max-Planck-Gesellschaft, München, 1999 (in German)

76. Ternashi T, Sugawara A, Shimizu T, Miyake M, Planar arrays of 1D gold nanoparticles on ridge-and-valley structured carbon. J Am Chem Soc **124**, 4210–4211 (2002)
77. Fahrner WR, Hilleringmann V, Horstmann J, Job R, Neitzert HC, Scheer H, Ulyashin A, Wieck A, Nanoprozessierung und Nanoelektronik, Fahrner WR (ed.), Fernuniversität Hagen, Germany, 2001, and many references therein (in German)

7 Ion-Track Applications

D. Fink, P.Yu. Apel, and R.H. Iyer

Just as low-energy ion irradiation of polymers has found a number of useful applications, also tracks of energetic heavy ions in polymers can be used in technology. In fact it took less than ten years after the detection of ion tracks (1958) for them to be applied (1967). Though there exist already quite a number of scientific applications, only a few industrial applications such as ion-track filters and field emitting flat panel displays have been commercialized.

In this chapter, we not only give an overview about existing applications and those that are currently under development, but we also try to give suggestions for future applications that have not yet been tackled.

7.1 Applications of Latent Tracks

It appears that, in comparison with etched tracks, the application potential of latent tracks is limited. At present three major possibilities show up:

- The exploitation of the material's radiochemical changes
- The exploitation of structural changes along latent tracks, with the consequence of modified transport and trapping properties along ion tracks
- The use of ion-induced phase transitions

The first possibility signifies use of the modified polymer itself as, e.g., a sensor. This has been successfully demonstrated recently by Chakane et al. [1]. The resistivity of Makrofol KG foils dropped after irradiation with energetic Ag ions in the presence of hydrogen gas from 400 to 320 MΩ, which corresponds to a sensitivity of $\sim$20%. When creating additional "irrigation channels" in the polymer by etch-track formation preceeding the ion irradiation, the hydrogen-sensing sensitivity was enhanced by up to 80%, however, at the expense of slower response. FTIR examinations showed that the CH rocking signal (488 cm^{-1}) and the –OH group signal (3508 cm^{-1}) increased, and the –OH–O– stretching line (3287 cm^{-1}) decreased after hydrogen exposure, however, this information alone is not yet sufficient to understand the underlying mechanism. Possibly the excess hydrogen is loosely bond to some dangling bonds, thus giving rise to protonic conductivity.

Another example of exploiting radiochemical changes in polymers along the ion tracks: polysilanes transform into silicon-carbide-like matter upon swift heavy-ion irradiation (see Vol. I, Sect. 8.4.3) [2]. Though the as-implanted ion-track zones show up as soft and deformable fibers in SEM images (Fig. 4.29), they apparently reorganize to $\sim$50-nm thick crystalline rods of good mechanical stability after prolonged thermal annealing at elevated temperatures (see Vol. I, Fig. 8.16) [3]. Therefore it has been suggested to use them (after appropriate doping) as future conducting cantilevers in AFM microscopy on specially cut silicon supports, just as carbon nanotubes can also be applied as AFM tips [4]. Apart from this possible application, the tiny SiC_x–Si transition zones might be useful for electronic applications in the case of previous appropriate doping of the components.

In connection with the second possibility, solution diffusion through irradiated membranes can be used for separation purposes. Here both the solubility and diffusivity of the penetrant are the decisive parameters to control the transport. Furthermore, it has been suggested to dope latent tracks for future polymeric nanoelectronics [5], to expand the old concept of low-energy polymer doping [6] for this purpose. The main idea is to preferentially dope the free volume along ion tracks with suitable materials that could give rise to the desired electronic properties. However, until now that concept has not yet left the stage of basic research.

Another possibility is to use swollen latent tracks filled with electrolytes for pulsed transport of electrical signals (see Fig. 3.20, Sect. 3.4.1) for biological applications, to exploit the similarity in their behavior with charge transport through biological membranes. Again, this concept has not yet left the experimental stage.

Swift heavy-ion-irradiated polymer foils are said to be used, however, already commercially as seals to protect the interior of cleanrooms or small electric motors against penetration of dust and water vapor from the ambient, though keeping their internal pressure under ambient conditions, due to the foils' ready permeability for lighter gas molecules from the ambient.

There exists also the possibility to let metal atoms or metallic clusters precipitate along latent tracks to obtain conducting nanowires for future electronic applications [7].

The last possibility is to exploit ion-induced phase transitions along swift heavy-ion tracks. Possibly this point does not apply to polymers, as it is at present not even known whether polymeric phase transitions such as the sol-gel transition are influenced by ion irradiation at all. It is, however, important for other carbonaceous matter such as fullerite or diamond that is transformed upon swift heavy-ion impact into a nongraphitic, sp2-rich conducting carbon phase within cylindrical extended regions (tracks) of some 50 nm width [8,9]. There are attempts to exploit this effect for field-emission displays [9]. As this confined sp2-rich phase is expected to have a more open structure than diamond itself, it is speculated that one might be able to preferentially dope tracks in diamond for electronic purposes [8].

7.2 Applications of Etched Tracks

Ion-track-etched membranes find a steadily increasing number of applications in frontier areas of science and technology [10–15]. There is now an increasing demand for these membranes because of the growing number of applications such as microfiltration, ultrafiltration, superclean environments, medical science and pharmaceutical industries, hospital treatment of burns and wounds, respiration protection, gas diffusion and separation, pervaporation, analytical sciences, and separation sciences. For many of these applications, high-performance microporous membranes with well-defined and uniform pore size and pore density are required. Some of the other important and desirable properties of the microporous membranes are their uniform thickness, their inertness to toxic environments, a high tensile strength, large porosity, large possible sizes, and nearly unit tortuosity (i.e., ratio of the length of the pore to the thickness of the membrane).

7.2.1 Etched-Track Membranes

The selectivity or rejection ability ϕ of a microporous membrane is the main parameter to describe the membrane's capability to prevent the passage of micro- or nanosized particles through its pores. It is defined as:

$$\Phi = (1 - C_{\mathrm{p}}/C_0)100\%, \tag{7.1}$$

where C_{p} is the permeate concentration and C_0 is the initial solution concentration. This selectivity is usually defined by a parameter $\lambda = d/D$, with d the particle size and D the pore size. The relationship $\phi(\lambda)$ allows one to choose the optimum membrane structure that results in maximum efficiency at given selectivity [16]. In ultra- and microfiltration, the flowthrough membrane is pressure driven (i.e., it is operated with a pressure difference applied across the membrane). The peculiarities of the physicochemical particle–pore surface interaction do not affect the selectivity significantly is this case. As a result, the relationship $\phi(\lambda)$ is identical for membranes of various types.

Different methods of pore-diameter determination yield different $\phi(\lambda)$ curves, the disagreement being the larger the narrower the pores. The reason for this data scattering was found to stem from the deviations of the real pore shapes from the idealized cylindrical form, as real tracks are more or less conical or hyperboloidal in shape, due to a finite etching speed. For instance, in paper [16] it was found for PET track-etch membranes that $D_{\mathrm{SEM}} > D_{\mathrm{HD}} > D_{\mathrm{BP}}$, with D_{SEM} being the diameter as determined from SEM observations, D_{HD} being the diameter resulting from hydrodynamics (i.e., from the viscous flow of a liquid through the pores), and D_{BP} being the pore diameter as obtained by the "bubble point" method [16]. Whereas D_{SEM} roughly corresponds to the largest diameter D_{max} of a conical track, D_{BP} rather refers to the smallest diameter D_{min}. All diameters D_{min}, D_{max}, and D_{HD} are correlated with each other in the given example [16] via:

$$3(D_{\max}^4/D_{\mathrm{HD}}^4)\beta^3 - \beta^2 - \beta - 1 = 0, \tag{7.2}$$

where $\beta = D_{\min}/D_{\max}$. Hence it appears to be advisable to use the "bubble point" method to obtain the most realistic $\phi(\lambda)$ relationship of a microporous filter membrane. With this measurement approach, the track membrane selectivity ϕ reaches 50% for $\lambda = 1.3$ and 100% for $\lambda = 2.0$. As also ITS essentially refers to the inner diameter $D_{\min}$ of conical tracks, this technique should be of equal use for $\phi(\lambda)$ determination [16]. However, (7.2) is not universally valid; note that other pore profiles are also possible [13] and DSEM can be smaller than DHD which is typical of some commercial TEMs. It has recently been proposed to use tracks with suitably etched nanopores as filters for hydrogen purification; permselectivities $(PH_2)/(PCO_2)$ of up to 4 were reported [17].

Track-etched membranes (TEMs) can be produced by fission-product beams or by high-energy heavy-ions from accelerators. Production using accelerated heavy ions is the preferred mode because (a) the accelerated ions are of stable nuclei and therefore radioactive contamination of the irradiated material is totally absent, in contrast to track production by fission products, (b) all accelerated ions have the same mass and well-defined energy that ensures high homogeneity of the pore size of the membranes, (c) the tracks produced in this way have excellent parallelism as the ion beams used usually have divergences of less than one degree, and (d) the flux of the accelerated ions as well as their energy can be easily varied to produce membranes of any desired pore density and of different thicknesses.

Track-etched membranes offer distinct advantages over conventional membranes due to their precisely determined structure. The well-defined pore characteristics make them eminently suitable for a wide variety of applications such as ultrafiltration ($< 0.1\,\mu$m) and microfiltration (0.1–1.5 μm) membranes for complete removal of viruses, bacteria, dust and colloidal particles from fluid media. This is apparent from the relative sizes of ions and molecules and the optimum ranges of pores sizes for different applications as illustrated below in Table 7.1 and Fig. 7.1 [18].

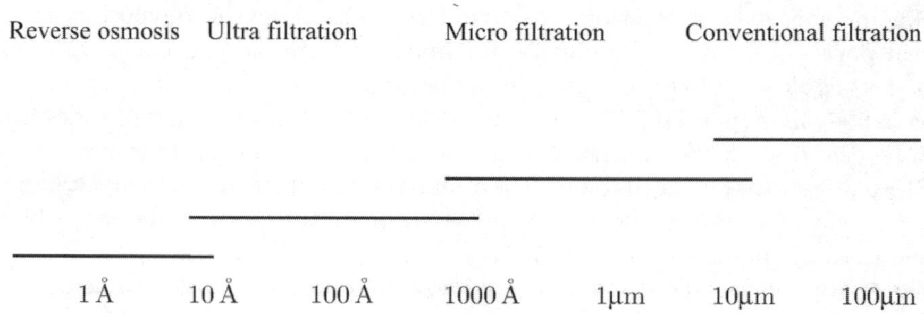

Fig. 7.1. Optimum ranges of pore size vs. applications

Table 7.1. The relative size of some ions and molecules

H_20	0.2	nm
Na^+	0.37	nm
Sucrose	1	nm
Haemoglobin	7	nm
Influenza virus	100	nm
Staphylococcus bacteria	1000	nm = 1 µm
Starch		10 µm

Therefore many countries have embarked upon setting up facilities for continuous production of nuclear-track membranes of the order of thousands of square meters per year [11]. The first track membranes using heavy-ion accelerators were produced in Dubna in the early 1970s and the first pilot plant constructed in 1980 [12]. Facilities for regular industrial production of these polymeric membranes of a wide range of thicknesses, pore densities and pore diameters are now available at the accelerators of JINR, Dubna [11–13], at the HMI, Berlin, at the CRC, Louvain-la-Neuve, and at the GANIL, Caen. Besides this, there are several sites especially in the United States where TEMs are produced both by accelerated ion beams (Brookhaven) and by fission-product beams.

Track-membrane production is meanwhile highly advanced. There exist systems to obtain not only parallel tracks but also tracks with other angular distributions [11–13]. In this way, the probability of overlapping of individual pores throughout the whole foil is drastically reduced and the membrane retention ability for filtering applications significantly increases. The theory of clustering of tracks that determines the quality of the microporous foils is given, for instance, in [11, 16].

For 1995, the market analysis for this type of foils gave $2 \times 10^5\,m^2$ for superclean elements production, $2 \times 10^5\,m^2$ for medical purposes, and $1 \times 10^5\,m^2$ for other applications such as ecological investigations, respiration protection, analytic purposes, etc.

A basic requirement for medical applications of microporous foils is the possibility to sterilize them. This is indeed demonstrated; such foils have been shown to be biosafe, having passed USP Class VI-121°C Plastics test [19], i.e. they can be autoclaved at 121°C or steam-sterilized without shrinkage. PET and PI are also stable to sterilization by γ-irradiation and chemical sterilization techniques.

The inherently hydrophilic nature of, e.g., PET can be modified to produce microporous membranes that have exceptional hydrophobic or oleophobic properties [19]. Recently, microporous PI and PET foils have also been laminated with other polymer foils to obtain flexible microchannels with porous walls with widths ranging from 30 to 500 µm, heights ranging from 5 to 20 µm, and lengths up to 6 cm, for advanced filtration purposes or medical applications [19,20]. Such composite membranes utilize polyester or PP media. Laminated membranes can be sealed by ultrasonic, heat, or RF sealing technologies.

Present commercial microporous products are offered not only as plain foils, but also as self-adhesive membranes, cut into various shapes, and delivered in roll form on release paper to meet the requirements of automated manufacturing lines [19].

Successful attempts have been undertaken to reduce the width of etched tracks for ultrafiltration applications by inserting into them Au tubules of defined inner diameter, to cleanly separate small molecules on the basis of their molecular size [21]. This may find applications for highly selective membranes for chemical separations.

Microporous polymer foils are also used as common backings for collecting aerosol particulate matter in environmental research due to their ability to absorb matter (e.g., aerosols) in their large open volume, and due to their low content of elements other than H, C, and O. Multielement analysis of the sampled material is then conveniently performed by proton-induced X-ray emission spectrometry (PIXE) [22].

7.2.2 Applications of Ion Tracks in Dosimetry and Related Fields

There exists a very broad range of applications in the field of dosimetry, covering, e.g., the art of determining the track-forming ion specie and its energy from track observations, techniques for fast visualization of track distributions and track counting, autoradiographic techniques, and techniques of lowering the detection limits of ionic radiation. Furthermore, track recording is commonplace for applications in medicine, biology, geology, mineralogy, uranium and oil prospection, space science, and many other fields. However, as there are excellent reviews and books available in these fields, we do not incorporate these topics in this book.

7.2.3 Applications of Etched Tracks in Lithography

The use of ion tracks enables one to form micro- and nanostructures that are inaccessible by X-ray and electron lithography. Since each individual particle creates a large defect density, only a small number of ions is necessary, generally less than some $10^9\,\mathrm{cm}^{-2}$. As such doses can be delivered within fractions of a second by heavy-ion accelerators, ion-track-based lithography offers a realistic alternative to existing synchroton-based deep lithography. To promote these efforts, a "European Network on Ion Track Technology" has been founded recently [23]. Two strategies of ion-track lithography are being followed. One is ion-track projection lithography, where stencil-mask technology developed from X-ray lithography is applied, to remove precisely defined zones along a prescribed direction using energetic ions. In this way, deep-lying plateaus with 6 nm roughness and 90° sidewalls with aspect ratios of up to 40 have been obtained [20].

The other approach uses the stochastically distributed individual ion tracks and combines them with thin-film lithography (thus combining nano- and microtechnology) for fabricating three-dimensional circuitry based on ion tracks as vertical, and evaporated connections as horizontal, structuring elements [23]. The applications described below in Sect. 7.2.11 belong to this topic. In such cases the number of nanowires N_{cont} connecting two contacts with area F_{cont} on the front and rear sides of the foil fluctuate around a mean value $N_{\text{cont}} = \Phi F_{\text{cont}}$, with its statistical variation s given by the stochastic (Poisson) distribution of the irradiation density. Clearly, F_{cont} and/or Φ (the areal nanowire density, equivalent to the ion-track density) must be tailored so that there is always at least one nanowire connecting two contacts. Experimentally determined nanowire distributions coincide reasonably well with theoretically calculated ones. For $N_{\text{cont}} = 10$, $\sigma = 53\%$ [24].

Lithography can be combined with localized thermal treatment of latent tracks that largely annihilates them, and thus prevents their etchability [25].

7.2.4 Applications of Etch Tracks for Galvanic Replicas

The fundaments of electroreplication of etched-track membranes have been outlined some time ago [26]. Etched tracks are filled with some metal, e.g., Cu, either in the form of needles (rods) or tubules, and the matrix polymer is then dissolved. The remaining metallic structure can be used for various applications, e.g., as cooling devices for microelectronics due to its large surface-to-volume ratio and due to its small-size microstructure, or as a dew-point sensor, as an antiwetting coating etc.[1] In the latter case, one exploits the mushroom-like structure of the replicas that emerges when the galvanic deposition is carried out beyond the point of filling up the track completely.

7.2.5 Applications of Etched Tracks in Separation Technology

Membrane separation techniques have already been used in some industrial applications like the desalination of brine, salt manufacturing from sea water and oxygen-enriched air. Another technique of importance is pervaporation, which has been recently used for the separation of a number of organic liquid mixtures. Specially prepared polymer foils were used for all these applications, but only now has it turned out that etched tracks may also find useful applications in separation technology [27–31]. In this field the requirements are, e.g., small thickness, high chemical inertness, high porosity, small pore size and, in the case of handling of radioactive waste, a high radiation resistance. The most widely used membrane supports are made commercially by solvent

[1] Schulz A, Fraktal GmbH Quedlinburg (2001), personal communication

cast methods, by exposing polymer films to consecutive steps of cold stretching, hot stretching and heat setting and by various other methods [32]. These chemical membranes have some disadvantages such as low tensile strength, nonuniform pore density, and large tortuosity (i.e., the ratio of pore lengths to the membrane thickness), but their porosity is around 60 to 85%. Their poor mechanical properties are usually counteracted by a relatively large membrane thickness (typically 150 μm or so). The search for alternatives for these chemical membranes soon arrived at etch-track membranes as supported liquid membranes (SLMs) for the carrier-mediated transport and removal of, e.g., toxic metal ions from radioactive or other industrial effluents [33].

In comparison with solid membranes, liquid membranes (LMs) have the great advantage that the diffusion coefficient of impurities through them (typically 10^{-7} to $10^{-6}\,\mathrm{cm^2\,s^{-1}}$ [28] exceeds that in solids by many orders of magnitude. As this allows a high throughput of matter it is advantageous for technological purposes. A further enhancement in throughput is obtained by using the most suitable carriers which are complexing agents dissolved in an appropriate solvent, specifically selected so as to have a high reversible selective reaction with the solute to be transported. Transport across various liquid-membrane configurations such as bulk liquid (BLM), emulsion liquid (ELM), and supported liquid (SLM) membranes are fast emerging as potential separation and concentration techniques particularly for the processing of dilute solutions, in the separation and recovery of metals of critical and strategic importance, in the decontamination of low- and high-level radioactive wastes [29], etc. In order to overcome the drawback that liquid membranes are not stable in shape, one can, e.g., impregnate the liquid into the pores of a porous solid (ILM, impregnated liquid membrane). Two common geometries for supported-liquid membranes are the flat sheet and the hollow fiber. Track-etched membranes belong to the most promising porous flat sheets for this purpose.

In a typical carrier-mediated liquid membrane, there are two possibilities of the selective transport of a solute **A** (e.g., a metal cation) from an aqueous solution (feed) through a LM with a water-immiscible complexing organic agent **C** (carrier) to another (strip) [34, 35]:

1. The coupled transport from left to right via the LM in the center, where two species **A** and **B** are transported simultaneously:

feed	LM	strip
$\mathbf{A} \rightarrow$	$\mathbf{CA} \rightarrow$	$\mathbf{A} \rightarrow$
$\leftarrow \mathbf{B}$	$\leftarrow \mathbf{CB}$	$\leftarrow \mathbf{B}$

and

2. the facilitated transport, where only one specie A is transported:

feed	LM	strip
A →	**AC** →	**A** →
	← **C**	

By complexing with the extractant, the solute atom (or ion, molecule) **A** is able to enter the organic phase in which it is otherwise insoluble. Concentration gradients within the membrane cause the carrier–metal complex to diffuse to the opposite side where **A** is released and eventually a counterion **B** (e.g., a proton) is picked up. By controlling the environment on each side of the membrane (e.g., the pH value), the formation of the carrier–metal complex can be controlled.

Diverse practical applications of such membranes in the recovery of metals from hydrometallurgical leach solutions, industrial effluents, and in the removal of valuable fission products, uranium, plutonium, americium, etc., from the nitric acid waste streams generated during the reprocessing of spent nuclear fuels have been reported in the literature [30].

Liquid membranes (LM) containing mobile complexing carriers have been shown to be effective and attractive for the selective separation and concentration of a solute, e.g., metal ions (e.g., alkali atoms, Fe, Cu, Zn, U, Am, or Pu), weak acids and bases, organic compounds (e.g., olefines or phenol), biologically important components (e.g., amines, amino acids, cholesterol, drugs etc.), gaseous mixtures (e.g., O_2, CO_2, CO, H_2S, or NO), etc., from dilute solutions [34]. These membranes offer distinct advantages over other conventional chemical-separation techniques, e.g., lower capital cost, low energy consumption and low solvent inventory that provides opportunities for economic use of even exotic and expensive reagents not attempted so far. High separation factors of the order of 10^2 to 10^4 typically are some of the other outstanding merits of this novel technique. Different LM systems have the same general configurations consisting of an aqueous and receiving phase separated by a membrane that usually consists of an organic lipophilic liquid. Judicious selection and modification of the constituents of the systems allows one to approach a desired level of selectivity in solute transport.

SLMs are readily made by immobilizing a liquid phase containing a complexing carrier (which is an extractant for the metal cation to be separated) in the pores of the microporous substrate by capillaric force. The immobilization is done simply by immersing the microporous substrate in the carrier for 8–10 h. The metal ions (solute) are then transported through the membrane from one side to the other where they are decomplexed. LM carrier materials frequently have concave geometrical molecular shapes to embed efficiently the solutes. In this sense, e.g., crown compounds have gained great importance.

In a recent study [33] Iyer showed that the transport efficiency of Pb^{++} ions, chosen as a typical divalent metal ion using D2EHPA as the carrier through 10-μm thick track-based microporous foils with typically 2–5% porosity is practically equivalent to that of the efficiency of conventional membranes with 160 μm thickness and 85% porosity. Since the pores of TEMs are well defined, they can be used as model supports in SLM studies. The unit tortuosity in the TEMs used ensures unambiguity in the interpretation of the data on the transport of metal ions. The comparison of permeability coefficient (P) obtained by using TEMs of 10 µm thickness and PTFE membrane of 160 µm thickness indicated that P decreases linearly with thickness [33,36]. This interesting result shows that the diffusion of the metal complex across the membrane is the rate-controlling step, rather than the extraction and stripping reaction kinetics. Therefore the smaller thickness of ion-track-based microporous foils more than compensates for their poor porosity.

In test experiments on, e.g., an actinide (Am^{241}) separation/recovery [33, 36] the etch-track membranes already proved their competitiveness by absorbing practically the whole amount of the representative trivalent actinide ion Am(III). However, the Am permeation through commercial filter foils was found to be still somewhat more rapid than that through etch-track membranes of the same area [33]. Recovery and purification of U^{233} present at levels of about 1000 ppm in 4 mole/l nitric acid waste solutions containing bulk amounts (40–45 g/l) of aluminum was done successfully using PTFE membranes (0.4 µm pore diameter, 160 µm thick, 85% porosity). Within 4 h, 80% of the uranium was recovered, without any trace of aluminum being absorbed [33].

Another important example of the potential application of SLM-based separation schemes in waste management was the production of pure, carrier-free Y^{90} for nuclear medicine and therapeutic uses from high-level radioactive waste solutions [33]. Using a SLM configuration, pure Y^{90} in MBq to GBq quantities could be separated [36].

In the case of galvanic Cu waste solution reprocessing, Cu could be enriched by factors between 300 and 3000 in comparison with simultaneously abundant Zn and Sn [26]. Meanwhile results are also available on the transport properties of Cd, Co, Hg, Ni, Pb, and Zn by D2EHPA and Cr by Aliquat-336 embedded in etch tracks and PTFE membranes [33]. Whereas excellent transport efficiency was found for Cd, Cr, Cu, Pb, and Zn, the transport was slower for Co and Ni, and poor for Hg. As ions, separated by 2-ethyl hexanol, also showed only poor extraction.

Ideally, it would be desirable to use an SLM system for continuous operation on a long-term basis without regenerating the membrane with fresh carrier. However, in practice the carrier held in the pores of the membrane phase gets depleted over a period of time depending on various factors such as:

1. Solubility of the carrier in the aqueous feed and strip solutions with which it is in contact
2. Losses due to diffusion across the membrane phase
3. Losses due to the very small but finite osmotic pressure differences between the feed/strip solutions

Using the TEM (thickness 10 µm, pore diameter 0.4 µm, porosity 2.75%) and using Pb^{++} as a test metal ion, it is established that while a freshly prepared SLM shows a transport efficiency of ~98%, the value decreased to ~85%, ~55% and ~30% respectively after continuous use for 4, 14 and 20 days without regeneration of the membrane with fresh carrier [37, 38].

In general, the effectiveness of a liquid-membrane process depends, among other factors, on a small membrane thickness for a short diffusion length and a high concentration of the carrier in the membrane phase. Track-etch membranes would emerge as a potential and essential component in the area of supported-liquid-membrane-based separation and/or recovery of environmentally toxic and strategically important metal ions (hydrometallurgical industry) from dilute solutions of low metal ion concentrations. Membranes with porosity in the range of 2–5% and with an ideal pore diameter of 0.4–1.0 μm and membranes of 20–25 μm thickness (which are naturally more robust and easier to handle in engineering-scale modules) would be sufficient for this application and can be easily manufactured. One problem associated with the industrial-scale use of SLM configurations especially with thin membranes (as in the case of TEMs) is their poor stability for long-term continuous operations without regeneration. However, considering the fact that most of the separations using SLMs are complete (close to 100%) in 6–24 h, the problem of stability can be circumvented by designing easily replaceable SLM modules containing large-area membranes that can be inserted into the process stream without interrupting the process. The use of thicker membranes would make handling of membranes easier and allow construction of robust flow modules [37].

Finally let us mention that biomimetic SLMs, e.g., lipid-impregnated track-etch PET membranes have been used as a model for biological membraness [39].

An effect that also might play a role in future liquid-membrane technology is that narrow etched tracks in PC and PET always exhibit some cation-permselective behavior due to the charging-up of their walls (see Sect. 3.4.2). Also narrow (< 2 nm) Au nanotubules embedded in etched tracks display a nearly ideal cation-permselective behavior, i.e., these membranes trransport cations but reject anions. The reason for this behavior is found in the strong selective adsorption of some anions (e.g., Cl^- in a KCL solution, but not F^- in KF solutions!) on Au. As a result, the Au tubules have an excess of negative charge on their inner surface that makes them impermeable to anions. Ion permselectivity can be controlled by changing the potential U applied to the Au nanotubules, from ideal cation ($U = -0.5$ V) to ideal anion ($U = +0.5$ V) permselectivity, which induces excess negative or positive charges on the walls of the Au tubules, respectively. The potential of zero charge (p.z.c.) where there is no excess charge on the metal, is found to be −4 mV [40, 41].

It might become of importance for future separation science and medical applications that recently an ion pump has been developed [42] on the basis of asymmetrically etched nanopores. Upon application of a low-frequency electric field across the nanopore-containing membrane, ions such as K^+ can be pumped against their concentration gradient (see Sect. 3.4.2).

7.2.6 Applications of Etched Tracks in Sensing Technology

A major field of future potential ion-track applications covers sensors. Sensors can, on the one hand, probe physical, chemical, and biological parameters such as temperature, pressure, liquid flowthrough, magnetic field strength, humidity, the presence of photons or more energetic particles, etc., and on the other hand determine the concentration of components or trace elements in gases or liquids, or biological activities. The sensor market is rapidly growing at 5.3% per year,[1] with a value worldwide of 27.8×10^9 Euro in 1998, and probably 46×10^9 Euro in 2008. At present, the car industry has the greatest use for sensors, followed by the product-processing industry, household and office-equipment production, construction, machine building and aereospace industries.

There are a number of possible approaches to realise ion-track-based sensors. One can either use the microporous foils just as porous windows of other sensoring devices, or one can incorporate the sensor material into the microporous foil itself. Whereas the first approach has already found some technological applications, the latter is partly still in its development stage.

Semiconductor gas sensors generally have to be operated at elevated temperatures, ranging from 150 to 700°C to obtain the speed and magnitude of response required for practical applications. Consequently, portable gas sensors of integrated sensing systems should have a good thermal insulation in order to keep the power consumption low and to protect the adjacent circuitry from the sensor's high temperature. One way followed earlier was to fabricate the sensing element with the integrated heater in thin-film form on a suspended thin membrane. This concept was realized for alcohol (and ether, methoxyflurane, halothane, propylene) sensors based on tin-oxide ceramics, the latter having been produced from corresponding organo-Sn. The sensors showed good response at sensing temperatures of about 250°C [43]. Similarly, a humidity sensor based on C_{60} is in development that allows regeneration by annealing to temperatures above 100°C, by means of miniaturized integrated heater elements [44]. An alternative concept suggested here is to incorporate both the sensor and the resistance-heating elements within individual ion tracks in PI as concentric tubules. In this way the sensor size could be reduced from some 50 μm to ~1 μm, with an accompanying impressive reduction in power consumption, good temperature insulation due to the polymer's low heat conduction, and good shielding of the sensitive sensor element within the interior of the protective polymer foil. This concept has not yet been realized.

Clark Cells

Clark cells are amperometric sensors to determine concentrations of cathodically reducible or anodically oxidatable compounds or elements. The basic philosophy is to separate the sensor volume from the volume to be probed by a membrane that is selectively permeable for the element to be measured, but impermeable to other components. This concept lets the sensor's working electrode rest in a selected electrolyte with stationary properties, independent of its actual environment.

The common electrochemical basis of many chemical and biochemical reactions is a change of the oxidation state, induced by charge transfer, according to:

$$\mathrm{O} + ne^- \leftrightarrow \mathrm{R}, \tag{7.3}$$

with O = oxidance, R = reductance, e = elementary charge, and n = number of electrons transferred per reaction. This charge transfer may take place at the sensor electrodes. If oxygen is to be probed, the latter is reduced at a cathode to hydroxide:

$$\mathrm{O_2 + 2H_2O + 4e^- \leftrightarrow 4OH^-}. \tag{7.4}$$

For Cl sensoring, the corresponding reduction reaction in acidic media appears as:

$$\mathrm{Cl_2 + 2e^- \leftrightarrow 2Cl^-}. \tag{7.5}$$

The electrode current is proportional to the amount of reduced material.

Contemporary commercial Clark sensor cells, usually built as several cm large compact glass tubes with molten-in Pt cathode wires, have the disadvantage of being quite pressure and motion sensitive due to changes in the membrane geometry. Therefore a redesign has been made on the basis of polymeric ion tracks.[1] The new ion-track-based construction principle enables one to reduce the cell volume by about two orders of magnitude, thus making microanalytical oxygen measurements possible. Due to a fixed connection of tubularly arranged Au cathodes with silicone membranes, the Clark cell operates practically pressure and vibration free, and the polymeric carrier foil enables mechanical flexibility of the sensor. A AgCl electrode, produced by a Ag anode in the KCl electrolyte, enables one to obtain an in situ reference voltage for well-calibrated sensing. A first such test sensor has been built (Fig. 7.2b), and it already enables oxygen measurements of a quality comparable with conventional Clark cells (Fig. 7.2a), at a much reduced price. Further optimization is on the way, and Clark sensors for other elements are expected to follow.

In particular, oxygen sensors have recently experienced a great breakthrough in biosensors, as many biochemically important materials react with the enzyme glucose oxidase towards H_2O_2 that is transformed to water by anodic oxidation. They can be realized in miniaturized form according to the above described redox equations by using Au nanotubules for this purpose[1] [19].

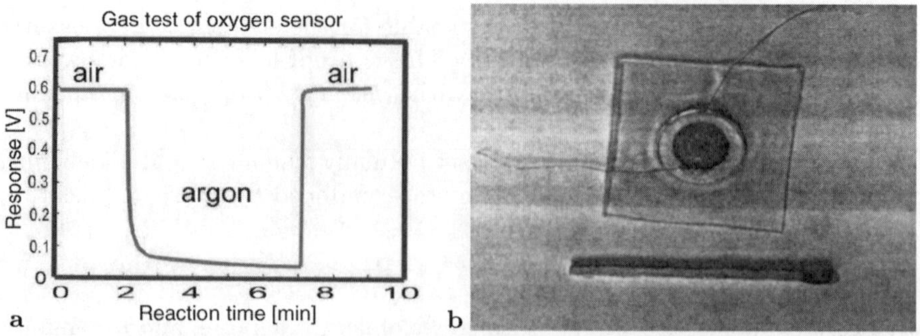

Fig. 7.2. Ion-track-based Clark sensor. (**a**) Typical response curve of oxygen in gas phase; (**b**) view of the first prototype, with a match below as the scale. The dark circular disk is the sensor element. Courtesy of SDK GmbH, Quedlinburg [1]

Sensors Integrated within Microporous Foils

The simplest concept is to fill the tracks with the corresponding sensing matter and to contact the tracks from both sides (Fig. 7.3a). In this case the chemicals to be detected have to diffuse through the polymeric bulk to the sensor. If, by contrast, the sensor is deposited in the shape of tubules, and if it is connected with concentric contacts (Fig. 7.3b), the matter to be probed has direct access to the sensor. Though being slightly more difficult to realize, this construction has a much faster response and greater sensitivity. One possibility to enable the concentric contact formation is to irradiate the foil with low-energy ions (with range $\ll$ foil thickness), so that the sample surfaces become saturated with dangling bonds within that range, and can subsequently bond the desired metals chemically. Another approach is the evaporation of the contact material at tilted angles.

The second concept was realized when PPy nanotubules embedded within a host polymer foil were doped with BF_4^- or ClO_4^-, PF_6^-. Then they vary in their electrical resistivity when exposed to a NH_3 atmosphere [45]. When

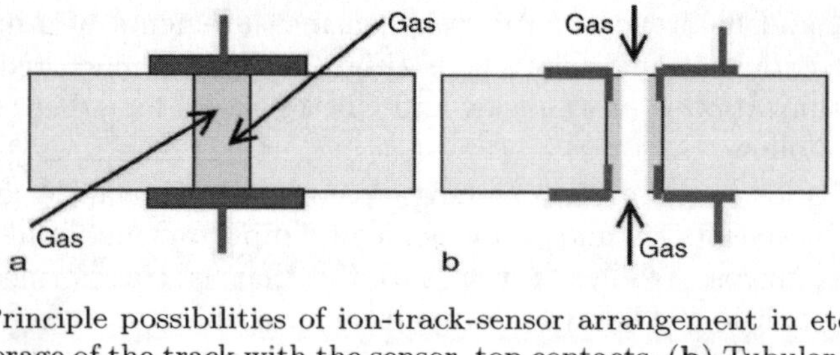

Fig. 7.3. Principle possibilities of ion-track-sensor arrangement in etched tracks. (**a**) Full coverage of the track with the sensor, top contacts. (**b**) Tubular track coverage with the sensor material, concentric tubular contacting. Whereas the chemicals to be probed (*arrows*) have direct access to the sensor in the latter arrangement, they have to diffuse through the polymer in the first configuration

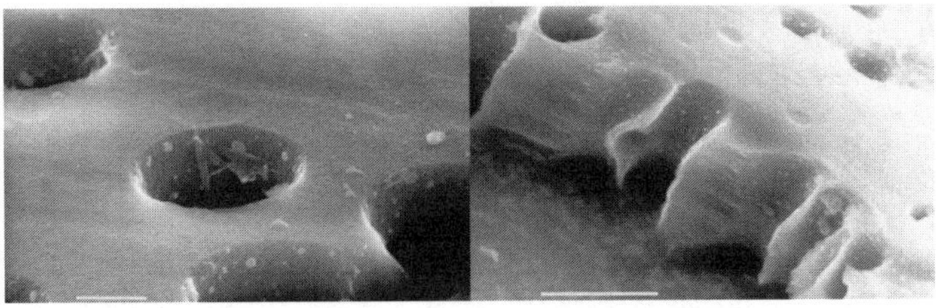

Fig. 7.4. Fullerite-covered etched tracks in a PET foil (4 layers deposited). *Left:* view from top, *right:* side view onto a fractured edge [47]. The C_{60} is seen to deposit as a smooth layer onto inner and outer polymer walls, with occasional larger-size precipitates

depositing Pd or Pt nanoparticles along tracks, the latter are expected to become hydrogen sensing; and phthalocyanine embedded within tracks should exhibit methane-sensing properties.

Sensors Based on Fullerite

The conductivity of fullerite (i.e., the solid state of fullerene, C_{60}) has a strong dependence on humidity, temperature, and pressure, and therefore it has been suggested to use it as a sensor material [46]. The C_{60} has been incorporated in etched tracks by letting it precipitate heterogeneously from a saturated solution in benzene or toluene onto a microporous foil, where it deposits in the form of very smooth thin layers, see Fig. 7.4. Some additional larger C_{60} precipitates are also found within the etched tracks. This deposition process can be repeated multiple times, with the C_{60} tubule resistivity gradually decreasing. However, for more than about 10 deposition cycles, the layer homogeneity deteriorates, resulting in less-efficient resistivity improvement per deposition step.

Figure 7.5a shows the temperature dependence of such tubules. Within the temperature interval from -200 to $+100°$C, the conductivity of these C_{60} tubules shows a strong negative temperature coefficient of resistance (TCR) with a drop of up to an order of magnitude per 10°C. The conductivity has an Arrhenius-type behavior with 1.77 ± 0.2 eV activation energy, the latter being typical for fullerene.

Another peculiarity that showed up in these first tests was that C_{60} strongly interacted with metals at slightly elevated temperatures. When using Ag contacts, the diffusion of the metal into C_{60} and its reaction towards the formation of $C_{60}Ag_{12.4}$ gave rise to another branch in the Arrhenius plot that deteriorated the high TCR of C_{60} somewhat, Fig. 7.5a [48]. Similar results are found for Pd, Ni, and Cu. Au becomes nonwetting at elevated temperatures, but Cr appears to be a stable contact material for that purpose so that it is recommended for future C_{60} sensor contacts. Figure 7.5b shows that C_{60} tubules can also be used as pressure sensors.

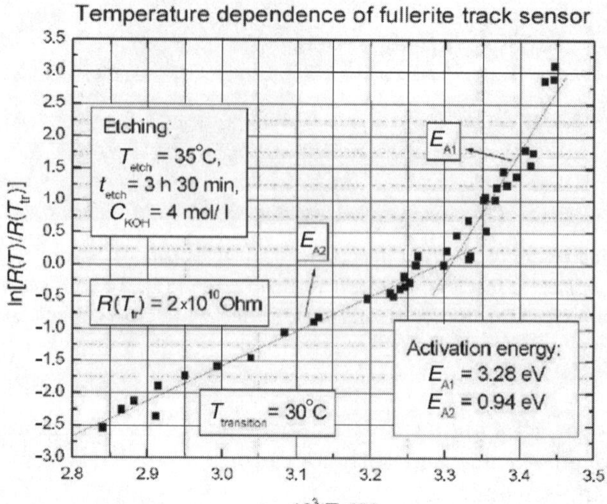

a

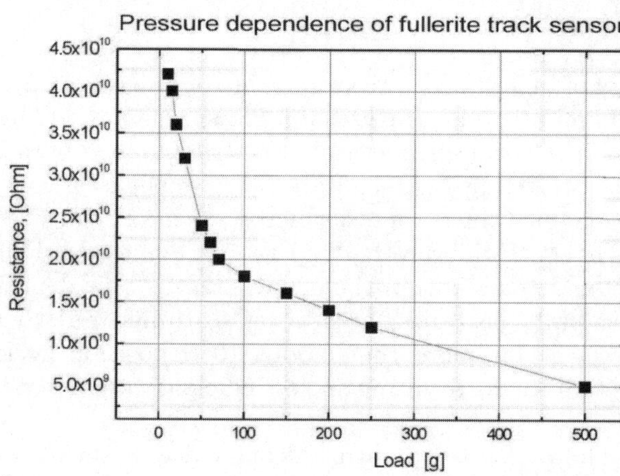

b

Fig. 7.5. Characterization of C_{60}-filled ion tracks. (**a**) Temperature dependence as seen in the Arrhenius plot, and (**b**) pressure dependence. Ag contacts used. Whereas the low-temperature regime reflects the behavior of pure fullerite, the high-temperature branch stems from $C_{60}Ag_{12.4}$ formed with the Ag contact. Such devices can be used as miniaturized pressure and/or temperature sensors

Optical Sensors

Such sensors have the advantage that they are immune against electromagnetic interference, and therefore can be deployed in areas where electrical-based sensors would fail or require expensive protection. One can also multiplex such optical sensors very simply. Up to now, this field is covered exclusively by optical-fiber sensors. They have already shown a great potential for chemical and medical sensing. One could think, however, also of incorporating optically active matter in etched tracks for sensing purposes, and of combining these sensors with optical fibers for transmission. This has not yet been realized.

7.2.7 Other Chemical Applications of Etched Tracks

The use of electrodes of nanoscopic dimensions (as produced by the template technique with etched tracks) offers the possibility to study the kinetics of heterogeneous electron-transfer reactions in electrochemistry, that are too fast to be studied at electrodes of conventional dimensions. Also, ultratrace levels of electroactive species that are up to three orders of magnitude below the usual level can be detected in this way.

Tracks could, in principle, also be applied usefully as microreactors in chemistry and pharmacology (see below), but apart from biomedical applications [49] (see below), no such work has yet been reported.

7.2.8 Biomedical Applications

It has always been mankind's dream to make blind men see, deaf men hear, and lame men walk again [50]. Only recently have the first experiments been conducted aimed at the future connection of artificial biosensors such as artificial retinas, ears, etc., with the human brain, or to use electrically stimulated muscles to regain the ability to walk or use destroyed hands again. Around 1995 the first studies began to let animal nerve cell communicate directly with a silicon chip [51, 52], and since then the first simple neuron networks were connected to silicon-based conventional electronics. Though all these attempts are still simple approaches in comparison to what God provides in nature, at least the general trend of research in this field has become obvious. A step forward towards future biomedical application will be to grow arrays of nerve cells on arrays of biocompatible electrical contacts embedded in nanoporous foils, for controlling, e.g., medical prostheses or other electromechanical devices. Complementary to this research are approaches that completely rely on inorganic sensor fields, e.g., on photoreceptor arrays of graded CdSe/CdTe heterostructures embedded in etched tracks [53]. Interestingly, the artificial photoreceptor elements can be made up to 5 times smaller than the natural retinal rod cells, which might have still unpredictable consequences.

Regular cell arrays are also of interest as platforms for future routine antibody tests. For all these purposes, ion-irradiated etched microporous foils may find interesting applications. The first such attempts have been performed [54], and one of them succeeded already to expand the lifetime of such transferred cells up to about a month – a time that is, in general, sufficient to study the cells' general behavior. It has turned out that, whereas individual isolated cells die rapidly, small cell assemblies exhibit much longer lifetimes. More precisely, the first cell layer in direct contact with the polymeric substrate sacrifices itself rapidly, to enable the survival of already about 60% in the next, and of an even higher cell percentage in the more distant layers. Osteoblasts (bone cells) have turned out to be most robust against cell transfer to artificial substrates [55].

Finally, foils of degradable polymers with dense layers of epithel cells grown upon them are under discussion as artificial skin for transplantations onto burnt areas and other lesions. Etched tracks in these foils might improve the cell adhesion and also enable a better exchange of gases and liquids. As it was observed that the presence of electric fields improves the healing process, the question has arisen whether there might exist biocompatible conducting polymers [56]. This question cannot be answered at present, nor is it known whether pronounced etched tracks can be produced in biodegradable polymers.

Biomaterials can be defined as the materials for medical and biological uses and divided into the biocompatible and biofunctional materials. The latter means that the materials – especially membranes – have a functionality to control the function of materials by information processing. For this sake they are also denoted (with still considerable exaggeration) as "intelligent materials". Specifically in relation with ion-track applications, one can relate here to the systems for spontaneous or automatized continuous drug release. The ultimate concept to be achieved here is the buildup of signal-responsive chemical-delivery systems (SRCDS) [57], by combining a biosensor with a drug reservoir inside a biocoating (protector), via suitable transducers, actuators, and eventually bioreactors, in an intelligent manner.

The most perfect example of these intelligent SRCDS systems is the natural neuron-synapse system [57]. Here information is delivered in the form of electrical pulse signals, then it is recognized by a receptor in a biomembrane where it causes the release of neurine transmitters from a reservoir cell; these are then captured by a receptor in the presynapse membrane and transformed from a chemical signal into an electrical pulse again. The present research aims at developing the first simplified SRCDS, according to the given natural example. As sensors, properly coated ISFETS (ion-sensitive field effect transistors) can be taken, and as actuators, electroresponsive or pH responsive gels can be used. The aim is the integration of all parts in compact membranes, with intelligent micropores taking over the task of actuators [57].

With micropores it may even become possible to produce biosensor arrays that may host in their pores a multitude of different sensors. The small size of etched tracks would enable an excellent spatial resolution; the polymeric nature of the base material can readily be rendered biocompatible, and the two sides of etched tracks in a thin foil would enable a straightforward throughput from the (e.g., optical, chemical, or pressure) signals on one side of the foil to the electrically stimulating pulse on the other side of the foil.

Another field of speculation is the future production of small-size radio-transmitting biocompatible sensors that could be incorporated in the respiratory or blood-circulating systems, for in situ control of chest or blood-pressure anomalities.

Finally, one can think of producing prostheses to regulate drug release inside the human body for long-time incorporation through tiny etched tracks [58]. Improving the tracks by grafting self-regulating gels (e.g., thermoresponsive gels, or gels that react to pressure, electric fields, pH value, etc.) onto the track walls, one can make the storage systems "intelligent" so that they supply the drug only in case of need. In fact, the first experiments of this kind have already been performed successfully [59, 60]. In particular, pores with diameters up to 10 nm are of interest for this purpose to produce haemocompatible membranes of high selectivity. For example, for insulin delivery, track diameters as small as 7 nm are required [61]. Hexagonally shaped supports are added in this case to prevent crack formation of the microporous foils that could have fatal consequences.

Another approach has been followed by Martin and coworkers [49]. They have shown that capped versions of PPy nanotubules can be loaded with enzymes to make an enzymatic bioreactor. A combination of electrochemical and chemical template-synthetical methods is used.

For medical-lab experiments it is useful to have homogeneous, flat, easily sterilizable, and transparent platforms for cell cultivation through which nutrition can be fed and waste products can be removed. Also here microporous foils have proven to be quite advantageous; they are already commercially available for this purpose. Low-fluence irradiated polymer membranes are also used commercially for chemical-pollution studies.

Another idea is to grow cells onto preselected spots on a petri dish that have been irradiated by energetic heavy ions at elevated fluences through masks, to improve the cell adhesion on the dish surface and to establish simultaneously an electrical contact from the dish interior to the outside world through the carbonized track zones, while maintaining sterility of the vessel inside. The first tests[2] have revealed that irradiated conducting zones of 0.5 mm diameter in PS and PP foils become brittle and easily crack due to the density differences between the polymeric host matrix (in this case polystyrene) and the irradiated carbonaceous spots. Possibly smaller spots with reduced material tensions do not fracture and thus enable the required biocompatible connections.

Still another approach for a future biosensor could be to deposit suitable fluorescent molecules – e.g., conjugated polymers – inside etched tracks (eventually firmly fixed by grafting), and subsequently to bond loosely to them specific molecules that, on the one hand, quench the fluorescence, and on the other hand exhibit well-tailored key structures that fit to specific antibodies. Once such an antibody arrives, the quenchant molecule will bond to

[2] Fink D, Klett R (1997), unpublished

it firmly and leave the fluorescent molecule. The renewed optical activity of the latter molecule acts as an indicator of the presence of specific antibodies with a sensitivity of typically 10^{-6} [62]. Subsequent exposure of the microporous foil with the fluorescent molecules to new quenchant molecules will deactivate the fluorescence so that the foil can be reused again. The advantage of the use of microporous foils in this case is that the biosensors can be used multiple times after regeneration, and that the transparent foils can be shaped as waveguides to enable the concentration of the fluorescent light in suitable photocells. This has not yet been done.

Development of Antibody Chips for Multianalyte Immuno Assays (MAIA)

A concept put forth by Prof. Ekins a decade ago [63] that could revolutionize laboratory diagnosis are multianalyte immuno assays (MAIA). This promises to be a cutting-edge technology in clinical chemistry. It will offer the advantage of estimating many analytes in one assay as compared to present assays where each assay can estimate only one analyte. By simultaneously carrying out all relevant tests for a given disease or associated diseases, MAIA can save valuable time for patient management. In short, an antibody chip MAIA can revolutionize immunoassay technology just as DNA chip is revolutionizing gene analysis.

At the core of MAIA is the "antibody chip": a small inert matrix on which a number of antibodies, each specific to a different analyte, is immobilized at spatially determined sites. Choosing a suitable matrix, applying the correct chemistry to bind the antibodies at a high density in specific locations is the key to developing "antibody chips". In a collaborative programme with the Biomedical group of the Bombay Atomic Research Center (BARC), a technique using TEMs (pore diameter 1 μm, pore density $10^7/\mathrm{cm}^2$, thickness 10–23 μm) was standardized to spot antibodies and to demonstrate the validity of this process. The first test experiments showed that PET-TEM films qualify as a good matrix for making antibody chips [64].

7.2.9 Applications of Etched Tracks in Novel Packing Materials

The light-absorbing n-type semiconducting transition metal oxides and chalcogenide compounds such as TiO_2, ZnO, SnO_2, $CaTiO_3$, Fe_2O_3, MoO_2, MoS_2, NbO_5, CdS, $KNbO_3$, $SrTiO_3$, CdSe, CdS, WO_3, and $Ti_xZr_{1-x}O_2$, and even SiC are photocatalysts that activate various chemical reactions by converting UV-radiation energy to chemical energy action, or to photoactivated catalysis. This can be applied to deodorization of gases, cleaning of gases from organic vapor, cleaning water of contaminants or treatment of waste water. For example, in green-house agriculture and the food industry it is highly important to reduce the content of ethylene (C_2H_4) gas in air, as the presence of only 50 ppb of ethylene is sufficient to kill plants and to initiate the premature flower, vegetable, and fruit aging process. The plant hormone ethylene is produced by all fruits, flowers, plants and vegetables, but cigarette smoke, plastics, silicone rubber, etc., are also ethylene sources. By photosemiconductors, the ethylene gas is permanently changed to harmless CO_2 and water via:

$$C_2H_4 + 3O_2 \longrightarrow \boxed{TiO_2 \text{ cat.}} \longrightarrow 2CO_2 + 2H_2O. \tag{7.6}$$

In waste-water treatment, alkanes, PCBs, alkenes, PAHs, alkynes, dioxins, ethers, furanes, aldehydes, pesticides, ketones, herbicides, alcohols, phenols, amine compounds, cyanides, amide compounds and esters have already been removed successfully.

Furthermore, photosemiconductors have some biological efficiency e.g. in cleaning and deodorization of gases of organic vapor, cleaning water, pasteurization, antifouling, treatment of waste water, the decomposition of pathogenes such as bacteria, viruses, dust mites, mold, spores, fungi and other micro-organisms – both airborne or in liquids, as in the control of algae growth. Therefore they have been suggested to be used, e.g., for food packaging, after appropriate embedding within a thin polymeric foil [65]. This idea could be applied for wrapping foils for food ("artificial egg shell"),[3] or as antimicrobial foils for sterilization purpose in medicine as plasters and bandages, to reduce the incidence of bacteria and viruses in wounds and to suppress the growth of fungi. Sunlight as well as indoor fluorescence UV lamps are sufficient to activate the antimicrobial efficiency.

Finally, the properties of strong UV absorption with simultaneous transparency for visible light makes thin layers of such photosemiconductors, embedded in a suitable carrier foil, useful to prevent sunburn of surfaces of fruits and vegetables that would otherwise damage the fruit tissue and lead to the development of disease symptoms. For all these applications, the technological interest centers on rutile (TiO_2), due to its high photocatalytic function, because its energy level is deeper than that of other photocatalytic materials. Consequently, the oxidising action of the positive holes generated by photoactivation is heightened. Furthermore, TiO_2 is relatively cheap and nontoxic.

TiO_2 membranes have already been created by transforming TiO_2 sol to gel and subsequently firing them. However, the brittleness, flaking and other deterioration of the emerging membranes prevented them from being applied. Therefore, Hulteen et al. [41] tried to incorporate TiO_2 fibrils in nanopores of polymers, and their action as a photocatalyst to decompose organic vapor molecules was studied. TiO_2 also has the function of an efficient nontoxic UV-light protectant for food. Many foods are known to contain unsaturated lipids and fats that develop off-tastes, off-colors, and oxidative rancidity during storage, e.g., by photo-oxidation in UV light of the C=C bonds of the unsaturated lipids.

[3] Tributsch H, Chapana Rojas J, Küppers U, Fink D, Petrov A (2002), unpublished work; patent pending (2003)

The photocatalytic reaction on the TiO_2 surface is thought to involve absorption of a UV photon by TiO_2 provided that the photon energy is equal to or larger than the bandgap energy of TiO_2, which is 3.2 eV (i.e., $\lambda < 400$ nm). Thereby an electron is promoted from the valence band into the conduction band, leaving a hole behind in the valence band. Some holes diffuse to the TiO_2 surface and react with –OH groups adsorbed there (which may stem from moisture or from water), forming hydroxyl and superoxide radicals, which subsequently leads to the formation of additional OH radicals and atomic oxygen. The radicals decompose organic molecules by oxidation towards CO_2 and H_2O because the potential energy of the radicals is greater than the bonding energy of almost all organic compounds. Only a small amount of UV irradiation from, e.g., fluorescent lamps is necessary for photocatalytic decomposition of nearly all contaminants in indoor and outdoor environments. The efficiency of this process is determined by the fraction of those electron–hole pairs that recombine rather than being photochemically active.

Template-synthesized TiO_2 structures dramatically increase the decomposition rate of organic matter by increasing the active TiO_2 area, though the expected yields are not reached – probably due to incomplete light illumination of the tiny fibrils [41]. As applications such as the electrolysis of water, decomposition of organic matter, and sterilization of germs occur on the TiO_2 surface it is desirable to increase the surface-area-to-volume ratio. Therefore a better approach rather than produce compact TiO_2 fibrils is to prepare tubules composed of individual TiO_2 nanoparticles, as this increases the overall reactive surface. Fine-grained particles of 1 to 100 nm diameter are most suitable for this purpose. There exist various methods to produce TiO_2 nanoparticles such as hydrolyzing Ti compounds or adding alkalines to Ti compounds and subsequently firing the products, or oxidizing Ti compounds in the vapor phase.

A disadvantage of this concept is that TiO_2 embedded in organic polymers also decomposes the host material. Therefore both the host polymer and the TiO_2 nanoparticles must be separated from each other by some nonoxidizable binder material as protective coating. Specifically for etched ion tracks in polymers, this signifies that the track walls must first be coated with some nonorganic material onto which, in a second step, TiO_2 colloids can be deposited. This binder material must be inert, such as SiO_2 sol, some siloxanes, Al_2O_3, colloidal SnO_2, etc. When dried or cured on the track walls, the deposited sols will convert, by cross-linking of the particles, to a network, usually a gel, of polysilicates, polyaluminates, etc.

The protective coating can also be a metal. Indeed, though the latter is not transparent it enhances the antimicrobial efficiency greatly: Combining the photosemiconductor particles with fine-grained metals ensures that each photosemiconductor and conductor (metal) particle in the composite constitutes of opposite electrodes of miniaturized electrochemical cells that promote electrolysis and catalytic activities. It is clear that the reactive ability of such a system is enhanced by ensuring that an aqueous environment is maintained around the sterilizing material. Furthermore, the conductor acts as a charge separator, impeding the recombination of conduction-band electrons and valence-band holes in the photosemiconductor. Accordingly, the antimicrobial action of simple photosemiconductors is greatly enhanced by the presence of metal.

Ag and Ni are preferred conductive coatings because they are resistant to oxidation, and Ag will form ions in solution that have known antimicrobial effects. Ag is particularly effective at killing bacteria and viruses whereas Ni is believed to be more effective against algae. Exposure to light is not required for the metals to act as sterilizing agents. Consequently, the photosemiconductor/metal combination exhibits its sterilizing activity even in darkness, although at a slower rate than if the photochemical cell had been activated. It is also contemplated to employ a combination of different particles in the foils to combat a whole spectrum of different micro-organisms simultaneously.

The antimicrobial action by the photosemiconductor/conductor film generally does not occur instantaneously. To counter this disadvantageous time constraint, one can either dope the photosemiconductor/conductor system with a metal of the platinum group such as Pt, Pd, Rh, Ru, Os, or Ir. In this way, an enhanced photocatalytic redox activity emerges so that the decomposition of contaminants adhering on the surface will be promoted.

Another, cheaper alternative is to combine the photosemiconductor/ conductor system with adsorbent particles such as apatites, zeolites, or activated carbon that trap the undesirable micro-organisms until they are eliminated by the antimicrobial agents generated by the electrochemical cells. These adsorbents also prevent the decomposition of OH^- groups that are essential to the trapping function. A preferred adsorbent is hydroxyapatite, a calcium phosphate ceramic, which is known to be particularly adept at adsorbing viruses.

From these considerations, the most promising structural concept for "artifical egg shells" emerges: The walls of the individual etched tracks in a suitable polymer foil (PET, PC, PI) are clad uniformly with, e.g., Ag by electroless plating, and thereupon TiO_2 nanoparticles are deposited, mixed with additions of a few per cent hydroxyapatite. Figure 7.6 shows SEM images of such tailored tubules. These and similar images reveal three interesting features. First, the higher the pH value of the TiO_2 colloidal solution used, the more often are tracks completely covered by TiO_2 platelets (not shown here) and thus made inaccessible, and many others show traces of broken covers. Hence applications require low-pH TiO_2 colloidal solutions for preparation.

The second observation is that the inner wall of the Ag tubules are well covered with a monolayer of TiO_2 particles. However, the particles on the foil surface do not touch each other to form a continuous layer but are distributed dispersedly on the shielding Ag layer, with gaps in between the particles. The final observation made is that particles on the foil surface are depleted around the etch tracks. The last two observations can be understood by surface

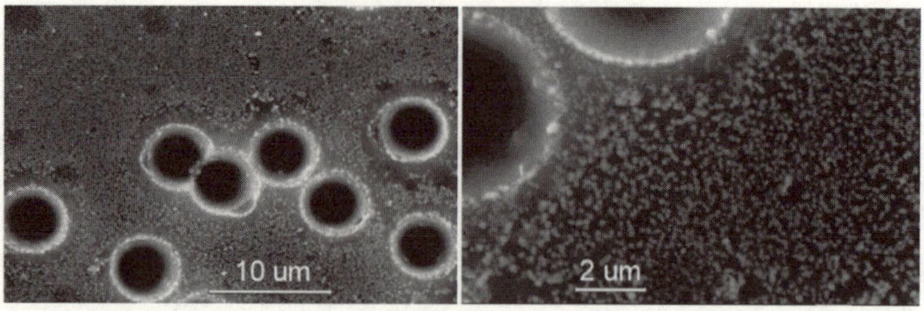

Fig. 7.6. Etch tracks in PI, coated with Ag and TiO_2 colloids (*white particles*). Views at different magnifications in the perpendicular direction

diffusion of the clusters on the microporous foil before drying. The particles will preferentially accumulate in zones of minimum potential energy that are the regions of highest surface curvature, i.e., the inner track walls. These tracks act as sinks for diffusing particles, whereas their vicinity is depleted of particles. So one has the favorite situation that the underlying physics makes the TiO_2 particles arrange exactly at the sites best suited for application.

What remains to do is to find the optimum opening angle of the etch-track cones, in order neither to transmit nor to reflect the impinging light, but to absorb it as homogeneously as possible on the etch-track walls over their whole lengths.

Ultrafine Au particles dispersed on transition metal oxide supports (such as TiO_2) have been found to catalyze important industrial reactions such as the oxidation of CO and partial oxidation of propylene to propylene oxide. Au_6 clusters can, e.g., be stabilized by phosphine (Ph) to yield $[Au_6(PPh_3)_6]$ $[BF_4]_2$ and stored in powder form until use. Aliquots of such a $\sim 10^{-4}$ mole/l Au_6L_6 colloidal solution (L = ligand) in CH_2Cl_2 can be deposited onto TiO_2 (immersed previously in acetone) to yield monolayers of single Au_6L_6 entities that show the catalytic activity. During the evaporation of the solvent, the $[BF_4]_2$ complex apparently evaporates; some 25 at.% carbon stemming from the phenyl groups and the solvent remain that do not affect the catalytic properties. Such Au_6L_6 clusters on TiO_2 particles, that have been deposited onto the inner walls of etched tracks, should enable one to obtain catalyst membranes of very high efficiency, due to the very large surface area enhancement by both the etched tracks and the TiO_2 nanoparticles [66].

7.2.10 Electronic Applications

Possible Production Strategies

Electronic applications of ion tracks are still at their beginning. The basic step is the production of conducting and/or semiconducting tracks. This can be accomplished either along latent tracks (see Sect. 7.1) or etched tracks. In the latter case, one can either use an electrolyte as the charge-carrying medium in which the specially shaped porous membranes are embedded, or one can insert a solid conducting or semiconducting material into etched tracks. In the latter case one can either deposit suitable material galvanically as massive wires, or via electrodeless deposition (ELD) or evaporation to obtain tubules [42]. A great advantage of the emerging single ion-track electronics (or: swift heavy-ion-track electronics (SITE) [5]) is that one does not require cleanroom conditions for most of the products to be prepared unless one combines SITE with lithography. Thus, even small companies will be able to produce future nanodevices with little investment of production capital.

A major limitation of all kinds of miniaturized electrical or electronic applications of ion tracks is the ability to dissipate the Joule heat that is generated as a result of the electrical current passing through the tracks. This problem is more severe in polymer-based electronics than in Si-based devices due to the much lower thermal conductivity of polymers. The answer to this problem is efficient cooling. For this purpose metallic replica of conical etched tracks have been developed[1] that enable an efficient cooling due to their large surface area, if put into intimate contact with the electronically active foils. Another possibility is to exploit the high porosity of the microporous foils themselves, including the electronically active tubules embedded therein, to enable laminar or convective flow of some insulating cooling liquid (e.g., an oil) through these pores. (However, high pressure differences are necessary to enable such flow, as the cooling liquid (or gas) throughput through pores with radius r scales with r^{-4}, according to the Hagen–Poiseuille law.)

Apart from this, one should take a polymer with high thermal stability such as PI as the microporous support membrane. This material is known to survive temperatures up to ~500°C without any chemical changes, up to ~600°C with limited carbonization, and up to ~800°C after full carbonization (in the latter two stages the pyrolyzed PI foil becomes conducting), while largely maintaining its shape (including etched tracks therein).

In principle, the whole zoo of electrical and electronic devices can be formed on the basis of ion tracks. Some of them will be treated subsequently. By combination of several galvanic and/or ELD steps with photoresist deposition, low-energy ion irradiation, and/or evaporation through masks, one should be able to produce various complex structures such as microinductances, microtransformers, nanobatteries, nanocondensors (Fig. 7.7), nanodiodes, nanotransistors, nano-LEDs, nanophotodiodes, and nanosensors [42]. Also, flexible 2D-device arrays (Fig. 7.8) should become feasible that could be applied, e.g., in field-emission displays. All these devices are in their development phase currently. The first test devices (see, e.g., Fig. 7.7) show that they are at least comparable – if not better – in their electrical characteristics.

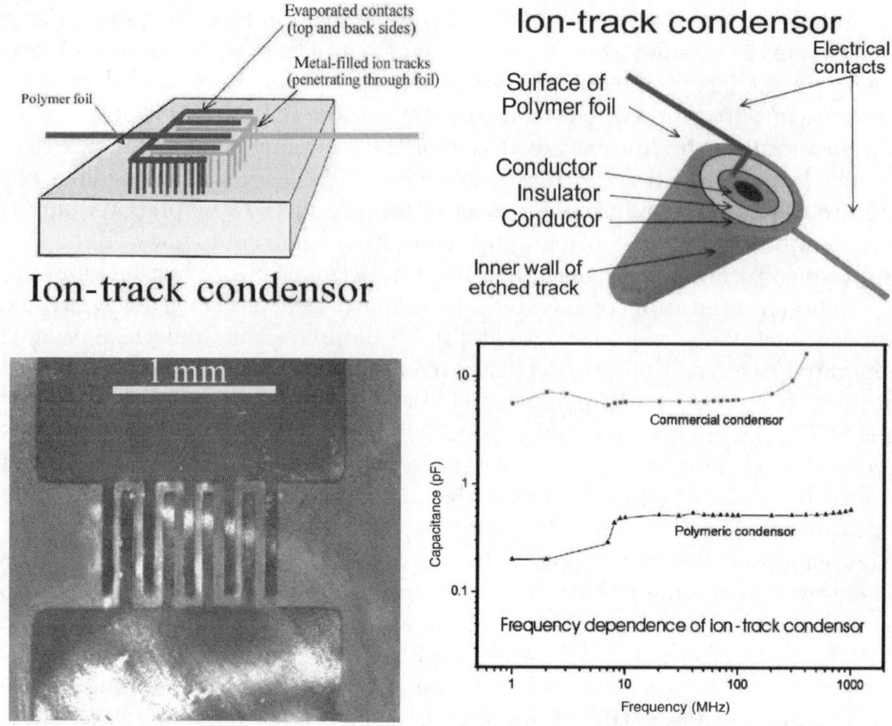

Fig. 7.7. *Top:* schematic concepts of ion-track-based condensors: *Left* a multi-track condensor; *right* a single ion-track condensor. *Below:* first, still macroscopic, ion-track condensor prototype and the frequency dependence of its capacity, as compared with a commercial 4.7-pF condensor

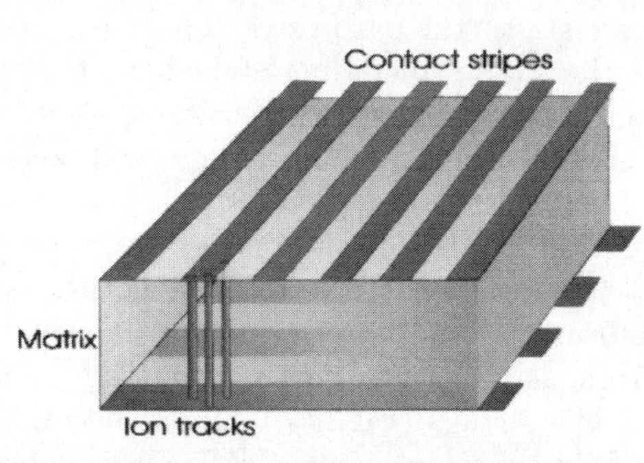

Fig. 7.8. Concept of two-dimensional arrays of ion-track-based devices, shown here for cylindrical tracks

New Tools for Microstructuring

At this point the reader's attention should be drawn to the emerging new C_{60}-containing polymers with exotic physicochemical properties (e.g., concerning the electrical conductivity) and good processability that might open new areas of yet unknown applications. As chemical polymerization reactions usually need processing temperatures below 100 to 200°C, the preparation of these new polymers could be performed in situ within etched tracks in, e.g., PI. Also in situ photopolymerization appears to be feasible. Similarly, new types of buckytube-containing polymers for aligned deposition along etched tracks are under discussion.

The reader should also keep in mind the possibility to include wetting/dewetting processes in the arsenal of tools for microstructuring of devices. If, e.g., certain regions of a microporous foil are hydrophobisized by, e.g., $C_{18}H_{37}SiCl_3$ through a mask, then dilute aqueous solutions of the PPy or Pani monomers will preferentially be adsorbed at these regions.

Schottky Diodes and Transistors

The first steps in this direction have already been performed. Metal/semiconductor structures (e.g., Cu/Se) have been produced by sequential galvanic deposition into etched tracks to produce arrays of nanometric Schottky barrier diodes [67, 68]. Other systems thus produced are Ni/CdSe/CdTe and Ni/ II-IV semiconductors [69]. Such structures show the phenomenon of resonant electron tunneling, where the current–voltage characteristics exhibits characteristic local maxima and minima at elevated voltages applied.

Recently a first ion-track-based field-effect-transistor-like structure was presented. Here a semiconducting material (CuCNS) was inserted galvanically into etched tracks within a specially prepared polymer foil. The latter was produced by joining both a metal-coated polymer foil and an uncoated polymer foil by glueing. Then the semiconducting wires were also contacted on both foil sides as additionally via the central metallic layer. In this way field-effect-like structures emerge where the current through the semiconducting rod can be controlled by the central gate electrode.[4] The device has the advantage of being robustly embedded in the polymeric foil that largely absorbs mechanical strain forces due to flexing, shearing, and tearing, so that the electronic parameters are hardly affected. It should also be possible to incorporate the recently produced static induction transistor (SIT) type using a layered structure of Au (drain) / organic iron phthalocyanine chloride (FePcCl) / Al (gate) FePcCl / Au (source) [70] into etched tracks, thus diminishing its dimensions considerably.

---

[4] Chen J, Könenkamp R (2001), personal communication

Quantum-Electronic Devices

It should be remembered that, when going to very small object sizes (typically below 10–40 nm in diameter [71], quantum effects show up. In connection with etched tracks, one can either produce such small objects by semiconductor or metal deposition within narrow etched tracks to form nanowires, or one can insert nanoparticles (of, e.g., CdS, PbS [72], Si and others) in colloidal form into larger etched tracks, to be deposited there.

Quantum confinement and single-electron charging effects in nanoparticles that act as, e.g., transistors offer the possibility of quantized shifts of the threshold voltage at room temperature that may lead to the development of multilevel logic devices. Metallic or semiconducting nanoparticles may be used for future nonvolatile memory devices. These nanoparticles have attracted a lot of attention in recent times both by theorists and experimentalists [73–75].

With decreasing wire radius, the valence band decreases and the conduction band increases in height, so that the overall bandgap increases. Quantum effects give rise to resonances in the current–voltage characteristics that could be exploited for specific electronic devices. Also, spontaneous magnetization of metal nanowires has been predicted [76].

Temperature Sensors

Thermoelectric nanodevices exploiting the Seebeck effect have been recently produced by pressure injection of molten Bi or Bi_2Te_2 into commercial porous Al_2O_3 films [71]. Though the thus-produced sensors are, in principle, capable of registering impinging heat pulses, their great disadvantage is the competing high thermal conductivity σ of the ceramic porous substrate that makes the devices slow and less sensitive in response. However, the thermal conductivity of nanostructures decreases with decreasing object size (e.g., $\sigma_{\mathrm{Bulk\ Bi}} = 8 \times 10^{-6}\,\mathrm{m^2\,s^{-1}}$, $\sigma_{\mathrm{Bi\ nanorods}} = 1.3 \times 10^{-6}\,\mathrm{m^2\,s^{-1}}$) [71], and nanoporous PI foils as substrates with their lower thermal conductivity should be capable of overcoming this problem.

Devices Exploiting Films of Metallic Nanoclusters

In the case that the electrodeless deposition (ELD) process of a metal or a semiconductor is interrupted before a consistent film has been deposited, the track-wall surface is covered with incoherent precipitations, see Sect. 6.2.3, Figs. 6.9 and 6.11a. The mechanism of conductivity along these precipitates depends then on the production parameters and the ambient (see Sect. 6.2.3): For distances of 1 nm and less between the metal clusters, electron hopping (tunneling) or percolation will dominate. For high ambient temperatures it is thermionic electron emission that governs the electrical current, and for high applied field strengths it is the field effect. If none of these conditions is fulfilled, the current is dominated by the conductivity of the polymeric substrate surface in between these metallic clusters. Usually it is the absorbed moisture and the remnant absorbed ions from the chemical deposition process that makes these polymer surfaces weakly conducting. Small variations in the intercluster distance – as experienced by pressure acting on the foils – lead to variations in the overall conductivity. Therefore these samples can be used as both sensitive humidity and pressure sensors. By covering the polymer surface with other specific sensoring materials, it might be possible to develop other sensors for different purposes.

These layers of intermittent metallic nanoclusters appear to exhibit still other interesting properties such as photoconductivity, so that one could think of exploiting this effect for the production of photoresistors.

Microbatteries

The first microbatteries have been prepared [77] with TiS_2/Au composite microtubules in polymers, counterelectrodes of metallic Li, and 3–7 vol.% solution of ethylene carbonate and diethyl carbonate with 1 **M** $LiClO_4$ as electrolyte. They have higher discharge capacities than conventional electrodes prepared from the same material. Furthermore, the intimate TiS_2/Au contact improves the low electrical conductivity of the Li^+-intercalated TiS_2 (and similar materials). Another attempt stems from Nishizawa et al. [78] who used 200-nm thick tubules of crystalline $LiMn_2O_4$ spinel as cathodes for rechargable Li batteries. Here, the spinel was produced by a thermal decomposition reaction of a $LiNO_3$ and $Mn(NO_3)_2$ solution embedded in Al_2O_3 ceramic foils at 500°C for 5 h in air. PPy-coated spinel nanotubules exhibited higher capacities than PPy-coated spinel films prepared under the same conditions. Instead of embedding these ceramics in alumina, one also could think at embedding them in PI, at slightly reduced reaction temperatures.

TiO_2 fibrils of 15 nm diameter have been shown to consist of bundles of highly crystalline TiO_2 structures with the rare anastase mineralogical orientation [001], i.e., with the c^*-axis being parallel to the fibril axis. In fact, only the anastase phase of TiO_2 not the rutile nor the brookite phases allow intercalation of Li^+ according to: $TiO_2 + X\,Li^+ + xe^- \rightarrow Li_xTiO_2$, which should therefore be applied for possible battery applications.

Light-Emitting Diodes (LEDs)

The idea to produce electronic devices by doping adequate polymers is not new; the principle feasibility of this approach was demonstrated as early as 1984 [6]. Recently, conjugated polymers have been used to produce thin-film semiconductor devices such as transistors [79], light-emitting diodes (LEDs) [80], electrochemically driven light-emitting cells (LECs) [81], and photodiodes [82]. Here, the polymer structure was modified chemically to control the energy gap, electron affinity, and ionization potential. For example, a dialkoxy derivate of poly(p-phenylenevinylene), PPV, was modified by adding a cyano group (yielding CN-PPV) to increase the ionization potential and electron affinity by $\sim 0.5\,\mathrm{eV}$, thus giving better electron injection and transport properties for LEDs [83], where the electron and hole currents are controlled by the interface formed between PPV and CN-PPV [84]. A similar compound, MEH-PPV, forms subgap absorption bands of 0.6 and 1.6 eV, characteristic of doped polymers of this type [85]. Meanwhile a lot of conjugated light-emitting polymers have been developed for any color [86] that could be useful to be integrated in etched ion tracks.

Photoinduced charge transfer is observed in blends of MEH-PPV and fullerite, C_{60} [87], and shows very efficient quenching (with factors up to 10^{-4}), as expected for intimate mixing of donor and acceptor species. Similarly, mixtures of MEH-PPV and CN-PPV showed quenching of around 2 to 5%. These mixtures were used to produce photovoltaic cells. The use of phase-separated mixtures as mentioned above has the advantage that both the photogenerated electrons and holes can be transported to the electrodes before recombination occurs. Collection of both carrier types is necessary for operation in a photovoltaic mode [82,88]. Similarly, the self-organizing property of polymer mixtures (i.e., the phase separation) was used in LEDs to obtain voltage-controlled color sources [89]. By inserting such mixtures into etched tracks, light-emitting or photovoltaic devices with nanometric-sized diameters can be obtained [90,91].

Nanometric LEDs can also be obtained by mounting an undoped self-carrying InP nanowire in between suitable n-doped and p-doped contacts. In such an arrangement, both the electrons and holes diffuse into the nanowire. They recombine in the center of this nanowire, or close to the p-type contact, and emit strongly polarized light thereupon [92]. Such an arrangement can also be considered as an example of ion-track-based LED designs where it is sufficient to deposit a suitable undoped material along the tracks, and both the n- and p-doped contacts onto front and back sides of the polymer foil. Then the light emission will occur in the track interior towards the polymeric foil that can be designed as a light guide.

Other Possible Optoelectronic Devices

Semiconducting chalcopyrite thin films such as Ag_2S that have been used in various optical and electronic devices, such as photoconducting cells, IR detectors, solar-selective coatings, and photovoltaic cells, could also be deposited chemically onto ion-track walls to produce miniaturized versions of these devices.

Another promising development that shows up in contemporary electronics, and that should find some application potential also in connection with ion tracks, is the use of organic semiconductor crystals. Materials of interest are anthracene, tetracene, pentacene, and α-6 thiophene. They form layered structures, resulting from the counterplay of van der Waals attraction and repulsion due to the Pauli principle. They form weak bonds with some tendency for localization, and have mobilities around $1\,\mathrm{cm^2/V\,s}$ (for comparison, classical semiconductors have mobilities of 100 to 1000 $cm^2/V\,s$). This makes them feasible for switching frequencies up to 100 kHz. Fullerene also belongs to the materials of interest. The first LEDs and FETs have been produced with these materials [93]. One can think of future two-dimensional display arrays of such structures, according to Fig. 7.8.

Crystals of 0.5 mm thickness and 1 to 10 mm length of polyacenes, oligothiophenes, phthalocyanines, and C_{60} have been produced by classical zone melting at 200° to 300°C in a hydrogen atmosphere, and the first superconducting test devices made of these organic crystals ($T_c \approx 2.3$–2.4 K) and fullerite ($T_c \geq 52$ K) have been reported [94]. Again it is expected that these structures can also be inserted into ion-track structures, to obtain new miniaturized devices.

Displays

Metallic wires can be used, e.g., as field-ion or electron emitters at low potential difference for large-area displays, high quantum-yield photocathodes, or energy-saving radar tubes. Microtube arrays may prove an alternative to electron-impact sources (with field emitters of typically some 10 µm diameter), by transmitting the gas molecules through the throat of sub-µm-sized tubules from the back side of the field-ion source and drawing them to their rim by dipolar electric-field forces produced by an intense field gradient. Again, the two-dimensional arrangement of these elements according to Fig. 7.8 appears feasible.

Microwave Generators and Filters

As nanowires or -tubules absorb certain wavelengths of electromagnetic radiation similarly to those used in microwave ovens and communications, an application potential for microwave filtering arises, to produce shields for microwave ovens and mobile phones [95]. Other possible applications may comprise high-power microwave generation, radiation- and temperature-insensitive electronics, and ultrafast computer technology making use of 1D quantum dots.

Alternative Materials

Apart from using polymers as carrier materials for latent or host tracks in electronic applications, it has been proposed to use diamond instead [8]. Though this proposition is fascinating due to the many properties that make diamond superior to any other electronic material, it is, however, difficult to realize. Even CVD-produced diamond layers are more expensive and much less flexible than polymer films, and thermal manipulations with diamond usually require temperatures of the order of 1000°C and more.

Electrolyte-based Electronics?

It has been reported in Chap. 3.4.2 that ion-track membranes with nanoscopic conical pores embedded in an electrolyte show a diode-like behavior, depending on both the concentration and pH value of the electrolyte [96–98]. On the one hand, this effect may have some biological interest as a model of so-called anomalous rectification found for a number of the cell membrane ionic channels having conical shapes. Furthermore, one might think of applications such as pH sensors due to the pH dependence of this effect, as electric valves for electrolytes, or as chemical valves for neutral molecules [98].

Apart from this direct application potential, the question arises whether it should not be possible to control this diode-like current behavior by some type of gate electrode so that transistor-like structures emerge. Once this has been accomplished, the way would be open for a new type of semiconductor-free electronics that could be fully based on electrolytes as the charge-carrying media, and on ion-track devices as the structural units to switch the ion currents, and to separate different electrolyte vessels from each other. Work has been resumed to realize this concept.

7.2.11 Magnetic Applications

Micromagnets and Microtransformers

The HMI research group is working on the production of micromagnets and microtransformers by combining deposition of metallic conductors along selected ion tracks inside a thin polymer foil with evaporation of appropriate conducting wires on the front and rear sides of this foil, Fig. 7.9a,b [99]. It has been shown [24] that the many parallel metallic nanowires that form the vertical connections behave just as one individual wire with correspondingly larger diameter. As it is easily possible to produce in this way coils with more or less quadratic cross-sections (according to both concepts (a) and (b), the quality factors of such magnets are supposed to be superior to thin-film magnets obtained by evaporation, as in the latter case cross-sectional ratios of 1:50 between height and width are common, which results in considerable inductive loss. By using conical instead of cylindrical etched tracks, the quadratic cross-section can be rounded somewhat, which reduces magnetic stray fields even more.

For low-frequency applications, coils should contain cores of ferromagnetic, and for higher frequencies of ferritic materials to provide good conducting paths for the magnetic flux. With increasing operating frequency, the size of the required magnetic components approaches zero. Therefore coreless coils were suggested for frequencies exceeding $\sim 10^{5-6}$ Hz. In fact, the first macroscopic magnets and transformers have meanwhile revealed the feasibility of this concept, and equivalent circuits have been established to describe their characteristics [100].

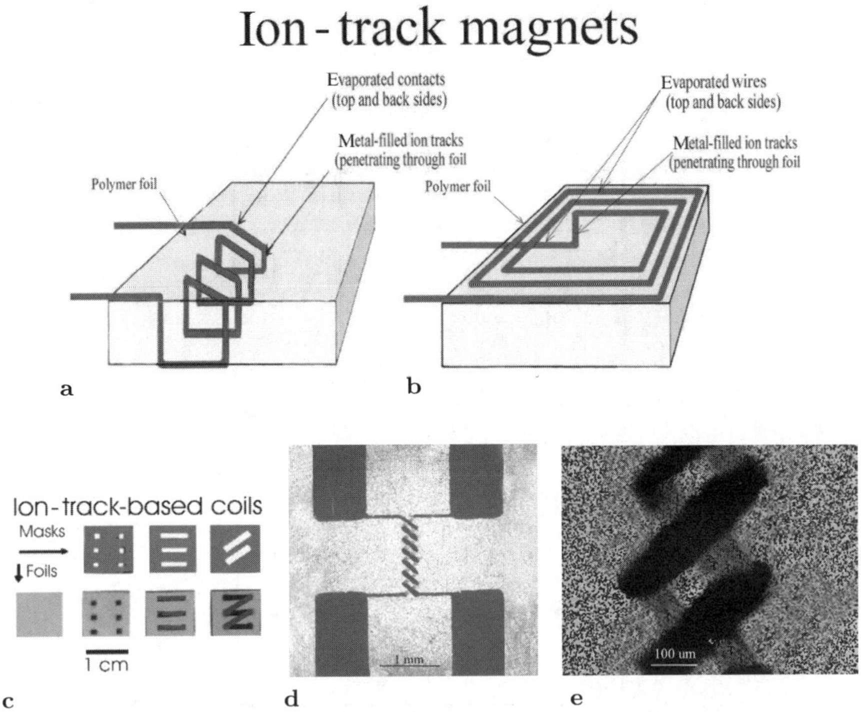

Fig. 7.9. Ion-track based micromagnets. (**a,b**) Principle sketch of the two possible concepts. Whereas in (**a**) many track-based feedthroughs are required, the concept (**b**) makes use of only one feedthrough.[6] A first macroscopic feasibility test device is displayed in (**c**). Shown are the masks used for the individual metal deposition steps (*top*) and the different stages of device preparation (*bottom*). *From left to right:* untreated microporous foil (PI, 23 μm thick, 10^7 etched tracks/cm^2)/foil with metal tubules along tracks/foil with conducting tubules and evaporated metallic connections between tubules on front and back sides. (**c,d**) First miniaturized ion-track-based transformer [99]. (**d**) *Grey:* PI microporous foil; *dark:* evaporated Au wires. (**e**) The individual etched tracks of the PI foil can be clearly recognized as small dark spots in the optical transmission microscope; the evaporated metallic zones show up as nontransparent areas

Such devices have resonant frequencies $\upsilon_{\mathrm{res}} = (2\pi(L_{\mathrm{eq}}C_{\mathrm{eq}})^{1/2})^{-1}$, with L_{eq} being the total leakage inductance and C_{eq} being the total device capacity. By addition of external condensors, υ_{res} can be tuned. The voltage gain of corresponding transformers (defined as the ratio of output voltage to input voltage) is low at low frequencies, attains its maximum at υ_{res}, and drops to zero beyond. Therefore the operational frequency of coreless transformers should be near but below υ_{res} for power transfer. They operate over a wide-

[5] This design is closely related to the coreless printed circuit boards (PCB) as described, e.g., in [100].

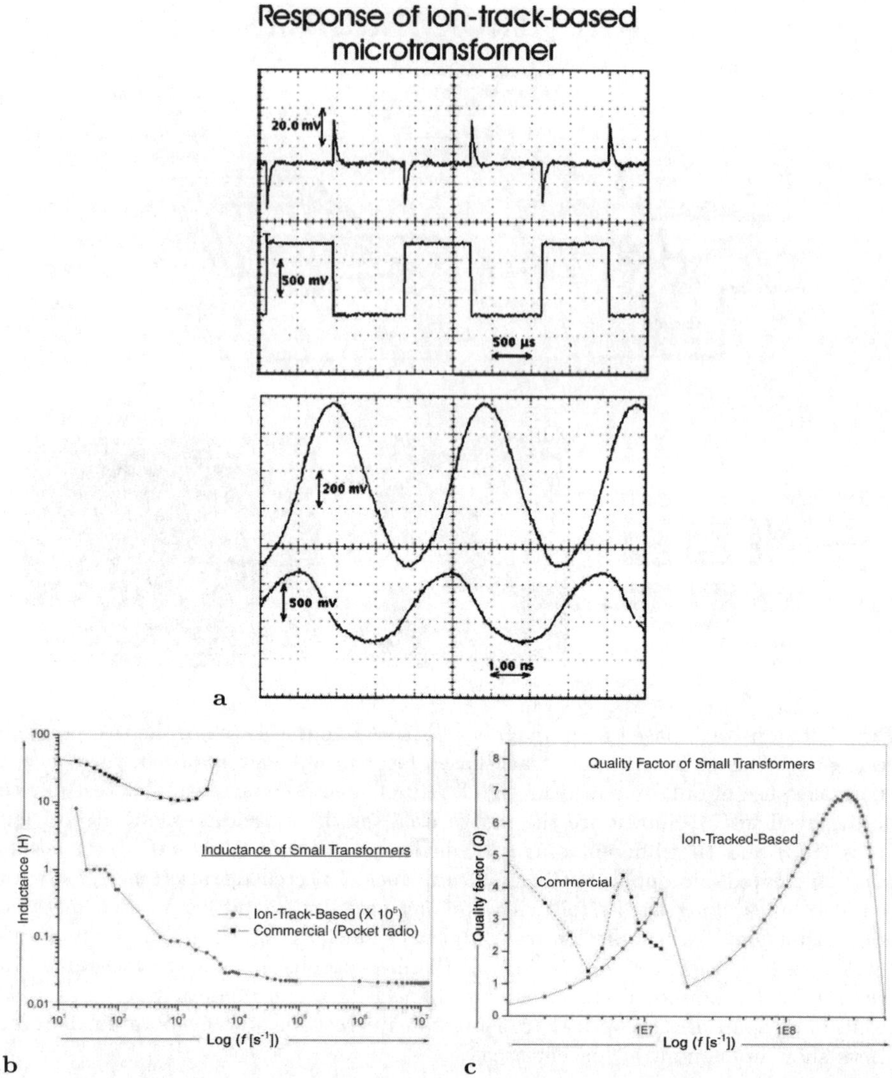

Fig. 7.10. Characterization of the first ion-track-based test transformer (still macroscopic, with size: 6 mm × 6 mm × 23 µm; 1 winding for primary and secondary coil each). (**a**) *top* a rectangular input signal (*below*) yields the expected differentiated response (*above*), and a sinusoidal input yields (*bottom*) a cosine response signal. Whereas the phase shift is 90° at 1 MHz (*top graph*), it breaks down at 1 GHz (*bottom graph*). (**b**) The inductance, and (**c**) the quality factor of the test transformer, as compared with a commercial small transformer from a pocket transistor radio

frequency range within which efficiencies exceeding 90% can be obtained. The maximum-impedance frequency υ_{mi} is slightly below υ_{res}. The voltage gain at υ_{mi} is high and the power requirement has its minimum there, so that such a transformer can also be used economically for signal enlargement.

At an operating frequency υ, such structures emit electromagnetic radiation perpendicularly to the coil plane at a wavelength $\lambda = c/\upsilon$, with $c =$ speed of light. The power P radiated from such devices drastically decreases with the emitted wave-length and increases with the coil radius $r : P \sim (r/\lambda)^4$. Therefore the criterion for a good loop radiator, $r \approx \lambda$, does not hold for small coils (for example, for coils with 5 mm size, $(r/\lambda)^4 \approx 10^{-16}$), which means that electromagnetic radiation from coreless magnets or transformers is negligible.

The first ion-track-based microtransformers have been produced meanwhile, Fig. 7.9c. Due to their large size their range of operation is still limited to the sub-GHz range, and due to their large height-to-width ratio and the limited number of windings their inductance is still quite poor. Nevertheless, they already allow reasonable signal transformation and have non-negligible quality factors (Fig. 7.10). Therefore it is expected that future miniaturized ion-track-based magnets and transformers with more windings and better aspect ratios (see, e.g., Fig. 7.9d,e) will show improved properties that will make them interesting for technical applications.

In principle, it should be possible to proceed with the miniaturization so far that coils with less than $\sim 10^{-6}\,\mathrm{cm}^3$ volume emerge, thus decreasing in size below even the smallest magnets ever produced by conventional lithographic techniques. It should also be possible to combine ion-track-based condensors and magnets in one device. By combining lithography and etched tracks, magnetic-field sensors based on magnetoresistive ion tracks have already been produced [101].

Magnetic Nanowires and Dots for Data Storage

Nanowires of Ni, Fe, FeCo, Fe-CoV, or Ni/Fe alloys produced by galvanic deposition inside etched tracks (see, e.g., [102]) (or porous Al_2O_3 membranes) show special magnetic properties. Upon formation, the magnetization tries to orient along an "easy" direction, e.g., a crystallographic axis. Whereas the magnetic relaxation of an individual nanowire correlates with $\{n_0 \exp(-a\,t)\}$, an assembly of magnetic wires relaxes according to $\{n_0 - S\ln(t)\}$ (n_0, a, $S =$ const., $t =$ time). Within arrays of magnetic nanowires (or other magnetic nanoparticles), domain formation takes place. If the magnetic particles are regularly arranged in a hexagonal lattice, ferromagnetic interaction was found, whereas a square arrangement favors antiferromagnetic interaction. Whereas large wires have a domain structure, smaller wires consist of single domains. With decreasing wire dimensions the hysteresis of the ferromagnets becomes narrower and steeper, i.e., the magnets become harder. Finally, a transition to superparamagnetic behavior (i.e., with no hysteresis) occurs.

The same behavior is also found in matter with a very dilute ($\leq 0.01\%$) concentration of magnetic particles.

Upon application of a reverse magnetic field, the magnetic orientation of discrete ferromagnetic nanoparticles (e.g., nanowires) switches to the reverse direction at a well-defined applied field strength, whereas a dense array of magnetic wires gives rise to a whole series of individual responses at different field strengths [103]. The research on such magnetic nanostructures ("magnetic dots"), including magnetic nanowires, is of technological importance in the search for improved mass-storage devices. It is expected that proper application of these magnetic dots will increase the present (2002) storage density of 4 (industrial application) to 100 (laboratory models) Gbyte/inch^2 by still one or two orders of magnitude. For this type of research, special magnetic microscopes (MFM, magnetic force microscope) with special magnetic dipole moment calibration techniques have been developed [104].

Heterogeneous alloys consisting of ferromagnetic grains with dimensions in the nm range surrounded by a nonmagnetic matrix (as realized by, e.g., Co-Cu alloys, or by magnetic nanowires in etched tracks) are of interest because they can exhibit giant magnetoresistance (GMR). The first magnetic sensors have been constructed with these structures, combining electrodeposited Ni tracks with evaporated contacts [105]. This approach may prove to be suitable where a low unit cost and a large response are more important than very high sensitivity to small field changes. Cu-Co heterogeneous multilayered nanowires produced by galvanic deposition in both PC etch tracks and Al_2O_3 nanopores show GMR, though until now still at sub-per cent abundance. Annealing at 400°C doubles this effect [106]. The reason for the GMR effect is explained by the very short electron mean free path in these structures. The GMR in individual magnetic nanowires is found to correlate inversely with their magnetothermopower [107]. Also, high-frequency studies are favored in magnetic nanostructures as their dimensions are smaller than the skin depth. These nanowires exhibit properties such as resonance-frequency tunability and zero-field ferromagnetic resonance absorption, which are very interesting for future applications.

Superconducting nanowires show an increase in the thermodynamic critical field H_c, due to the small transversal dimension leading to an incomplete Meissner effect. A nonzero resistance occurs below T_c in these one-dimensional superconductors due to fluctuations of the superconducting order parameter (phase slips). The destruction of superconductivity as observed in the current–voltage characteristics may be explained by the formation of phase-slip centers [108].

7.2.12 Optical Applications

Arrays of micromirrors might become of interest in future image processing of small-sized displays. They can be produced quite easily by irradiating swift heavy ions through suitable masks. The radicals thus formed in the

a b

Fig. 7.11. Illustration of metal deposition onto a polymer, using ion-induced activation. Example: irradiation of $10^{12}\,cm^{-2}$ 300-MeV Xe^{21+} onto PI through a mesh with 50 µm × 50 µm openings, and subsequent Cu deposition in an alkaline chemical deposition bath. (**a**) Optical microscopy image, focused to front object plane. *Dark:* deposited Cu; *bright:* no Cu deposition. (**b**) Slight defocusing of the object reveals in the microscope that the Cu films on the individual irradiated spots are not planar but curved, thus reflecting the incoming light to individual microfoci, i.e., that an array of micromirrors has been formed in this way [7]

irradiated volume can, on the one hand, promote etching of very deep patterns, and, on the other hand, simultaneously serve as nucleation centers for chemical deposition of metals (see Sect. 6.2.3), Fig. 7.11a. As etching and the metal film deposition within the transmitted areas can be tailored so that yields not planar but curved surfaces, optical micromirror arrays may emerge, Fig. 7.11b.

Nanoscopic metals as produced by template techniques show special optical properties. Thus Au (or Ag) fibrils in etched tracks (or Al_2O_3 pores) can become red purple or blue, depending on their size, similarly as is known from, e.g., Au colloidal suspensions. The colors result from the plasmon resonance band on the nanometal that corresponds to the wavelength of light that induces the largest electric field on the nanometal particle. For wavelengths larger than the fibril diameter dipole excitation, and for wavelengths of the order of the fibril diameter, multiple excitation plays a role [109]. Applications of this effect for X ray, EUV, and optical filters are under discussion [110–112]. The effective medium theory predicts even optically transparent metallic structures for electrically disconnected metal particles that are small compared with the wavelength of the incoming light.

Nanorods of high-bandgap materials such as ZnO can find applications as optoelectronic switches in the blue wavelength region (45 to 100 nm). Such structures have hitherto been grown only from the gas phase by deposition of $Zn(C_5H_7O_2)_2$ onto 200°C hot Si or Al_2O_3 crystalline substrates, and subsequent decomposition of the organometal at 500°C [113], but deposition of

[7] Fink D, Petrov A, Berdinsky A (2001), unpublished results

corresponding sols within etched ion tracks in PI at ambient temperature, with subsequent burning at about 400–430°C appears to be a promising alternative.

Erbium-doped SiO_2 colloids, if embedded in polymers, have been shown to exhibit up to 80% quantum efficiency for luminescence [114]. Such structures could, in principle, also be embedded in tracks in polymers to yield 2-dimensional luminescent displays. The advantage of producing nanocrystals of a given material is that a high optical gap of a bulk material can be decreased, which enhances its technical applicability.

7.2.13 Mechanical Applications?

Apart from the use in lithographic techniques for micromechanics, tracks have not yet been applied for mechanical applications to form, e.g., new types of actuators. In principle, there might arise the possibility to incorporate memory-shape alloys within tracks, e.g., by evaporation. They could then be used to force the microporous carrier foils to bend into different preselected shapes and thus act as actuators.

7.2.14 Statistical or Spatially Defined Ion Impact?

An important point that has hitherto restricted the applicability and acceptance of ion-track technology by industry is the unpredictable nature of statistical ion impact on the sample materials. Whereas the positions of ion incidence on a polymer foil do not play any role for many applications such as filtering purposes, this point becomes crucial when thinking of producing sophisticated electronic, optic, or biological devices with resolutions in the μm- or even nm-scale range. For this purpose, highly elaborate systems have been developed that deliver sequential individual ion impacts within a well-defined pattern with sub-μm resolution [115]. Though these setups will doubtlessly gain great merits for special applications with high demand for precision, they are obsolete for cheap mass production, due to excessive time and cost of production.

Therefore it appears that, when considering ion-track applications in larger scale, a principle change in the conceptual approach towards technological applications may be necessary, as concerns the statistical ion impact on the targets. Whereas in classical lithography each irradiated area of a mask receives a well-defined and reproducible dose of, e.g., UV light, and hence every prepared sample will be a useful product, this will not be the case in electronics based on individual ion tracks. Here it may well happen that, depending on statistical coincidences, occasionally in given functional regions ion tracks are completely missing or catastrophically overlapping, both of which render these products useless. So it may well be possible that some fraction of the produced products will have to be abandoned after careful product control.

Though this philosophy appears unacceptable for the "classical" semiconductor industry, it should be kept in mind that exactly this approach is God's strategy adopted for nature. Billions of extremely complex functional organisms of a given specie are produced, and subsequently subjected to the rigorous test of survival, during which a certain number of organisms fails. Taking into account the enormous degree of sophistication of each of these entities, this strategy might appear at first glance as a waste of energy and resources, however, the long success story of biological life on earth shows that this appears not to be the case. As the net overall biological investment in both useful and useless organisms is apparently much lower than the overall gain – the specie's guaranteed survival – the strategy appears nevertheless to be economical!

The same applies to ion-track technology: Provided that the price for an individual ion-track-based device can be made low enough to afford discarding, e.g., 10% of all products and nevertheless to make profit, this strategy will pay.[8] In fact, the success of, e.g., the first ion-track-based oxygen Clark sensors (see Sect. 7.2.6) as compared with classical sensors, that lowered the price of production by at least an order of magnitude, indicates that this goal appears to be realistic. In comparison with the price per transistor in silicon-based electronics, the price of an individual track in a polymer foil is much lower already (of the order of 10^{-12} Euro) – and thus eventually will justify the adoptation of the "biological strategy" for the emerging swift heavy-ion-track technology.

References

1. Chakane S, Singh F, Avasthi DK, Bhoraskar SV, Usefulness of nuclear tracks in polymers for hydrogen gas sensing. 21st Int. Conf. on Nuclear Tracks in Solids, New Delhi, 21–25 Oct., 2002
2. Herden V, Das Verhalten von lichtinduzierten Ladungsträgern in Polysilanen unter besonderer Berücksichtigung von Dotierungen und strahlenchemischer Vernetzung. PhD. Thesis, Technical University Berlin, No. **D83** (2001) (in German)
3. Fink D, Petrov A, Stolterfoht N, Wilhelm M, Hoffmann V, Richter A, Behar M, Farenzena L, Papaleo R, Hirata K, Kobayashi Y, Chadderton LT, Schulz A, Fahrner WR, Creation of nanoscale objects by swift heavy ion track manipulations. Proc Intl Symp on Material Chemistry in Nuclear Environment, Tsukuba, Japan, March 13–15, 2002. To be published as a JAERI report (2003)
4. Cheung CL, Hafne JH, Lieber CM, Carbon nanotube atomic force microscopy tips: Direct growth by chemical vapor deposition and application to high-resolution imaging. PNAS **97**, 3809–3813 (2000)

[8] These thoughts were developed during an informal discussion initiated by K. Hjort on the occasion of the 5th Intl. Symposium on Swift Heavy Ions in Matter, Giardini Naxos, May 22–25, 2002

5. Fink D, Klett R, Latent Tracks in polymers for future use in nanoelectronics, an overview about the present state-of-the-art. Braz J Phys **25**, 54–75 (1995)
6. Koshida N, Wachi Y, Application of ion implantation for doping of polyacetylene films. Appl Phys Lett **45**, 436–439 (1984)
7. Biswas A, Awasthi DK, Kanzow J, Ding SJ, Fink D, Gupta R, Zaporojtchenko V, Faupel F, Nanostructural modifications in Au cluster arrays distributed in Teflon AF layers upon MeV heavy ion impact. To be published (2003)
8. Fink D, Chadderton LT, Cruz SA, Fahrner WR, Hnatowicz V, te Kaat EH, Melnikov AA, Varichenko VS, Zaitsev AM, Ion tracks in condensed carbonaceous matter. Radiat Eff Defects Solids **126**, 247–250 (1993)
9. Krauser J, Weidinger A, Zollondz JH, Schultrich B, Hofsäss H, Ronning C, Conducting ion tracks for field emission. Proc of the Workshop on European Network on Ion Track Technology, Caen, France, 24–26. Feb. 2002
10. Starosta W, Buczkowski M (eds.), Proceedings of the 2nd meeting on "Particle track membranes and their applications", Institute of Nuclear Chemistry and Technology, Dec. 2–6, 1991 Warsaw, Poland. INIS-mf-14419
11. Starosta W, Buczkowski M (eds.), Proceedings of the 3rd International Conference on "Particle track membranes and their applications", Oct. 26–30, 1993, Jachranka, Poland (1993)
12. "Flerov GN, Apel PYu, Didyk AYu, Kuznetsov VI, Oganessian RTs, The use of heavy-ion accelerators for the production of track membranes. Atomic energy **67**, 274–280 (1989) (in Russian)
13. Apel PYu, Track etching technique in membrane technology. Radiat Meas **34**, 559–566 (2001)
14. Yamazaki IM, Patterson R, Geraldo LP, J Membrane Sci **118**, 239 (1996); and: Umeuchi S, Nishimura Y, Yamazaki IM, Murakami H, Yamashita M, Electrical field effects on adsorption and fluorescence spectra of pyrene doped in a PMMA polymer film. Thin Solid Films **311**, 239–245 (1997)
15. Spohr R, Nuclear track irradiations at GSI. Nucl Tracks **4**, 101–108 (1980)
16. Spohr R, Ion Tracks and Microtechnology. Principles and Applications. Vieweg & Sohn, Braunschweig, 1990
17. Vijay YK; Acharya NK, Wate S, Avasthi DK, Nanofilter for hydrogen purification. Int J Hydrogen Energy, in press (2003)
18. Iyer RH, Dwivedi KK, Four decades of nuclear tracks research in India. Invited talk IT-1, ICNTS-21, October 21–25, New Delhi, India, to be published in Radiat Meas, 2003
19. RoTrac® Capillary pore membranes. Information leaflet of Oxyphen GmbH Dresden, Rossendorfer Technologiezentrum, Bautzener Landstr. 45, 01454 Grosserkmannsdorf, Germany, 2001
20. Bertsch A, Benninger F, Munnik F, Mikhailov S, Renaud P, High aspect ratio patterning of polymers by ion beam LIGA. Proc of the Workshop on European Network on Ion Track Technology, Caen, France, 24–26. Feb. 2002
21. Jirage KB, Hulteen JC, Martin CR, Nanotubule-based molecular-filtration membranes. Science **278**, 655–658 (1997)
22. Halder H, Menzel N, Hietel B, Wittmaak K, A new analysis chamber with a rotating target holder for total-sample PIXE analysis of aerosol deposits collected in berner impactors. Nucl Instrum Methods **B150**, 90–95 (1999)

23. Hjort K, The European network on ion track technology. Proc of the Workshop on European Network on Ion Track Technology, Caen, France, 24–26. Feb. 2002
24. Lindeberg M, Hjort K, Statistics of flexible circuitry based on ion track defined nanowire cluster links. Proc of the Workshop on European Network on Ion Track Technology, Caen, France, 24–26. Feb. 2002
25. Dehaye F, Balanzat E, Ferain E, Legras R, FTIR spectroscopy study of chemical modifications induced by swift heavy ion irradiation in bisphenol A polycarbonate. Proc of the Workshop on European Network on Ion Track Technology, Caen, France, 24–26. Feb. 2002
26. Vetter J, Spohr R, Application of ion track membranes for preparation of metallic microstructures. Nucl Instrum Methods **B79**, 691–694 (1993)
27. Franken Feld JW, Li NN, Recent advances in liquid membrane technology. In: Handbook of Separation Process Technology, John Wiley, NY, Chap. 19, pp. 840, 1987
28. Ohara PA, Bohrer MP, Supported liquid membranes for copper transport. J Membrane Sci **44**, 273–287 (1989)
29. Pandey AK, Iyer RH, Track-etched membranes in separation sciences (invited talk). Proc of the Xth Natl. Conf. On Solid State Nuclear Track Detectors, Dept of Physics, Kurkshetra University, Kurkushetra, Oct. 3–5, pp. 143–150, 1996
30. Shukla JP, Kumar A, Singh RK, Iyer RH, Separation of radiotoxic actinides from re-processing wastes with liquid membranes. In: ACS Symposium series 642, RA Bartsch, JD Way (eds.), ACS, Washington DC, Chap. 27, pp. 391–408, 1996
31. Sudowe R, Pemzhorn RD, Vater P, Abu Jaber S, Brandt R, Filters with small holes ($d < 1\,\mu$m) as a tool to separate gases. Radiat Meas **28** 811–816 (1997)
32. Baker RW, Roman IC, Lonsdale HK, Liquid membranes for the production of oxygen-enriched air. I. Introduction and passive liquid membranes. J Membrane Sci **31**, 15–29 (1987)
33. Iyer RH,Bhagwat PV,Surendran P, Nair JP, Production of track-etch membranes using the BARC-TIFR heavy ion accelerator and their applications (invited talk by Iyer RH). Proc of the XIth Natl Symp on Solid State Nuclear Track Detectors, Dept. of Physics, Guru Nanak Dev University, Amritsar, India, October 12–14, pp. 103–114, 1998; and
Iyer RH, Track-etched membranes: preparation and applications in separation science – an overview (invited talk). Proc of the XIIth Natl Symp on Solid State Nuclear Track Detectors, D.A.V. College, Jalandhar, India, October 29–31, pp. 9-22. 2001
34. Noble RD, Way JD (eds.), Liquid membrane Technology: An overview. In: Liquid Membranes, Theory and Applications, American Chemical Society, Sympos. Series 347, Washington DC, pp. 1–27, 1987
35. Danesi PR, Separation of metal species by supported liquid membranes. Separation Science Tech **19**, 857–894 (1984)
36. Pandey AK, Gautam MM, Shukla JP, Iyer RH, Effect of pore characteristics on carrier-facilitated transport of Am (lll) across track-etched membranes. J Membrane Science, **190**, 9–20 (2001)

37. Iyer RH, Separation and recovery of radioactive and non-radioactive toxic trace elements from aqueous industrial effluents. To be published in a special issue on "Frontier areas in biodegradation/bioremediation" in Indian Journal of Experimental Biology, 2003
38. Ramanujan A, Atchuthan PV, Opalakrishnan V, Dhami PS, Kannan R, Balu K, Iyer RH, J Radio Anal Nucl Chem, communicated (2000)
39. Lev AA, Korchev YuE, Rostovtseva TK, Bashford CL, Pasternak CA, Lipid impregnated nuclear filters as a model for studies of surface conductance and channel phenomena. Proc 11th School on Biophysics of Membrane Transport, Wroclaw, Poland, May 4–13, 321–349 (1992)
40. Martin CR, Nishizawa M, Jirage K, Kang M, Lee SB, Controlling ion-transport selectivity in gold nanotubule membranes. Adv Mater **13**, 1351–1362 (2001)
41. Hulteen JC, Martin CR, Template synthesis of nanoparticles in nanoporous membranes. Chap. 10. In: Nanoparticles and Nanostructured Films, Fendler JH (ed.), Wiley-VCH, Weinheim, pp. 235–262 (1998)
42. Siwy Z, Fulinski, Fabrication of a synthetic nanopore ion pump. Phys Rev Lett **89**, 198103-1–198103-4 (2002)
43. Micheli AL, Chang SC, Hicks DB, Tin oxide gas sensoring microsensors from metalorganic deposited CMOD thin films. Ceram Eng Sci Proc **8**, 1095–1105 (1987);
and: Xu JJ, Shaikh AS, Vest RW, Indium tin oxide films from metallorganic precursors. Thin Solid Films **161**, 273–280 (1988)
44. Berdinsky A, Fink D, Petrov A, Fullerene humidity sensors. To be presented at the Ul-san Conf, Corea, June 2003
45. Ferain E, Legras R, Track etched templates use for nanoscale material synthesis. Presented at the 5th Intl. Symposium on "Swift Heavy Ions in Matter", May 22–25, 2002, Giardini Naxos, Italy, contribution P8–78
46. Berdinsky AS, Shevtsov YuV, Okotrub AV, Tombin SV, Chadderton LT, Fink D, Lee JM, Sensor properties of fullerene films and fullerene compounds with iodine. Chem Sustain Dev **8**, 141–146 (2000)
47. Berdinsky AS, Fink D, Petrov A, Chadderton LT, Krasnoshtanov SM, Rylova ES, The effect of external mechanical stress on the fullerite conductivity. 3rd Sibirian Russian Workshop and Tutorial on Electron Devices and Materials (EDM), Erlagol, Altai, Vol. 1, pp. 40–44 (2002)
48. Berdinsky A, Fink D, Müller M, Petrov A, Chadderton LT, Apel PYu, Formation and conductive properties of miniaturized fullerite sensors. Proc MRS Boston, Fall meeting, 27.11.–1.12. 2001; Contribution Y4.7 (2001)
49. Parthasarathy RV, Martin CR, Synthesis of polymeric microcapsule arrays and their use for enzyme immobilization. Nature **369**, 298–301 (1994)
50. The Holy Bible. See, e.g., Jes. 32,2; 42,18; Matth. 9, 27–31; Mark. 7,31–37; John 5,1–15
51. Weber C, Mayer KM, Neurotechnologie: e-mail ins Hirn. Focus **26**, 130–134 (2000) (in German)
52. Puntle C, Biochip mit menschlicher Zelle. Focus **10**, 184 (2000) (in German)
53. Klein JD, Herrick RD II., Palmer D, Sailor MJ, Brumlik CJ, Martin CR, Electrochemical fabrication of cadmium chalcogenide microdiode arrays. Chem Mater **5**, 902–904 (1993)
54. Chin VI, Ozkan M, Bhatia S, Rapid three-dimensional arraying of single cells. Proc MRS Boston Fall Meeting, 27.11.–1.12. 2001, Contribution Y5.8

55. Dalby M, Johnstone H, Gallagher J, Wood MA, McGhee K, Casey B, Affrossman S, Wilkinson CDW, Curtis ASG, Nanometric patterns for tissue engineering: fabrication and in vitro biocompatibility. Proc MRS Boston Fall Meeting, 27.11.–1.12. 2001; Contribution Q5.1
56. Krsko P, Stuccio C, Libera M, E-beam patterned poly(ethylene glycol) for spatial control of surface bioactivity. Proc MRS Boston Fall Meeting, 27.11.–1.12.2001; Contribution Q5.3
57. Kaetsu IK, Biocompatible and biofunctional membranes by means of radiation techniques. Nucl Instrum Methods **B105**, 294–301 (1995)
58. Renardy M, Planck H, Trauter J, Zschocke P, Siebers U, Zecorn T, Federlin K. In: Clinical Implant Material, G Heime, U Soltész, AJC Lee (eds.), Adv Biomater **9**, p. 633 (1990)
59. Tamada M, Yoshida M, Asano M, Omichi H, Kakakai R, Spohr R, Vetter J, Thermoresponse of ion track pores in copolymer films of methacryloyl-L-alanine methyl ester and diethylene glycol bisallyl carbonate. Polymer **33**, 3169–3172 (1992)
60. Shtanko NI, Lequieu W, Du Prez F, Goethals E, Preparation and properties of thermoresponsive track membranes. Proc of the Workshop on European Network on Ion Track Technology, Caen, France, 24–26. Feb. 2002
61. Desai TA, Nanoporous microfabricated membranes: From diagnostics to drug delivery. Proc MRS Boston Fall Meeting, 27.11.–1.12. 2001; Contribution Y5.5
62. Wang D, Bazan GC, Heeger AJ, Biosensors from conjugated polyelectrolytes: Energy transfer complexes. Proc MRS Boston Fall Meeting, 27.11.–1.12. 2001; Contribution BB10.37
63. Ekins RP, Chu FW, Multi analyte microspot immunoassay-microanalytical "compact disk" of the future, Clin Chem **37**, 1955–1967 (1991)
64. Rajan MGR, Bharati Gupta, Immobilisation of antibody spots on glass surfaces-potential use as antibody chips for multianalyte assay. Paper presented at the 9th Asia Pacific Conference on Clinical Biochemist, New Delhi, March 2002;
and: Rajan MGR, Bharati Gupta, Iyer RH, Samuel AM, Track-etched membranes – a novel substrate for "anti-body chips" for multi analyte immunoassays. To be presented in Nuclear and Radio-Chemistry Symposium, BARC, Mumbai, India, February 10–13, 2003
65. The citations in this book are extracts from the following patents: Kimura et al., US patent 6,228,480, May 8, 2001; Lim et al., US patent 6,187,696, Feb. 13, 2001; Simmons Jr. et al., US patent 6,154,311, Nov. 28, 2000; Katsumata, US patent 6,012,267, Jan 11, 2000; Sakurada et al., US Patent 6,004,667, Dec. 21, 1999; Overcash et al., US patent 5,981,011, Nov 9, 1999; Geddes et al., US patent 5,560,945, Oct 1, 1966; Glenn et al., US patent 6,069,112, May 30, 2000; Hayakawa et al., US patent 6,013,372, Jan 11, 2000
66. Chusuei CC, LaI X, Davis KA, Bowers EK, Fackler JP, Goodman DW, A nanoscale model catalyst preparation: Solution deposition of phosphine-stabilized gold clusters onto a planar TiO_2 (110) support. Langmuir **17**, 4113–4117 (2001), and many references therein
67. Chakarvarti SK, Vetter J, Microfabrication of metal-semiconductor heterostructures and tubules using nuclear track filters. J Micromech Microeng **3**, 57–59 (1993)

68. Biswas A, Avasthi DK, Singh BK, Lotha S, Singh JP, Fink D, Yadav BK, Bhattacharya B, Bose SK, Resonant tunneling in single quantum well heterostructure junction of electrodeposited metal semiconductor nanostructures using nuclear track filters. Nucl Instrum Methods **B151**, 84–88 (1999)
69. Klein JD, Herrick RD, Palmer D, Sailor MJ, Electrochemical fabrication of cadmium chalcogenide microdiode arrays. Chem Mater **5**, 902–904 (1993)
70. Rajesh KR, Menon CS, Investigation of SIT characteristics of iodine doped FePcCl thin films. ID47, Intl. Workshop on Nanostructures for Electronics and Optics, Dresden, 18–21 Aug. 2002
71. Borca-Tasciuc DA, Chen G, Borshchecsky A, Fleurial JP, Ryan MA, Lin YM, Dresselhaus MS, Thermal characterisation of nanowire array in α-Al_2O_3 matrix. Proc MRS Boston, Fall meeting, 27.11.–1.12. 2001, Contribution V2.7
72. Gorer S, Hodes G, Quantum size effects in the study of chemical solution deposition mechanisms of semiconductor films. J Phys Chem **98**, 5338–5346 (1994)
73. Yanson AI, Yanson KI, van Ruitenbeek JM, Observation of shell structure in sodium nanowires, Nature **400**, 144–146 (1999)
74. Zabala N, Puska MJ, Nieminen RM, Electronic structure of cylindrical simple-metal nanowires in the stabilitzed jellium model. Phys Rev **B59**, 12652–12660 (1999)
75. Puska MJ, Ogando E, Zabala N, Shell and supershell structures of nanowires: A quantum-mechanical analysis. Phys Rev **B64**, 033401-1–033401-4 (2001)
76. Zabala N, Puska MJ, Nieminen RM, Spontaneous magnetization of simple metal nanowires. Phys Rev **B80**, 3336–3339 (1998)
77. Cepak VM, Hulteen JC, Che G, Jirage KB, Lakshmi BB, Fisher ER, Martin CR, Chemical strategies for template syntheses of composite micro- and nanostructures. Chem Mater **9**, 1065–1067 (1997)
78. Nishizawa M, Mukai K, Kuwabata S, Martin CR, Yoneyama H, Template Synthesis of polypyrrole-coated spinel $LiMn_2O_4$ nanotubules and their properties as cathode active materials for lithium batteries. J Electrochem Soc **144**, 1923–1926 (1997)
79. Burroghes JH, Jones CA, Friend RH, New semiconductor device physics in polymer diodes and transistors. Nature **335**, 137–141 (1988)
80. Burroghes JH, Bradley DCC, Brown AR, Marks RN, Mackay MK, Friend RH, Burn PL, Holmes AB, Light-emitting diodes based on conjugated polymers. Nature **347**, 539–541 (1990)
81. Pei Q, Yu G, Zhang C, Yang Y, Heeger AJ, Polymer light-emitting electrochemical cells. Science **269**, 1086–1088 (1995)
82. Halls JJM, Walsh CA, Greenham NC, Marseglia EA, Friend RH, Moratti SC, Holmes AB, Efficient photodiodes from interpenetrating polymer networks. Nature **376**, 498–500 (1995)
83. Moratti SC, Bradley DDC, Cervini R, Friend RH, Greenham NC, Holmes AB, Light-emitting polymer LEDs. Proc SPIE – The Intl. Soc. for Optical Engineering, Ser. 2144, 108–114 (1994)
84. Greenham NC, Moratti SC, Bradley DDC, Friend RH, Holmes AB, Efficient light-emitting diodes based on polymers with high electron affinities. Nature **365**, 628–630 (1993)
85. Schlenoff JB, Obfrut J, Karasz FS, Transport, magnetic, and optical properties of electrochemically doped poly (1,4-dimethyloxyphenylene vinylene). Phys Rev **B40**, 11822–11833 (1989)

86. Rees ID, Robinson KL, Holmes AB, Towns CR, O'Dell R, Recent developments in light-emitting polymers. MRS Bull **27**, 451–455 (2002)
87. Morita S, Zakhidov AA, Yoshino K, Doping effect of buckminsterfullerene in conducting polymer: change of absorption spectrum and quenching of luminescence. Solid State Commun **82**, 249–252 (1992)
88. O'Regan B, Gratzel M, A low-cost, high-efficiency solar cell based on dye-sensitized colloidal TiO_2 films. Nature **353**, 737–740 (1991)
89. Berggren M, Inganäs O, Gustavsson G, Rasmusson J, Andersson MR, Hjertberg T, Wennerström O, Light-emitting diodes with variable colors from polymer blends. Nature **372**, 444–446 (1994)
90. Berggren M, Inganäs O, Micrometer- and nanometer-sized polymeric light-emitting diodes. Science **267**, 1479–1481 (1995)
91. Berggren M, Inganäs O Micrometer- and nanometer-sized polymeric light-emitting diodes. Science **267**, 1479–1481 (1995)
92. Gudiksen MS, Semiconductor nanowires as optoelectronic building blocks: from fundamental physics to devices. Proc MRS Boston Fall Meeting, 7.1.–1.12. 2001, Contribution V8.8
93. Schön JH, Dodabalapur A, Kloc Ch, Batlogg B, A light-emitting field-effect transistor. Science **290**, 963–965 (2000)
94. Schön JH, Organic semiconductor crystals. Seminar presented at the HMI Berlin on 5.2.2001
95. Pool R, Physicists tackle theory, tubes, and temperature. Science **247**, 1410–1412 (1990)
96. Baur D, Apel PYu, Korchev YE, Müller C, Siwy Z, Spohr H, Spohr R, Surface gel in ion track etching – observations and consequences. Proc of the Workshop on European Network on Ion Track Technology, Caen, France, 24–26.Feb. 2002
97. Apel PYu, Tuchkin SV, Gotlib VA, Lebedeva NE, Lev AA, Rectification property of conical track etched pores and an explanation of the change of rectifying direction at some conditions. Proc of the Workshop on European Network on Ion Track Technology, Caen, France, 24–26.Feb. 2002
98. Apel PYu, Korchev YuE, Siwy Z, Spohr R, Yoshida M, Diode-like single-ion track membrane prepared by electro-stopping. Nucl Instrum Methods **B184**, 337–346 (2001), and references therein
99. Petrov A, Production of micro- and nanoelectrotechnic devices by help of ion tracks in insulators. PhD. Thesis, Fernuniversität Hagen (2004)
100. Hui SY, Tang SC, Chung H, Coreless pinted crcuit board (PCB) transformers – fundamental characteristics and application potential. IEEE Circuits and Systems Society Newsletters 3–47 (2000)
101. Hjort K, The European network on ion track technology. Presented at the 5th Intl. Symposium on "Swift Heavy Ions in Matter", Giardini Naxos, Italy, May 22–25, 2002
102. Shao I, Searson PC, Cammarata RC, Electrochemical deposition of FeCo and FeCoV thin films and nanowire arrays. MRS Proc 27.11.–1.12. 2001, Contribution M5.4
103. Farrell D, Yamamuro S, Majetich S, Magnetic interactions in self-assembled nanoparticle arrays. Proc MRS Boston Fall Meeting, 27.11.–1.12. 2001, Contribution Y9.1

104. Carl A, Kirsch S, Wassermann, Large scale periodic magnetic nanostructures fabricated by optical interference lithography. Proc MRS Boston Fall Meeting, 27.11.–1.12. 2001, Contribution Y9.2
105. Lindeberg M, Gravier L, Ansermet JP, Hjort K, Processing magnetic field sensors based on magnetoresistive ion track defined nanowire cluster links. Proc of the Workshop on European Network on Ion Track Technology, Caen, France, 24.–26. Feb. 2002
106. Fedosyuk VM, Schwarzacher W, Kasyutich OI, Yi G, Heterogeneous alloy nanowired by electrodeposition. Phys Low-Dim. Struct. **11/12**, 61–68 (1999)
107. Gravier L, Ohgai T, Fabian A, Wade T, Wegrowe JE, Ansermet JP, Magnetoresistance and magnetothermopower of magnetic nanowires. Proc of the Workshop on European Network on Ion Track Technology, Caen, France, 24.–26. Feb. 2002
108. Pireaux L, Properties of magnetic and superconducting nanowires. Proc of the Workshop on European Network on Ion Track Technology, Caen, France, 24.–26. Feb. 2002
109. Kröll M, Laurent S, Blau WJ, Nonlinear optical properties of metallic nanowires. Proc MRS Boston Fall Meeting, 27.11.–1.12. 2001, Contribution V12.9
110. Mitrofanov AV, Pudonin FA, Apel PYu, Gromova TI, The ultraviolet transmittance of porous VUV and X ray diffraction filters. Nucl Insrum Methods **A282**, 347–351 (1991)
111. Mitrofanov AV, and Apel PYu, Optical properties of nuclear track filters. Proc of the 2nd Intern Workshop on Solid State Detectors and Their Applications, Dubna, 24–26 March 1992. Dubna, JINR E3-93-61, 164–168 (1993)
112. Mitrofanov A, Pudonin F, Starodubtsev N, Zhitnik I, Spectral filtration of XUV-radiation with the help of thin-film filters in the TEREK-C solar telescope and the RES-C spectrohelograph. Proc of SPIE 3406, 35–44 (1998)
113. Liu SC, Wu JJ, Growth of highly oriented ZnO nanowhiskers by chemical vapour deposition. Proc MRS Boston Fall Meeting, 27.11.–1.12. 2001, Contribution V2.4
114. Polman A (2001), Exciting lanthanide implanted photonic materials. Presented at the 11th Intl. Conf. on Radiation Effects in Insulators, Lisbon, Sept. 3–7, 2001, unpublished
115. Cholewa M, Fischer BE, Heiß M, Preparatory experiments for a second-generation radiation-biological single-hit facility. Presented at the 5th Intl. Symposium on "Swift Heavy Ions in Matter", May 22–25, 2002, Giardini Naxos, Italy

Part III

Conclusion

8 Concluding Remarks

D. Fink

After having summarized the physical and chemical aspects of ion irradiation of polymers in the first volume of this book, in the second volume the emphasis was put onto transport processes in, and technical applications of, ion-irradiated polymeric matter. As in Volume I it can be concluded here that a general understanding has been obtained, but that many details are still missing. Similarly as in the first volume, we therefore compile here a list of open questions that have not yet been answered satisfactorily up to now, and of yet unrealized possible concepts for future ion-track applications. This will hopefully stimulate readers to perform their own research work, to fill in the remaining gaps, and to encourage them to realize new technical concepts that make use of ion impact into polymers.

Transport Processes in Polymers and Latent Tracks

Let us consider how much the charge-carrier production in polymers differs for UV, γ, electron and ion irradiation.

Can one tailor the ion irradiation of polymers so that nuclear or electronic defects dominate in acting as trapping centers for a given penetrant? Depending on the answer, one might think of doping polymers with mobile metal atoms (or clusters) at low concentration, and then letting these metal atoms diffuse until they precipitate at the traps. In this way, one would be able to create various buried conducting 3D structures (e.g., metallic nanowires, coils, etc.) within a polymer, by writing the corresponding patterns with an ion nanobeam into a polymer.

More experiments should be undertaken on the migration of penetrants in C_{60}^{+}-irradiated tracks in polymers.

How do clustering processes of penetrants in polymers occur in closer detail? Systematic experimental studies should be compared with theoretical models that should be elaborated especially for ion-irradiated polymers. In particular, one should determine at which condensation/evaporation ratio the inverse Ostwald ripening sets in for clusters in polymers.

How is electrical charge transported – and eventually trapped – in irradiated polymers? The present answers are still partly speculative and therefore not yet fully satisfying.

There have hardly been any TDS experiments undertaken on defects and trapped impurities in ion tracks.

Quite a number of experiments were already performed on the penetration into, and the doping of latent ion tracks. Can applied electric fields influence the mobility and distribution of electrolytes within latent tracks? It should be determined how, in detail, the anion of a given cationic marker influences the mobility of the latter, and vice versa. More systems with different anionic and cationic mobilities of liquid penetrants in latent tracks, should be examined to establish some systemacy. Can one eventually produce galvanically metallic structures within latent tracks?

Treeing is a well-known deterioration effect in polymers occuring, e.g., in high-voltage cables. How is the electrical breakthrough behavior influenced by ion irradiation? Are the treeing structures eventually aligned along the ion tracks in this case?

There should be more study in detail of the oscillatory behavior of electrolyte charge transport through very thin pores.

Likewise, there should be more study in detail of the directional dependence of electrolyte charge transport through conical pores. Can this transport be switched by external fields so that transistor-like structures emerge?

Ion-Track Etching

Though the etching of ion tracks with liquid etchants has been examined extensively, almost nothing is known about the possibility to etch tracks with reactive gases, e.g., with ozone. Such examinations appear promising as, e.g., plasma etching with reactive atomic oxygen has already brought very good results in other fields.

Only a selected number of polymers has been examined up to now concerning their behavior upon ion-track etching. More polymers should be studied, essentially conducting polymers, synthetic and "living" biopolymers, biodegradable polymers, and naturally produced polymers (such as starch foil, chitin, etc.). Large-area microporous foils of silicone rubber should be produced, to add this material to the list of commercially available microporous thin foils.

Etching studies have been largely restricted to cases where the etching speed in tracks is higher than in the bulk, to produce pores. In polymers where ion irradiation produces preferentially cross-linking instead of chain scissioning (e.g., in the case of polysilanes, many organic rubbers, etc.), the etching speed appears to be reduced along the tracks so that one will produce protruding structures such as hillocks, instead of holes. These cases should be studied in detail, to yield information about new possibilities to produce structured polymer surfaces, e.g., for field-emission devices.

The ITS technique, hitherto successfully applied to etched tracks, should be expanded to also study the above-mentioned cases of reduced etching along tracks.

Transport Processes in Microporous Foils and Etched Tracks

The experiments on the transmission of highly charged ions through etched tracks ("guiding effect") should be expanded to the transmission of electrons and positrons.

Ion transmission through tracks should also be studied for tracks of shapes other than cylindrical, and with the inner track walls covered by magnetic materials, semiconductors and/or nanoclusters.

The first experiments on the possibility of ion focusing by such appropriately bent microporous foils should be undertaken. It should be examined as to how far ion acceleration along etched tracks is possible, eventually after embedding small metallic Einzel lenses therein.

The fact that the presence of etched tracks in a polymer foil enlarges its total surface by factors of up to 100 or 1000 also signifies that the total near-surface volume fraction of this sample increases considerably. Taking into account de Gennes' experiments that point to, e.g., enhanced near-surface self-mobility and reduced T_g, one should expect a dependence of these quantities on the degree of etching of a microporous foil, from the moment when the intertrack distances becomes lower than twice the molecular chain length. No such experiment has ever been made.

Metal–Polymer Interaction

Reliable data on metal diffusion in polymers are scarce and the diffusion mechanism is not well understood. Therefore, more systematic experiments should be performed on the metal–polymer interaction. Specifically, it should be determined whether the inverse Ostwald ripening process can also be verified in these systems.

Also, systematic depth-profile measurements should be undertaken, combined with microscopic cluster population examinations, to correlate the cases of seemingly negative effective diffusion coefficients and of deep penetration with the corresponding cluster rearrangements. For this purpose the metals should be brought into contact with the polymers by both evaporation or ion implantation at sufficiently high fluences.

Furthermore, the correlation between the polymer surface energy, metal/polymer interface energy, and metal surface energy should be applied to metal diffusion in void-rich polymers to see how much the presence of the metal might influence the polymer's void distribution.

Interfaces

Examinations on the behavior of interdiffusion of ion-irradiated and pristine polymers should be undertaken.

Interfaces of two adjacent polymers always show some surface segregation that should influence the impurity migration through these interfaces. Corresponding systematic experiments should be undertaken, for unirradiated as well as for ion-irradiated interfaces.

Ion irradiation should be used for fine tuning of polymer multilayer structures that have been formed by surface-directed spinodal decomposition.

Ion-Track Manipulations

More small-angle scattering studies of galvanically or chemically embedded metallic nanowires/tubules in etched tracks should be performed. For this purpose, synchroton sources will also be helpful.

The wide spectrum of deposition of organometals in tracks should be studied, as well as the possibility to deposit II/IV and III/V compounds by electrodeless or galvanic deposition.

Ceramic thin foils should be produced by high-fluence irradiation of polysilanes. Similarly, *SiC nanocrystal rod* formation should be perfected for nanoscopic electronic and mechanical applications.

Ion-Track Applications

Fiber Bragg Gratings for Sensing

Highly sensitive optical-fiber sensors have been created, e.g., for hydrophones in undersea surveillance that operate with interferometers, the latter allowing detection of changes as small as 1 pm. In particular, fiber Bragg gratings (FGBs) have gained importance in this field. These are sensors in which a spatial modulation has been imposed on the refractive index of the core of a single-mode fiber, by exposing the fiber to two coherent intense UV beams; their interference produces a permanent axially varying intensity pattern that is recorded by the fiber. Strain that modifies the grating can be recorded with high accuracy. The same construction principle could also be applied to ion-irradiated waveguides for dedicated sensor devices, especially taking into account that ion-beam patterning can have a better resolution than UV patterning.

The *high-fluence swift heavy-ion microlithography*, meanwhile already applied in first studies, should be elaborated further. The advantage in comparsion with conventional lithography is the possibility to build up deep-lying structures as thick as some 50 to 100 µm, however, at the expense of strong mask corrosion by electronic sputtering, and of crack formation along the carbonized modified polymeric zones. Studies should be undertaken into how to minimize these effects.

Catalytics

The deposition of suitable organometals within etched tracks offers an ideal way to produce advanced catalysts. For example, the deposition of Pd-betadiketonate in tracks should produce porous Pd layers with catalytic activity on the etched track walls.

Electronics

Conducting nanowires or other (semi)conducting structures should be produced via trapping of metals or metallic clusters along tracks.

Biomedicine

Platforms for medical experiments should be built up on the basis of latent and conducting tracks in biocompatible thin polymeric foils, in which individual cells can thus be contacted via the tracks, to record their action potentials.

Sensors

The first ion-track based sensors have already been produced and are now being offered commercially; more of them are under development. They cover fields from physical properties up to biological information. In particular, more sensors for medicine are worth developing, e.g., for monitoring electrolytes in blood such as K, Na, and Ca, for determining blood gases, for measuring the blood pH value, for determining antibodies against various diseases, and for detecting D-dimers (the latter protein is released when a blood clot breaks down). A blood pH value below 7.4 indicates the presence of dead body cells, e.g., after a stroke, due to their decay towards lactic acid and other products [1]. For the latter case one can incorporate, e.g., suitable dyes in polymers. It was shown above that such dyes can be readily grafted to polymers after ion irradiation.

Miniaturized semiconductor gas sensors combining concentric sensing and heating elements in the same etched track appear worthy of development.

Ion-Track Technology for Counterterrorism

A very wide new market has opened up after the attack against the World Trade Center in New York on Sept. 11, 2001. Ion-track technology for counterterrorism is an emerging major R & D area that should be taken serious from now on. Let us cite J. Wadsworth from Lawrence Livermore Labs in this connection [2] (emphasized by the author): "In the longer term, advanced detection systems will require advances in materials science such as *smart membranes* and surfaces for *biosensing* in liquids and gases, *microfabrication* and *microelectromechanical systems technologies*, and *chip-based architectures* for rapid multiplexed detection. Advanced diagnostics and treatments will ultimately rely on high-resolution protein–protein and protein–DNA imaging and *high-throughput* methods for rapid determination of protein structures."

Ion-track-based polymeric sensors appear to find most open reception as biosensors. The above-cited "smart membranes" can readily be realized on the base of microporous track foils. What is also required in this connection is advanced "biodetection instrumentation, pioneering the instrumentation and 'ruggedization"' [2]. As an example, Wadsworth gives "the biological aerosol sentry and information system (BASIS), which is a mobile field laboratory that collects samples and utilizes polymerase chain-reaction techniques to identify DNA". In fact, aerosol collection is already now a favorite domain of microporous track foils. Cladding the inner walls of such aerosol filters with adequate biological sensor material is hence a possible strategy to realize such nanoscopically dimensioned devices in cheap mass production.

Additionally to such biosensors to detect biologically hazardous materials in air and water, one requires new types of sensors to detect illicit chemical-weapons production. What is needed is "to include materials-science-based ideas such as novel chemical sensors, miniature air vehicles, ... and ultra-lightweight optical systems" [2] in the arsenal of antiterror fighting. Again, ion-track-based devices or device components will have good opportunities for application, being extremely lightweight and small.

Actuators

Ionic-conductive polymers such as PEO can convert electrical energy into mechanical energy. This is highly interesting for medical applications as these soft polymers can mimic biological responses. The actuator motion arises from the cation attraction to the anode relative to the polymer host, which creates a pressure gradient leading to deformation. This deformation is reversible with the direction of the electric field and displays stable operation after more than 10^7 cycles in air. The actuation can be increased by the use of the plasticizer poly(ethylene glycol) (PEG) that decreases the crystallinity of the medium, leading to increased ion mobility. In this way, present PEO-PEG actuators have achieved 2% strain at applied voltages of 1–3 V, a frequency of 0.5 Hz, and a power density of ~ 50 J/kg, which is half that of biological muscles [3]. As the polymeric crystallinity can also be reduced by ion beams, ion irradiation at moderate fluences may be another helpful means to improve the actuator response further and/or to tailor the actuator shape precisely by irradiation through adequate masks.

With this outlook we close the overview of the second volume of this book on ion irradiation of polymers.

References

1. Baldini F, Mignani AG, Optical-fiber medical sensors. Mater Res Soc Spring Meeting, April 3, 2000, San Francisco, MRS Bull **27**, 383–387 (2002)
2. Wadsworth J, Science and Technology and Counterterrorism. Mater Res Soc Spring Meeting, April 3, 2000, San Francisco, MRS Bull **27**, 348–352 (2002)
3. Brinkman K, Low voltage achieves large strain in electroreactive polymer. Mater Res Soc Spring Meeting, April 3, 2000, San Francisco, MRS Bull **27**, 352 (2002)

Index

Springer Series in

MATERIALS SCIENCE

Editors: R. Hull R. M. Osgood, Jr. J. Parisi H. Warlimont

10 **Computer Simulation of Ion-Solid Interactions**
By W. Eckstein

11 **Mechanisms of High Temperature Superconductivity**
Editors: H. Kamimura and A. Oshiyama

12 **Dislocation Dynamics and Plasticity**
By T. Suzuki, S. Takeuchi, and H. Yoshinaga

13 **Semiconductor Silicon**
Materials Science and Technology
Editors: G. Harbeke and M. J. Schulz

14 **Graphite Intercalation Compounds I**
Structure and Dynamics
Editors: H. Zabel and S. A. Solin

15 **Crystal Chemistry of High-T_c Superconducting Copper Oxides**
By B. Raveau, C. Michel, M. Hervieu, and D. Groult

16 **Hydrogen in Semiconductors**
By S. J. Pearton, M. Stavola, and J. W. Corbett

17 **Ordering at Surfaces and Interfaces**
Editors: A. Yoshimori, T. Shinjo, and H. Watanabe

18 **Graphite Intercalation Compounds II**
Editors: S. A. Solin and H. Zabel

19 **Laser-Assisted Microtechnology**
By S. M. Metev and V. P. Veiko
2nd Edition

20 **Microcluster Physics**
By S. Sugano and H. Koizumi
2nd Edition

21 **The Metal-Hydrogen System**
By Y. Fukai

22 **Ion Implantation in Diamond, Graphite and Related Materials**
By M. S. Dresselhaus and R. Kalish

23 **The Real Structure of High-T_c Superconductors**
Editor: V. Sh. Shekhtman

24 **Metal Impurities in Silicon-Device Fabrication**
By K. Graff 2nd Edition

25 **Optical Properties of Metal Clusters**
By U. Kreibig and M. Vollmer

26 **Gas Source Molecular Beam Epitaxy**
Growth and Properties of Phosphorus Containing III–V Heterostructures
By M. B. Panish and H. Temkin

27 **Physics of New Materials**
Editor: F. E. Fujita 2nd Edition

28 **Laser Ablation**
Principles and Applications
Editor: J. C. Miller

29 **Elements of Rapid Solidification**
Fundamentals and Applications
Editor: M. A. Otooni

30 **Process Technology for Semiconductor Lasers**
Crystal Growth and Microprocesses
By K. Iga and S. Kinoshita

31 **Nanostructures and Quantum Effects**
By H. Sakaki and H. Noge

32 **Nitride Semiconductors and Devices**
By H. Morkoç

33 **Supercarbon**
Synthesis, Properties and Applications
Editors: S. Yoshimura and R. P. H. Chang

34 **Computational Materials Design**
Editor: T. Saito

35 **Macromolecular Science and Engineering**
New Aspects
Editor: Y. Tanabe

36 **Ceramics**
Mechanical Properties, Failure Behaviour, Materials Selection
By D. Munz and T. Fett

37 **Technology and Applications of Amorphous Silicon**
Editor: R. A. Street

38 **Fullerene Polymers and Fullerene Polymer Composites**
Editors: P. C. Eklund and A. M. Rao

Springer Series in

MATERIALS SCIENCE

Editors: R. Hull R. M. Osgood, Jr. J. Parisi H. Warlimont

39 **Semiconducting Silicides**
Editor: V. E. Borisenko

40 **Reference Materials in Analytical Chemistry**
A Guide for Selection and Use
Editor: A. Zschunke

41 **Organic Electronic Materials**
Conjugated Polymers and Low Molecular Weight Organic Solids
Editors: R. Farchioni and G. Grosso

42 **Raman Scattering in Materials Science**
Editors: W. H. Weber and R. Merlin

43 **The Atomistic Nature of Crystal Growth**
By B. Mutaftschiev

44 **Thermodynamic Basis of Crystal Growth**
P–T–X Phase Equilibrium and Non-Stoichiometry
By J. Greenberg

45 **Thermoelectrics**
Basic Principles and New Materials Developments
By G. S. Nolas, J. Sharp, and H. J. Goldsmid

46 **Fundamental Aspects of Silicon Oxidation**
Editor: Y. J. Chabal

47 **Disorder and Order in Strongly Nonstoichiometric Compounds**
Transition Metal Carbides, Nitrides and Oxides
By A. I. Gusev, A. A. Rempel, and A. J. Magerl

48 **The Glass Transition**
Relaxation Dynamics in Liquids and Disordered Materials
By E. Donth

49 **Alkali Halides**
A Handbook of Physical Properties
By D. B. Sirdeshmukh, L. Sirdeshmukh, and K. G. Subhadra

50 **High-Resolution Imaging and Spectrometry of Materials**
Editors: F. Ernst and M. Rühle

51 **Point Defects in Semiconductors and Insulators**
Determination of Atomic and Electronic Structure from Paramagnetic Hyperfine Interactions
By J.-M. Spaeth and H. Overhof

52 **Polymer Films with Embedded Metal Nanoparticles**
By A. Heilmann

53 **Nanocrystalline Ceramics**
Synthesis and Structure
By M. Winterer

54 **Electronic Structure and Magnetism of Complex Materials**
Editors: D.J. Singh and D. A. Papaconstantopoulos

55 **Quasicrystals**
An Introduction to Structure, Physical Properties and Applications
Editors: J.-B. Suck, M. Schreiber, and P. Häussler

56 **SiO_2 in Si Microdevices**
By M. Itsumi

57 **Radiation Effects in Advanced Semiconductor Materials and Devices**
By C. Claeys and E. Simoen

58 **Functional Thin Films and Functional Materials**
New Concepts and Technologies
Editor: D. Shi

59 **Dielectric Properties of Porous Media**
By S.O. Gladkov

60 **Organic Photovoltaics**
Concepts and Realization
Editors: C. Brabec, V. Dyakonov, J. Parisi and N. Sariciftci